From Perception to Pleasure

The Neuroscience of Music and Why We Love It

Robert Zatorre

UNIVERSITY PRESS

OXFORD
UNIVERSITY PRESS

Oxford University Press is a department of the University of Oxford. It furthers
the University's objective of excellence in research, scholarship, and education
by publishing worldwide. Oxford is a registered trade mark of Oxford University
Press in the UK and certain other countries.

Published in the United States of America by Oxford University Press
198 Madison Avenue, New York, NY 10016, United States of America.

Library of Congress Cataloging-in-Publication Data
Names: Zatorre, Robert, author.
Title: From perception to pleasure : the neuroscience of music and
why we love it / Robert Zatorre.
Description: New York : Oxford University Press, 2024. |
Includes bibliographical references and index. |
Identifiers: LCCN 2023020457 (print) | LCCN 2023020458 (ebook) |
ISBN 9780197558287 (hardback) | ISBN 9780197558300 (epub) | ISBN 9780197558317
Subjects: LCSH: Music—Psychological aspects. | Musical perception. |
Cognitive neuroscience. | Pleasure.
Classification: LCC ML3830 .Z27 2023 (print) | LCC ML3830 (ebook) |
DDC 781.1/1—dc23/eng/20230613
LC record available at https://lccn.loc.gov/2023020457
LC ebook record available at https://lccn.loc.gov/2023020458

DOI:10.1093/ oso/9780197558287.001.0001

To my family, with love and gratitude.

From Perception to Pleasure

Contents

Preface

People often ask me how I got into the study of music and the brain. There were several factors that led me down that path, but perhaps the most relevant were some transformative musical experiences. One such epiphany occurred in my early adolescence, when somebody lent me a record (we had vinyl back then) of music by Béla Bartók (it was his 1936 composition *Music for Strings, Percussion, and Celesta*). I had barely ever heard of Bartók and had little idea of music beyond the Beatles, the Doors, and the Moody Blues (and the Tangos that my parents listened to, but that I deemed old-fashioned at that age). But when I first listened to this soft, hypnotic, layered and, in places, relentlessly rhythmic music, I was transfixed: I had chills, palpitations, and feelings I couldn't even describe because I had not experienced them before. I had no idea that music could have such effects. That, and other similar experiences, motivated me to learn music. After a few years, I figured out that I was a merely competent, but far from outstanding, musician; yet the more I learned about music, the more I became intrigued to understand how and why music works the way it does. So in university I turned to science, which had always been important to me—even as a child—for potential answers. I soon discovered that the field seeking answers to such questions was still in its infancy, but that provided an opening to try to figure out a few things. I've been at it ever since. And, although there's still plenty left to figure out, this book is the outcome of what I've learned.

We fast-forward to just a few years ago, when I first sat down to write this book. Initially, it meandered through lots of interesting terrain; but I eventually lost interest in pursuing it because it wasn't leading anywhere. I was also simply having too much fun in the lab to devote the necessary time to it. It took some more reading of science books that I enjoyed, and many discussions with my colleagues and students (especially the many long chats with my wife and collaborator, Virginia Penhune), for me to understand that a good book needs to have a good narrative. The delay between my initial attempt and the one that led to this book was well-timed for two reasons. First, because a lot of new research on music, auditory circuitry, pleasure, reward, and emotion had been coming out from many labs, including our own, and a coherent, important story about the neural basis of music and its enjoyment was emerging that I felt needed telling. The second factor was the silver lining offered by the pandemic: with labs closed and nowhere else to go, I had the time and opportunity needed to write.

In putting this book together, I had a choice between writing a straightforward academic book, or something geared more to the general public. In the end I leaned more toward the former, but the book is a bit of a hybrid; I wanted to make it reasonably readable by interested, nonspecialist readers, but without sacrificing detailed content. For that to work, I included occasional anecdotes and some personal angles, in addition to the facts and figures. This approach was encouraged by the two best pieces of advice I received from two good scientific friends, based on their own experiences writing wonderful science books: Adrian Owen told me that I should write the book that I wanted to write, not the one that others expected me to. In other words, it should be a story that I wanted to tell. Ani Patel advised me to write an academic book, but one that would also incorporate my own experiences in the discovery and interpretation of the findings. In other words, I should give the background and history behind the studies and the people who carried them out. So there's a bit of a scientific memoir embedded in there along with all the graphs and brains.

A consequence of these decisions is that this book is not a textbook; it is not meant to cover exhaustively all the relevant research areas that could have been discussed. There are many truly fascinating topics that I decided are largely out of scope for the main concepts I wish to develop. That's why for example the topic of brain plasticity, and how neural circuits become optimized for musical processing with training or experience, is only briefly mentioned in a few sections. Similarly, questions about genetic contributions to musical skill, and how they may interact with experience-dependent plasticity, will have to wait for future appropriate attention. Another important domain that only gets occasional mention is the development of musical processes, over a lifetime, and how training interacts with age to yield advantages for early training. Nor do I have much to say about the evolutionary aspects of musical functions, fascinating though that question may be. And, although I do describe certain pertinent studies of neurological disorders and touch briefly upon potential clinical applications toward the end, this book is primarily about the neural circuitry that underlies musical perception and pleasure—and not about dysfunctions.

The wonderful thing about science is that it's not static. It keeps moving. But it's also incremental. There's a funny sense some people have that studies more than a few years old are somehow *passé* because they didn't use the latest fancy gadget or analysis. Yet, what we know today was built on solid findings from the past; so another important goal I had in mind was to give some context, based on knowledge from earlier work, that I hope will make the origins of some of our current ideas clearer, without neglecting the many really exciting novel findings. In general, I have opted to present and explain a lot of empirical data because any hypotheses or theories we can build have to be built on that foundation. As Brenda Milner used to tell me, "Theories come and go, but data always stay," by which she did not mean that theories were not important, but rather that theories and interpretations must change in order to explain the accumulation of both old and new empirical knowledge (that is, assuming you've done your experiments right). So I've taken that advice to heart.

So the book you are now contemplating, dear reader, is the outcome of this effort. It will be for you to judge how successful the endeavor turned out to be, of course, but I hope that you will find the ideas I've put forward intriguing and that they will provide a platform upon which future work can continue to build.

Introduction

As neither the enjoyment nor the capacity of producing musical notes are faculties of the least use to man in reference to his daily habits of life, they must be ranked amongst the most mysterious with which he is endowed.
—Charles Darwin, *Descent of Man* (1871)

Our species has been making music most likely for as long as we've been human. It seems to be an indelible part of us. The oldest known musical instruments date back to the Upper Paleolithic period, some 40,000 years ago. Among the most intriguing of these are delicate bone flutes, seen in Figure 1.1, found in what is now southern Germany (Conard et al. 2009). These discoveries testify to the advanced technology that our ancestors applied to create music: the finger holes are carefully bevelled to allow the musician's fingers to make a tight seal; and the distances between the holes appear to have been precisely measured, perhaps to correspond to a specific musical scale. This time period corresponds to the last glaciation episode in the Northern Hemisphere—life could not have been easy for people living at that time. Yet time, energy, and the skills of craftworkers were expended for making abstract sounds "of the least use . . . to daily habits of life." So, music must have been very meaningful and important for them. Why would that be?

That question leads to another, more personal one for me: why should a neuroscientist study this apparently mysterious capacity we humans have for creating and enjoying music? It's a question I was often asked early in my career when there was little in the way of an established discipline to draw from. The question took two forms. My colleagues in the sciences would sometimes express skepticism about the value of understanding the psychological and neural bases of music, in part on the grounds that there were more important things to study (language, vision, memory). My colleagues in the humanities, on the other hand, would sometimes object to the idea that an artistic endeavor such as music should even be the subject of scientific study in the first place, perhaps reacting against a perceived reductionist agenda on the part of scientists that would diminish the status of music.

Happily, in the intervening years, both camps have largely (though not universally) come around. Indeed, researchers in language or memory and scholars in musicology or

From Perception to Pleasure. Robert Zatorre, Oxford University Press. © Oxford University Press 2024.
DOI: 10.1093/oso/9780197558287.003.0001

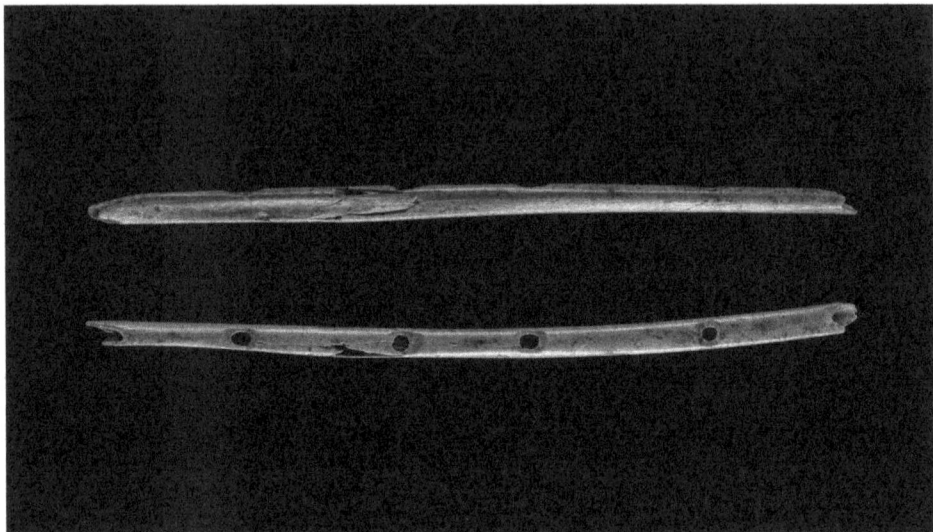

FIGURE 1.1 Paleolithic flute. This instrument, made from the wing bone of a vulture, dates from the Upper Paleolithic period, approximately 40,000 years ago. It is 22 cm long but is broken at one end, so it was likely longer originally and may have contained more than the remaining five finger holes. Reproduced with permission from (Conard et al. 2009).

music theory frequently have become enthusiastic collaborators and contributors to the goals of understanding the neural basis of music. Thanks to the support received from many quarters and to the growth of a worldwide community of scientists and scholars dedicated to the enterprise of music neuroscience, an enormous amount of progress has been made, which I draw on extensively in this book. Why have so many finally recognized the interest of understanding music in the brain?

The short answer, I think, is written—chiseled into cement in fact, as seen in Figure 1.2—on the front wall of the building I work in, the Montreal Neurological Institute. "The problem of neurology is to understand man himself" proudly proclaimed Wilder Penfield, the founder of our Institute. In slightly more updated language, we might rephrase this statement to say that neuroscience strives to understand what makes us human. I have always found it remarkable that Penfield, a neurosurgeon who built the institute together with its hospital to treat neurological diseases, nonetheless had a grander vision in mind. Studying the nervous system will certainly lead to cures for terrible disorders and lessen suffering, but it will also teach us something much greater, he seems to be saying to us: it will tell us what it means to be human. I believe that the study of music and the brain fits in very well with this larger quest. Music is a ubiquitous, essential aspect of humanity—I would go so far as to say that it is a type of species-specific communication system. That is why our ancient ancestors practiced it and why we continue to do so today. As such, it seems to me that if we are to truly understand the full nature of human experience, we cannot leave music out of it.

The slightly longer and more prosaic answer to why the study of music neuroscience has blossomed is that because of its complexity and multidimensional nature, music

LA NEUROLOGIE CHERCHE À COMPRENDRE L'HOMME LUI-MÊME

THE PROBLEM OF NEUROLOGY IS TO UNDERSTAND MAN HIMSELF

WILDER PENFIELD 1891-1976

FIGURE 1.2 Penfield quote. Panel mounted on the front wall of the Montreal Neurological Institute and Hospital, displaying the statement of its founder, Dr Wilder Penfield. Photograph by the author.

provides a remarkable window onto the most advanced aspects of human cognition. As such, compared to other research domains, it allows us a different and complementary perspective to better understand these functions.

Consider a simple, everyday scenario: you are waiting for an elevator in the lobby of a busy building when someone walks by whistling a tune. You recognize it, and as you imagine the rest of the tune in your mind, perhaps you gently sway or tap your foot to it. It brings a smile to your face because not only is it a beautiful tune but it also reminds you of your vacation last summer, when it first became popular. Nothing extraordinary about any of that. Except that the apparent simplicity of everything that happens in those few seconds belies the remarkable computations going on in your brain.

First, the sound entering your ear contains not only the acoustical features corresponding to the tune, but also a multitude of other random noises all mixed in; so your auditory system must segregate the relevant sound from the background. You then need to hold the sounds in your mind for a few moments—long enough to be able to extract the pitches and durations, register the relationships between successive sounds, and recognize the tune. Recognition is itself remarkable because you may never have heard the song whistled before—instead you only heard it sung, in a different key, and at a faster tempo. Yet, your perceptual and memory systems instantly sort all that out. The ability to imagine the tune also calls on memory—information stored somewhere in the brain is retrieved and played out internally, generating an experience akin to the original perceptual event. Your motor system also becomes engaged, requiring that sound impulses become transformed into commands to the muscles. Yet the movements are not simply timed to the start of each tone, but rather to the beat of the music, an abstract quality related to its metrical structure. Emotions may arise in you, due to the structure of the rhythm and melody, leading to an aesthetic appreciation of the melody's beauty. The music's twists and turns form built-in surprises that give you a frisson of pleasure. And finally, associations with past personal events may further enhance the affective experience.

A lot of this book is based on unraveling many of these aspects of musical cognition, and especially how they are instantiated in neural circuitry. A critical point is that music is not just about sending and receiving acoustical signals; it also leads to complex emotional states of enjoyment, pleasure, and much else. The main goal of this book therefore is to attempt to answer the overarching question of why we humans seem to love music so much from a cognitive neuroscience perspective. Different disciplines of study might approach this question differently: a sociologist, an anthropologist, a music historian, or a music theorist might all give equally valid, though maybe divergent, answers. But I believe neuroscience can offer a unique and essential perspective. Even if historical or musicological factors are obviously important at some level of explanation, they are all ultimately expressed though the activity of our brains. Music exists because people make it and perceive it and respond to it, and none of that could happen if our brains were not perfectly adapted to those tasks. Therefore, to address why humans make and enjoy music, I shall propose an answer—or a first approximation at least—emphasizing how neural circuitry enables musically relevant processes to happen and how certain neurobiological mechanisms in our brains generate the pleasure that we feel from it. And by so doing, my hope is that we will gain novel insights into the many complex cognitive functions that music depends upon, while also understanding music itself at a deeper level.

At its simplest, the thesis I shall defend follows from the title of this book: from perception to pleasure. Musical pleasure, I argue, arises from the functional interactions between two distributed and complex neural systems: the perceptual/motor/cognitive system on the one hand and the reward system on the other. The former is instantiated in various corticocortical loops within each cerebral hemisphere, that can be roughly subsumed under the rubrics of auditory ventral and dorsal pathways. The latter consists of subcortical structures in the midbrain and basal forebrain, along with orbitofrontal and other cortical regions. Each system is specialized for particular kinds of processing. The perceptual/motor system is responsible for generating representations of the environment so that we can act upon it; the reward system, in contrast, is responsible for assigning value to the stimuli in the environment or to the outcome of our actions, so that we will know how to act (approach, avoid, consume, examine, etc.) To follow the argument through, it will thus be necessary to understand how the various components of these systems work and how the two systems are connected to each other.

Before we get into these details in subsequent chapters, there are a few general issues that need to be introduced, as they are central to the arguments I shall develop. First, there are questions regarding how neuroscience findings from other species may be understood in the context of music processing. Then, I will discuss how two important concepts in contemporary cognitive science and neuroscience, namely, statistical learning and predictive coding, pertain to the main questions of this book.

Comparative Neuroscience

How should we interpret neuroscience findings from nonhuman species in relation to music? We can start by looking at the auditory system. A wealth of anatomical and physiological knowledge about auditory processing pathways has been acquired over the past 50 years, much of it coming from animal studies, especially nonhuman primates such as monkeys. The problem is that the very abilities that I would argue are most characteristic of human auditory cognition—music and speech—are hard to study in other species, since they neither speak nor play music.

In fact, the auditory cognitive capabilities of monkeys, in particular, are often quite limited compared to their highly developed visual skills, which is likely related to differences in the way the auditory system is organized in that species (Fritz et al. 2005, Balezeau et al. 2020). This situation is problematic because monkeys are most often used in neuroscience studies, as it is an animal that is fairly close to us in its neural architecture. It's a bit like studying bird flight if your best available experimental model is a chicken—it would still be useful, but likely incomplete. Whatever neural features may be particular to humans that allow complex functions like music to emerge spontaneously and universally are therefore difficult to understand fully via this approach.

I do not intend to say that other species do not exhibit behaviors that are relevant to understanding music or that they are not valuable to study (for an excellent discussion of this point, see Honing 2019). Indeed, there is a wealth of evidence that many different animals show some skills which may be thought of as analogous, if not necessarily homologous, to human music-related skills, such as rhythmic entrainment for example (Wilson and Cook 2016, Patel 2021). Meanwhile other skills, such as relative pitch, seem to be mostly absent in other animals (Bregman et al. 2016, Elie and Theunissen 2016, Shannon 2016). This topic is largely beyond our scope in this book, but further cross-species study will be very important—particularly when paired with comparative neuroscience—so we can better understand the specialized neural circuits in human beings that allow particular perceptual and motor abilities to emerge.

I prefer to think about the comparative issue in terms of biological adaptation rather than in terms of a simplistic human superiority concept. An axiom of evolutionary theory is that an organism's specialized features reflect its optimal adaptation to the environment it lives in. Bats provide an excellent example of a specialized auditory system: their brains are highly adapted for echolocation, allowing them to navigate and seek out food in flight (Covey 2005). If we humans had to survive by shouting around in the dark to find insects to eat, it would not work out so well. On the other hand, we are highly cognitive creatures who live in complex organized societies. We are full of thoughts, memories, and desires; we experience joys, loves, fears, and other emotions; and we want constantly to communicate these internal states to others. We depend, then, on the generation, manipulation, and transmission of abstract, information-bearing patterns to survive and thrive. In this context, our musical and linguistic abilities can be seen as specializations or adaptations (see further discussion in Chapter 5) that permit complex systems of communication via vocal and other auditory signals. And so, in this book, I take the approach of describing the functional and structural neural specializations we possess that allow music to happen, and that enable music to express, communicate, and regulate affective states.

Keeping all these considerations in mind, there is an enormous amount to learn from comparative neuroscience that is relevant for our goals. For one thing, there is a good deal of evolutionary convergence across species in certain structures compared to others. In particular, subcortical structures, including many portions of the reward system and of the auditory brainstem, are relatively conserved across humans, monkeys, and rodents, which allows us to be more certain about cross-species parallels in those structures. Conversely, there is more divergence in the functional organization of auditory cortex across these species (Malmierca and Hackett 2010). Anatomical (Petrides 2014) and functional (Rocchi et al. 2021) connectivity of auditory cortex in humans does find many parallels in the macaque brain. But since the kinds of detailed tract-tracing that can be performed in animals are not possible in humans, and the kind of cognitive tasks that can be done with humans are often difficult with animals, much remains unknown about exactly what features of anatomical and functional organization distinguish us from our simian neighbors and thus allow our special auditory abilities.

The few direct comparative studies that exist of human cortical activity patterns versus that of nonhuman promates in response to sound tend to show that functional organization is broadly similar in terms of how low-level acoustical features are encoded (Schönwiesner et al. 2014), which is good news for us. But those studies also show that there are nonetheless salient differences in particular aspects of processing, especially those that are most relevant for music and speech (Erb et al. 2019, Landemard et al. 2021), including lateralization of functional connections (Rocchi et al. 2021). Therefore, in what follows, I will make use of as much information as possible from other species, keeping in mind, however, that not all aspects of neural organization are identical.

There is another valuable reason to consider cross-species comparisons: they can yield insight into the universality of neural mechanisms that might be relevant for aspects of music processing. There is significant controversy in music cognition about the extent to which a given phenomenon that is evident in people from one culture may or may not be representative of what happens in other cultures. Since most music-related research, with some exceptions, has been carried out with Western music and with listeners enculturated in this tradition, it is a very fair question. As such, if we want to be conservative, our conclusions should largely be limited to Western music unless there is direct evidence to the contrary. Direct evidence would come from comparing the behavioral and neural responses of people with experience in non-Western musical systems. In this book I will mention such research when it exists and is relevant, although my goal is not to try to resolve this complex question, which is currently under active investigation (Jacoby et al. 2020).

But comparative analysis can help to address these issues indirectly because if a given phenomenon—take as an example, the representation of harmonic and octave pitch relationships that we shall discuss in Chapter 2—is observed in both monkeys and humans, then one may conclude that the neural processes subserving that function were already present in our common evolutionary ancestor some 25 million years ago (Gibbs et al. 2007).

It can, therefore, be assumed to be a fundamental property of human auditory processing, antedating the influences of culture. Any given culture may of course adapt, modify, or even ignore harmonic or octave relationships entirely (Jacoby et al. 2019), and that is very interesting to know about because it reveals the interaction between culture and biology. But if similar neural representations also exist in other species, it would be hard to argue that the way these tonal relationships are processed merely reflects an entirely arbitrary cultural artifact.

Statistical Learning

Statistical learning constitutes another very relevant concept that will recur throughout the rest of this book. In its most basic form, the term refers to the ability to detect and learn regularities that exist in the pattern of events in the environment. More specifically, it is often used to describe learning of the statistical relationships between events, that is, the degree to which one event predicts the occurrence of another. In a famous study that opened the door into this research domain, my friend Jenny Saffran, now at the University of Wisconsin, demonstrated that young infants were able to learn the contingencies between pairs of adjacent speech syllables presented in an unbroken stream (Saffran et al. 1996). In other words, they were able to learn that when sound X occurs, it is more often followed by sound Y than by sound Z. This is called statistical learning, because the contingencies are probabilistic, just like in the real world: when you hear one shoe drop, it is usually—but not always—followed by the other shoe. Hence the expression "waiting for the other shoe to drop," which implies prediction and expectancy.

Statistical learning seems to be a very general phenomenon and has been documented not only in auditory but also visual modalities (e.g., [Fiser and Aslin 2002], and in several different animal species as well, e.g., [Meyer and Olson 2011]). As such, it is not surprising that it also occurs for tonal sequences, both in infants and adults (Saffran et al. 1999). This learning reflects the fact that in real music certain relationships are statistically more common than others. For example, as David Huron (2001) showed, at the most basic level, small musical intervals occur much more often than large ones in melodies drawn from many different cultures, across different continents, as shown in Figure 1.3. So the nervous system learns that, given a certain pitch, it is very likely that it will be followed by another tone close in pitch (say, one whole tone) rather than by a large leap (such as a sixth); larger intervals do occur, but are more surprising on a purely statistical basis.

More recent work on the phenomenon of statistical learning (for a review, see Saffran 2020) has emphasized more complex properties, including learning of long-range (nonadjacent) relationships that are critical for higher-order learning, for example of syntax, both in language and in music (Creel et al. 2004). An important conclusion from this body of work is that statistical learning is at least one important mechanism by which mental representations, or models, of patterns in the world are formed. This concept will also be important in the rest of this book since we are interested in understanding how these internal models are supported by the brain's functional architecture and how they generate responses to sounds that conform or not to the expectations.

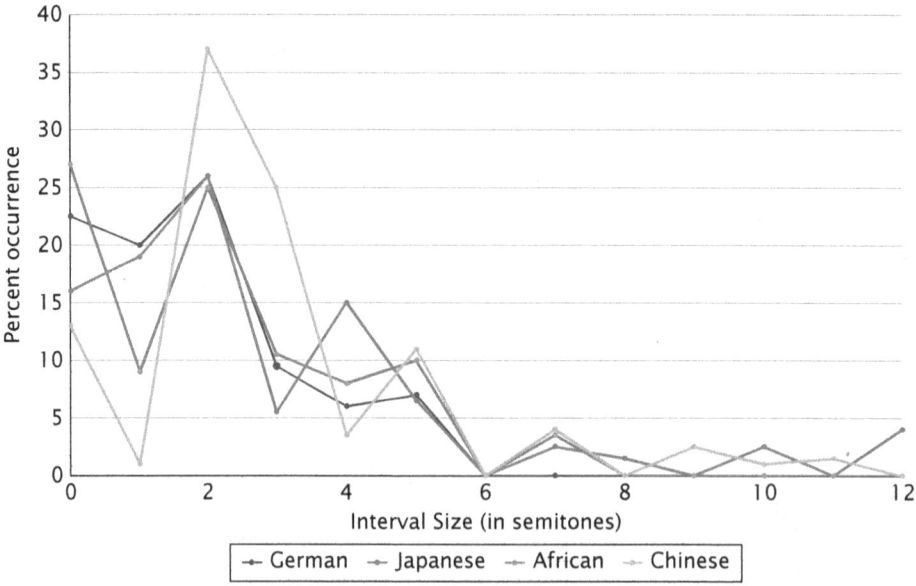

FIGURE 1.3 Distribution of melodic intervals. The graph shows the proportion of occurrence of certain musical intervals across four different geographic locations/musical cultures. Although some differences are clearly seen across groups, there is a general tendency for small musical intervals to predominate, with larger intervals becoming progressively less common. Adapted with permission from (Huron 2001).

Statistical learning may also be considered within the broader context of implicit learning, which has particular importance in the acquisition of musical knowledge (Rohrmeier and Rebuschat 2012). Considerable evidence indicates that listeners learn about the structure of the music that they are exposed to throughout life without needing any explicit training (Bigand and Poulin-Charronnat 2006) and without necessarily even being consciously aware of what they have learned (Rohrmeier and Rebuschat 2012). This means that we can study the neural mechanisms underlying important musical cognitive processes by studying listeners without formal musical training, but who have already internalized a lot of the relevant knowledge to be able to perceive and enjoy music via environmental exposure. However, this should not be taken to mean that musical training or experience has no effect on how musical representations are formed and maintained—training does make a difference—so that topic will also be visited at several points in the rest of this book.

An important distinction should be made between the outcome of learning based on statistical regularities (as would happen over a lifetime of listening to music within a certain style), as opposed to explicit knowledge of the pattern of sounds within a specific piece of music (e.g., as would happen when you hear a popular song so often that you know every part of it, or when a soloist has memorized a concert piece perfectly for a performance). These two kinds of learning, statistical and explicit, are of interest in the context of understanding our perceptual and emotional responses to music, because they generate two different kinds of expectancies. In the case of learning statistical relationships, so-called schematic expectancies are formed, which are based on general schemas, so that one predicts that Y will follow X because that's how it usually happens. In the case of explicit learning of a specific piece of music, so-called veridical expectancies are formed (Bharucha 1994, Huron 2006), which may not always match schematic knowledge (e.g., one has learned that in a particular passage Z, not Y, follows X).

Schematic expectancies are especially important for responding to music because they form the basis for making predictions about upcoming events in a very general way. They are based on a set of learned contingencies which can be applied to any musical piece, even a new one never before heard, as long as it is written within the musical system that has been previously learned. In the case of Western music, for instance, the regularities, or rules, of melodic structure, harmonic progressions, rhythmic timing, and metrical structure would all be internalized by listeners enculturated in that system (while listeners from other cultures of course would learn the relevant rules of their own system). Veridical expectancies may not always conform to the most commonly expected statistical relationships. It is, very often, the violations from the most likely outcomes that yield surprises and reward-related responses, as we shall discuss in depth in Chapters 7 and 8.

The interplay between the two kinds of expectancies is endlessly fascinating, as composers and musicians know very well. This relationship accounts, in part, for the fact that very well-known pieces of music can continue to surprise and give us pleasure—despite the fact that we know exactly what's coming—a conundrum which seems to have puzzled the philosopher Wittgenstein (Wittgenstein 1966). The answer is that even if we know some particular piece of music inside out and hence can accurately predict what will happen next, we still get some level of surprise when the music unfolds in a way which does not meet our schematic expectancies. Experimentally, it has been demonstrated that schematic expectancies tend to be resistant to being overcome by veridical ones, even after repetition (Tillmann and Bigand 2010), showing that they are indeed quite powerful.

An interesting aspect of statistical learning is that it manifests itself very early in life, as shown by the infant studies cited above. This precocious ability makes sense biologically since learning the properties of one's environment quickly most likely provides advantages for survival and gives a scaffold to learn other aspects of communication. Yet, there is also flexibility in the system such that, even in adulthood, exposure to a new set of statistical contingencies can result in adjustment of a listener's internal model. For example, adults can learn to perceive microtonal scales after brief exposure to those sounds (Leung and Dean 2018), and likewise, adult listeners show sensitivity to a novel musical scale structure after only half an hour of passive listening to such a scale (Loui et al. 2010). Importantly, for considerations of musical pleasure, manipulation of the properties of the novel musical scales can lead not only to learning but also to changes in preference (Loui et al. 2010), showing that there is a link between the formation of an internal model and liking of patterns. Notwithstanding the point made above, that schematic expectancies tend to be stable, such models need not be completely static either, especially with exposure to a very different system than the established one, which may result in adaptation of the existing system or perhaps formation of a parallel representation.

In terms of the neural mechanisms involved in statistical learning, it has recently been shown by my colleagues at the University of Barcelona that statistical learning of language-related sounds is strongly linked to the reward system (specifically the striatum, which we will learn more about in Chapter 6) because it depends on encoding of predictions and prediction errors (Orpella et al. 2021). This mechanism is essentially the same one that applies to music, as we shall discuss in detail in Part II of this book. It seems likely then that a similar process would underlie the learning of tonal relationships. This finding is especially interesting for one of the central arguments in this book: that the anticipatory processes enabled by learning of statistical contingencies are also dependent on processing within the reward system. It also provides a direct link between pattern learning and predictive coding, which is the next concept to consider in this introduction.

Predictive Coding

Traditional descriptions of how perception occurs emphasized how successively more complex representations of stimuli are evident as one ascends the hierarchy of a neural pathway. This feedforward account was based on the fact that, early in a sensory pathway, neurons usually encode basic features (e.g., edges in primary visual areas, or frequencies of sinusoidal tones in primary auditory regions), while later structures encode more complex and abstract patterns (e.g., shapes or faces in higher-order visual regions, tone combinations or melodies in higher-order auditory areas). Although the data upon which these models were built were certainly correct, this class of bottom-up models was incomplete and could not account for many perceptual phenomena.

A more recent conceptual development is provided by the theory of predictive coding, which has become a very popular way to understand perception and brain function more generally. (For a global review of this multifaceted topic, see [Friston 2010]; for a discussion of how the model applies to auditory neuroscience see [Heilbron and Chait 2018]; and for ideas about its relevance to music, see [Koelsch et al. 2019]). In a nutshell, this theory differs from the conventional feedforward understanding by proposing that there also exist very critical feedback, descending signals that propagate from higher to lower levels, in addition to the feedforward sensory signals. I should not really say that this idea is recent because, in fact, Hermann von Helmholtz developed a similar concept based on visual illusions in a treatise originally published in 1867 (von Helmholtz 1925), where he proposed that perception depends upon a process of "inductive conclusions," or "unconscious inference," such that the brain makes some kind of inference about how to interpret the sensory data it receives. Of course, modern versions of predictive coding are far more detailed and mathematically formulated; but nothing, it seems, is really that new under the sun.

More specifically, the idea behind predictive coding is that whereas the ascending signals transmit sensory data, the descending signals transmit *predictions* about what those sensory signals will be and when they should occur. Those predictions can be based on either local features of what events are occurring at the moment or in the very recent past, or on accumulated knowledge of common patterns—most likely obtained via statistical learning over protracted time periods—which allows the formation of stable internal models. This abstract knowledge of how events usually unfold generates expectancies, which are the basis for the predictive signals. When those predictions arrive at the lower stage of the hierarchy, they are compared to the actual input; if there is a difference between them, a *prediction error* signal is generated. This error signal is very important because it indicates a discrepancy between the external world (the stimulus—or at least its representation at the input level of the nervous system) and the internal world (the model, housed at some hierarchically higher level of the nervous system), thus serving as a learning signal to update the model.

A nontechnical, if somewhat simplified, way to think about predictive coding is to consider the metaphor of a ship being piloted through a tricky passage. There is a hierarchy on board, with the sailors at the bottom, the pilot in the middle, and the captain at the top. The sailors' jobs are to monitor the instruments and keep a lookout for what's out at sea. The pilot has a model (maps, prior experience, etc.) of what is needed to navigate, and has instructed the sailors about what to expect—currents, presence of rocks, winds, and so forth (i.e., the prediction signal). Similarly, the captain has given the pilot the overall plan (i.e., a high-level model) of the ship's ultimate destination and timeline, so that the global expectations of what should happen are laid out. The sailors do not need to send information continuously up to the next level, to the pilot, because they have already been told what to expect; they only need to monitor carefully and report if there is a discrepancy. If the current shifts from what was forecast, for instance, then an error signal is sent upstairs, causing the pilot to make an adjustment (i.e., update the model) to take that into account. The pilot does not need to inform the captain of such minor modifications because they don't change the global plan regarding the ship's destination (i.e., there is no discrepancy at that higher level).

But if something more dramatic happens, let's say an unexpected iceberg or a pirate ship appears, then the error signal needs to be propagated further, all the way to the captain. The captain knows the ship's mission and can therefore make the adjustment at this more abstract level in case of a threat that requires a complete change of approach to achieve the original goal. Such examples would constitute worse-than-expected outcomes; but there could also be unexpected positive outcomes: if the sailors spot a treasure chest full of gold floating nearby, that would certainly be unexpected, and they would certainly inform the bridge about their good fortune so that appropriate action could be taken to retrieve it.

A very rough analogy could be made between the three levels just described and an auditory neural hierarchy, with lower structures (e.g., auditory brainstem/midbrain nuclei), middle structures (auditory cortex), and high-level structures (frontal cortex), as seen in Figure 1.4. The auditory cortex receives inputs from lower levels and then generates predictions about what to expect next, which are sent to lower-level structures; these latter are inhibited if the sound environment is stable or repetitive, so as not to continuously send redundant information. But when the environment changes (e.g., in a musical context, a new and unexpected chord appears), then a prediction error occurs (indexed by mismatch responses, about which we will have more to say in Chapters 2 and 3). But it is only at the highest level of frontal cortex, where long-term goals and action plans are formulated, that changes to more global patterns (e.g., melodic phrases or sections) are processed.

There is one more important nuance about predictive coding, which is that the process is very dependent on how reliable the prediction error signals are. Solid information is weighted more heavily, whereas highly variable signals provide uncertain information

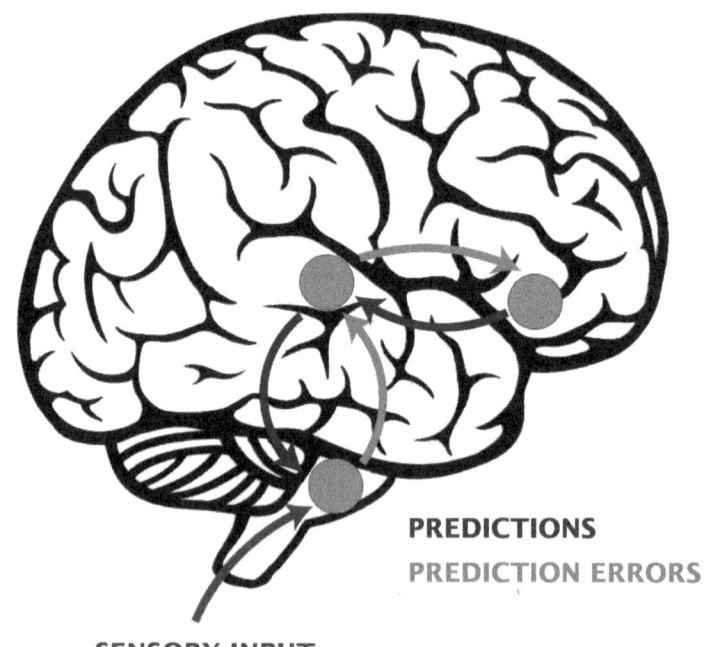

PREDICTIONS

PREDICTION ERRORS

SENSORY INPUT

FIGURE 1.4 Schematic depiction of predictive coding. Each blue dot represents a locus of processing within the auditory hierarchy, starting with the brainstem, then auditory cortex, and then inferior frontal cortex. Orange arrows indicate conceptually how afferent (bottom-up) sensory input (green arrow) propagates to higher levels of the hierarchy and signal prediction errors when an unexpected change has occurred. Purple arrows indicate efferent (top-down) outputs descending from higher to lower levels and representing prediction signals that modulate lower levels. Figure by F. Ahmed, used with permission

Conard, N. J., Malina, M., and Münzel, S. C. (2009). New flutes document the earliest musical tradition in southwestern Germany. *Nature* **460**(7256): 737–740.

Huron, D. (2001). Tone and voice: A derivation of the rules of voice-leading from perceptual principles. *Music Perception* **19**: 1–64.

and hence are less valuable to update the internal model. In our nautical example, if the sailor who reports something unusual out on the horizon is known to be habitually drunk and unreliable, the pilot may become skeptical and opt to ignore information from that source. Or the environment itself may lead to uncertainty: if it's very foggy and windy, the instruments may not yield valid results, for example. Similarly, in the brain, a noisy signal (e.g., a sound partly masked by acoustic noise) will not generate as much updating (or mismatch response) compared to a clear signal. This phenomenon is known as precision weighting, meaning that inputs are weighted according to their variability. In general, environments that have more entropy—in the sense of greater randomness or less structure—will generate less precise signals, leading to more poorly formulated models. This aspect of the theory turns out to have interesting implications for our understanding of music, because musical environments can also have greater or lesser degrees of predictable structure, leading to different kinds of responses to deviations from expectations, as we shall discuss in Chapter 8.

It is often stated in predictive coding accounts of behavior that the goal is to reduce prediction errors as much as possible. Thus, the ideal would be to anticipate every up-coming event and therefore be prepared for them, thereby enhancing an organism's fitness or survival. But this idea is a bit too simplistic because if such a strategy were the only one implemented, it could lead to getting stuck in a local minimum. In other words, just be-cause you can anticipate your environment perfectly does not mean that you have found the ideal environment to optimize your access to reward or avoid threat. For that, you may need to explore outside of your local comfort zone, where everything is predictable, and take a chance to find something better. Boredom is a strong motivator of behavior.

This conundrum is referred to as the exploration–exploitation trade-off and has a long tradition in various fields of psychology (Mehlhorn et al. 2015). Several accounts have been proposed to reconcile predictive coding theory with the drive for novelty and surprise (Schwartenbeck et al. 2013). But one that is particularly relevant for the central claims of this book is the idea that seeking information is itself a form of reducing pre-diction error, because it allows the accumulation of knowledge and the global reduction of uncertainty. This concept gives us curiosity: the drive to obtain information even if its use is not immediately apparent. In music cognition, this curiosity leads to preferences for music that is not too boringly predictable, but not too chaotic either, as we shall see in Chapter 8.

Furthermore, this curiosity factor interacts with individual differences in personality traits: some people are more naturally curious and open to experience than others. In ev-olutionary theory, not every member of a group needs to have the same level of curiosity, leading some individuals to focus more on exploitation while others focus more on explo-ration (a good strategy for survival, perhaps, since exploration entails possibility of greater reward, but also puts the explorer at greater risk—you might get eaten by a lion while you're out exploring). These differences are also manifested in music, where individual personality traits lead to different reward reactivity and are linked to preferences for more novel versus more familiar music, as we will discuss in Chapters 8 and 9.

There will be much more to say about predictive coding throughout this book; but es-pecially in Part II, where I will draw a link between predictions and whether or not they are fulfilled with their corresponding affective responses. The entire mechanism that allows pre-dictive coding in the auditory perceptual domain resides largely within the corticocortical loops described in Part I of the book. But there is a parallel type of predictive coding going on in the reward system (discussed in detail in Chapter 6), where the value of a stimulus is determined, and a *reward prediction error* is generated within subcortical reward structures, when the stimulus that is received is either better or worse than the prediction (Schultz 2017). In this form of predictive coding, therefore, the error signal is valenced—it has a positive or negative sign—unlike the sensory prediction errors. One of the main ideas that will be developed in Part II, especially in Chapter 8, is that sensory prediction signals ori-ginating in the auditory cortical networks propagate to the reward system, where they gen-erate valenced reward prediction signals. And it is this latter reward-related activity that is responsible for our strong pleasure responses to music.

The Plan of this Book

The plan of the book follows from the principal premise that to understand musical pleasure, we need to understand both the perceptual/cognitive system as well as the reward system. In keeping with that agenda, Part I: Perception deals with the first of these two systems, and Part II: Pleasure deals with the second one. The next three chapters after the present introduction will explore in detail the anatomical structure and functional properties of the auditory pathways, especially as they relate to music-relevant processes. A central tenet of neuroscience is that structure and function are intimately linked. So, understanding how these systems work requires some knowledge of their connectivity patterns and functional characteristics. Chapter 2 will deal with the inputs and outputs between the subcortical auditory structures and auditory cortex and with the basic processes carried out within the cortex, such as segregation of sound sources, pitch representation, and so on. Chapter 3 will introduce the auditory ventral processing pathway and its properties, including in particular working memory retention, and will discuss amusia as a disorder linked to this pathway. Chapter 4 will focus on the dorsal processing pathways, their role in working memory manipulation, and especially the important link between auditory and motor systems. Chapter 5 finishes Part I with a discussion of the specialization of the two hemispheres for musically relevant processing.

Part II of the book starts with Chapter 6, which details the structural and functional properties of the reward system outside of the context of music, emphasizing mechanisms of prediction and prediction error. Chapter 7 presents the empirical evidence that music engages the reward system and discusses the role of dopamine circuits. Chapter 8 integrates much of the prior information to develop a model for why music generates pleasure, based on concepts of predictions and expectancies, and the connections between auditory pathways and reward structures. Chapter 9 then follows up on these concepts to explore the role of the reward system more broadly in music-induced emotion and emotional regulation. Finally, a short coda brings us to a close.

PART I

Perception

2

Early Sound Processing
The Auditory Cortex, Its Inputs, and Functions

Every sound you hear—the startling trumpet blasts at the opening of Mahler's *Symphony No. 5*; your lover's voice picked out from across the room; the insistent call of the whippoor-will on a summer night—they are all carried by air molecules impinging on your eardrum. In other words, all the input to the auditory brain is unidimensional, consisting of minis-cule increases or decreases of air pressure which push or pull that tiny membrane in or out. Our entire universe of sound must therefore pass through the eye of this needle, posing a significant computational problem for the auditory nervous system. Worse yet, sounds made simultaneously by different sources in the environment (a commonplace occurrence in general, but in music in particular) are all summed up together by the time they reach your ear: they are entangled in the air vibrations, so that the auditory brain must disen-tangle them somehow.

The situation is nicely captured by a metaphor developed by my late McGill colleague Al Bregman: imagine you are sitting on a beach, but that you are only looking at the move-ment of the water in two channels dug into the sand, and not looking out at the sea itself. The sea, in this metaphor, is analogous to the air, while the movement in the two channels corresponds to the movement of the two eardrums, as seen in Figure 2.1.

From these waves you must figure out if there are boats, people, or fish in the water; their number, location, and movement; and also how choppy the water is. Not an easy task! Yet our auditory system solves the analogous problem every moment of every day. If you're in a park, say, you can easily tell that there are two people talking to your left, wind up in the trees, and someone playing a saxophone in the distance. To figure all this out, and much more, the auditory system makes use of a complex web of structures and processes, which we will discuss very selectively in this chapter, starting with a brief overview of the relevant brainstem nuclei and then focusing on the relevant structural and functional features of the auditory cortex itself.

From Perception to Pleasure. Robert Zatorre, Oxford University Press. © Oxford University Press 2024.
DOI: 10.1093/oso/9780197558287.003.0002

FIGURE 2.1 Bregman's scene analysis analogy. The figure depicts an analogy developed by Albert Bregman (Bregman 1994) to illustrate the difficulty that the auditory system faces in order to segregate and identify events in the environment. The analogy hinges on the idea that a person is on the beach but cannot look out at the ocean. She can look only at how waves cause movement of two pieces of cloth that cover two narrow channels dug into the sand. From observation of these movements, she must decide how many objects there are in the water and their identity, where they are located, if they are moving, if so in what direction, and if it's windy or calm. It appears to be an impossible problem, but the situation is strictly analogous to what the auditory system faces: the waves generated by the moving objects on the water are akin to sound waves generated in the air; the two channels correspond to the ear canals, and the cloth represents the movement of the ear drum. That movement is the sole input available to the auditory system, from which all other details must be derived. CC-BY, anonymous artist.

Pathways to Auditory Cortex: Inputs and Subcortical Nuclei

The front end to the auditory nervous system is the cochlea, the delicate coiled structure ("cochlea" comes from the Latin for shell) deep inside the temporal bone of the skull, which receives the vibrations passed on to it from the outside world via the eardrum and the tiny bones of the middle ear. The remarkable and complex biomechanics of the cochlea (Manley et al. 2017) are well outside our scope; but suffice it to say that the array of fine hair cells mounted along the basilar membrane (a thin layer of tissue lining the inside of the cochlea) are responsible for the transformation of vibratory motion into neural impulses. These signals then enter the central nervous system via the eighth cranial nerve that connects each ear to the corresponding cochlear nucleus, the first brainstem way station on the long journey to the cortex.

An important principle of neural coding emerges already at this very initial stage of processing: frequency representation. Because of various features of the cochlea's structure and organization, sounds coming into the ear are sent along to the brainstem, already partly separated according to their frequency, akin to how a prism separates white light into different colors. Thus, neurons in the cochlear nucleus respond to different frequencies of sound in an orderly manner according to their topographic position in an array of neurons, a principle known as tonotopy. This tonotopic organization is preserved at each stage of the ascending auditory nervous system all the way to the cortex, and we will have more to say about it below.

One of the most important functions of the subcortical structures is to combine and compare sounds arriving from the two ears, which occurs in the first location, where inputs from the left and right cochlear nuclei interact, in a structure called the superior olivary complex, which allows computation of spatial information. Although this topic will not be covered here, there is a wealth of neurophysiological literature, as well as human brain imaging studies, on the binaural computations that are essential for spatial hearing, which not only depends on subcortical but also on cortical contributions (van der Heijden et al. 2019), as will also be touched upon in Chapter 4.

Impulses from the cochlear nucleus are routed to the different nuclei of the brainstem via a network of ascending, descending, and crossing fibers, which richly interconnect the system (Irvine 2012) going through the olivary complex, inferior colliculus (in the midbrain), and after one final synapse in the medial geniculate body of the thalamus, reaching the auditory cortex (Figure 2.2A). The fibers from the level of the colliculus and upward can be classified into two broad categories, described fifty years ago by anatomist Ann Graybiel (Graybiel 1973) as belonging to a primary or lemniscal system, which contains the most sharply tonotopically tuned neurons, with short response latencies that project to core cortical regions (Malmierca 2015), and an auxiliary or non-lemniscal system parallel to the first but containing more broadly tuned neurons, innervating different nuclei within each structure (Figure 2.2B).

The lemniscal system's job is basically to provide fast and accurate information about the stimuli in the environment. The non-lemniscal system, on the other hand, has a more integrative function, as it projects more diffusely to several brain structures beyond the auditory cortex, including the amygdala, orbitofrontal cortex, and insula. It also innervates structures in the reticular activating system in the brainstem (Reese et al. 1995), which is likely why a thunderclap can wake you from sleep or why a cymbal burst can rouse you in the middle of a symphony.

The functional contributions of each of the structures in the auditory pathway are only partly understood, and the interactions between them add complexity to the story. Of greatest relevance for us, is that far from passively providing inputs to higher levels, there is a great deal of processing that goes on within subcortical nuclei and between the different levels; indeed the entire subcortical-cortical system should be seen as an integrated whole, contributing to many processes which are essential for higher-order aspects of perception (Coffey et al. 2019b). Among the many reasons for this conclusion is that the number of fibers descending from cortex to midbrain and brainstem nuclei is massive, considerably greater even than the number of ascending fibers (Winer 2005), indicating that the entire system works as a network with two-way traffic. The descending connections from cortex to

(A)

(B)

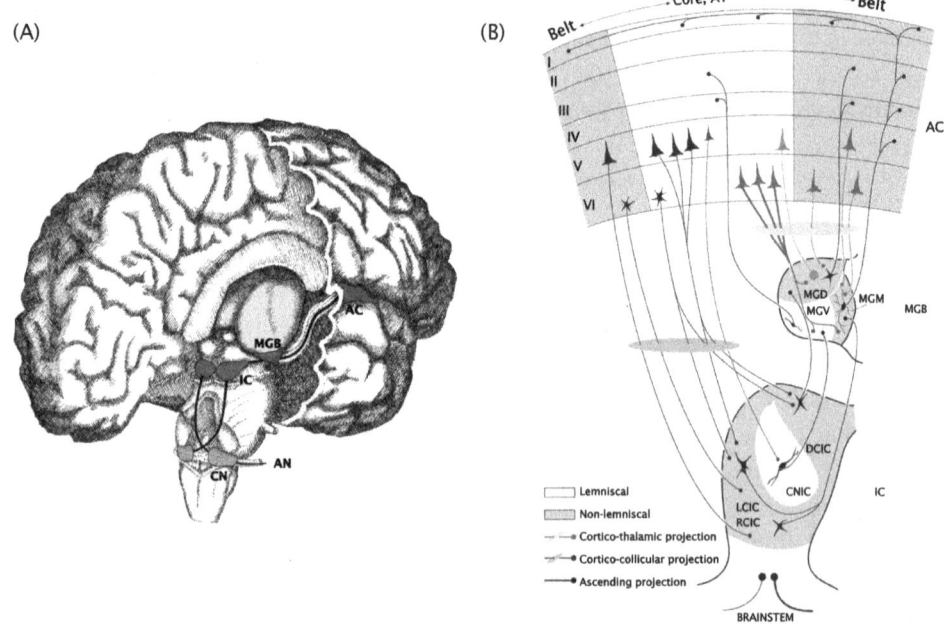

FIGURE 2.2 Auditory subcortical pathways. A: Simplified diagram of major subcortical auditory nuclei and their anatomical position within the brain. Input from the cochlea enters the central nervous system via the auditory nerve (AN) and synapses at the cochlear nucleus (CN). After connecting through the superior olivary complex (not shown), fibers ascend contralaterally to the inferior colliculus (IC) of the midbrain. From there, connections proceed to the medial geniculate body (MGB) within the thalamus and thence to the auditory cortex (AC). Prominent efferent outputs also descend from higher to lower levels of the auditory hierarchy. Drawing by E.B.J. Coffey used with permission.

B: Connectivity diagram schematically illustrating lemniscal (shown in white) and nonlemniscal (shown in gray) subdivisions of the IC, MGB, and cortex, and their ascending and descending projections. Black lines indicate ascending projections from the brainstem to the IC, MGB, and cortex; orange lines indicate major cortico-thalamic connections and purple lines major cortico-collicular projections. Descending projections largely correspond to the nonlemniscal system, which provides modulatory top-down signals to hierarchically lower regions.

From (Malmierca et al. 2015). Reproduced under the terms of the Creative Commons Attribution License (CC BY).

thalamus and inferior colliculus belong largely to the non-lemniscal pathway (Saldaña et al. 1996), with implications for the functional roles of these pathways in mediating top-down control signals (Figure 2.2B).

Sensitivity to Repetition: Sensory-Specific Adaptation and Novelty Detection

Neural responses often show a decline over time to repeated presentation of the same stimulus. This phenomenon of fatigue or adaptation can be found at almost all levels of the nervous system and can be accounted for via physiological mechanisms. But the adaptation can also be specific to a given feature of the stimulus rather than being general. For example, in the auditory system, habituation will occur if the identical tone is presented repeatedly; but if its frequency is altered, keeping other aspects of the stimulus the same, the neuronal response will recover (Condon and Weinberger 1991). A response can even be elicited by the absence of an expected sound. An everyday example of this effect is provided by the refrigerator in your kitchen: you may not notice that it's running because it generates a constant, low-level hum; but when it stops, you suddenly become aware of it because the background sound has changed, generating a brain signal that elicits attention. This phenomenon is known as sensory-specific adaptation, and the reason it becomes important is that it provides a mechanism for selectively responding to novelty in the environment.

Ethologically speaking, there is no point in responding to every stimulus that may impinge on us because there are too many of them, and they do not all require the same level of processing; an organism that cannot distinguish irrelevant background events from novel, potentially important ones, may not survive for long. Hence, to the extent that sensory-specific adaptation provides a means for detecting a change in ongoing inputs, this ability becomes very important indeed. In the context of music, change, surprise, and expectancies play a major role, as I shall argue throughout this book. Therefore, it behooves us to understand some of the basic physiology that contributes to these phenomena from the earliest levels of auditory processing.

Sensory-specific adaptation was first described as a phenomenon particular to auditory cortex in animal neurophysiology studies (Ulanovsky et al. 2003), where it was proposed to be the basis for the human mismatch response (to be dealt with in detail in Chapter 3). But stimulus-specific adaptation turns out to be present in subcortical nuclei too. Research in several animal species from Manuel Malmierca's group at the University of Salamanca has documented the existence of sensory-specific adaptation, both in the inferior colliculus (Malmierca et al. 2009), and in the medial geniculate of the thalamus (Anderson et al. 2009)—the two subcortical stations before reaching the cortex. Thus, some kind of mechanism for deviance detection seems to operate from very early stages in the ascending auditory pathway, underlining the fact that the so-called relay stations do not merely transmit information from earlier stations up to later ones but rather there is active processing going on in each structure.

What is the nature of these novelty detection processes at each level? Are they all redundantly doing the same thing? Some clever experiments have recently clarified how these processes differ from lower to higher neural centers (Parras et al. 2017); but for a better understanding, it is important to consider two different aspects of how novelty can arise. The simplest case is when a repeated sound changes to a new sound (or stops altogether). In such cases, novelty detection can proceed merely based on suppression of activity by the repeated stimulus and release from this inhibition upon arrival of the new stimulus. However, a response to a new sound can also occur at a higher level of processing because it is not predicted—not just because it differs from what was heard before.

As an example, consider a simple ascending or descending musical scale in which each tone is presented only once: every subsequent event is perfectly predictable once a few tones have been presented and the pattern established, even though each tone is not identical to its predecessor; but if the scale pattern is broken by repeating a tone, then that would constitute novelty (even though it is a repeat) and engender a response. When these kinds of stimulus sequences are compared to each other, a fascinating pattern emerges: at lower levels, such as the inferior colliculus, the simpler kind of novelty mechanism predominates based on adaptation, or repetition suppression; but as one ascends to the cortex, those kinds of responses are less prominent, and instead, the more complex form of prediction response is found such that there is sensitivity to changes in patterns rather than mere repetition (Parras et al. 2017). Note that such pattern sensitivity requires a certain degree of integration over time, a topic we will cover in Chapter 3.

A recent human study with high-field, functional MRI (fMRI) has added to the knowledge drawn from these animal studies. Alejandro Tabas and colleagues from Katerina von Kriegstein's lab in Germany, showed that if the listener knows in advance which sound out of a sequence is going to change, then responses to the change are inhibited (Tabas et al. 2020). This finding shows that a simple adaptation idea cannot explain everything about the effect, since the degree of suppression is greater as the listener is more certain about when the change will occur. This effect was detected in the inferior colliculus and medial geniculate, which may reflect the top-down influence of cortical modulation onto earlier structures.

It is also important to consider the different anatomical connections between subcortical and cortical auditory structures because the simpler kinds of suppression responses are generated primarily in the lemniscal pathway, whereas more abstract types of prediction-based responses are mostly found to emerge from the non-lemniscal pathway. Recalling that the non-lemniscal pathway accounts for most of the descending influences from cortex to subcortex leads to the conclusion that the more abstract prediction-type responses are associated with higher regions in the anatomical hierarchy. These findings taken together lead to an interesting model of subcortical-cortical hierarchical processing of novelty, which is largely in keeping with models of predictive coding (Carbajal and Malmierca 2018).

This theoretical model proposes that there is a cascade of ascending and descending influences such that the ascending lemniscal pathway, with its high-fidelity and rapid encoding of sensory features, is responsible for sending stimulus information up to higher levels and can therefore signal basic acoustical changes when they occur. This response to change could thus be considered as a prediction error to be used by higher-order structures that model aspects of the stimulus sequence. Conversely, the descending influences could be thought of as control or predictive signals which influence responses of earlier regions and inhibit their activity when a stimulus is predictable, based on the more abstract information represented at higher levels. Thus, the entire integrated cortical-subcortical system is responsible for these predictive processes. Carbajal and Malmierca (2018) put it well: ". . . this system of backward and forward connections between stations looks like the perfect network to host the top-down flow of predictions and the bottom-up transmission of prediction errors." Importantly in this hierarchical system, the cortical representations are the most abstract and, hence, of particular interest to us in understanding musically relevant processes with all their cognitive complexity. Therefore, we turn our attention now to the auditory cortex and its properties.

Auditory Cortex

A Bit of History

By the late 19th century, David Ferrier and John Hughlings-Jackson in England had already identified the approximate location of auditory cortex in dogs, monkeys, and humans based on lesions and even basic brain stimulation methods (Heffner 1987, Hogan and Kaiboriboon 2004), as seen in Figure 2.3.

A few decades later, careful anatomical work on cellular architecture, notably by Brodmann in Germany and von Economo in Austria (for a recent review, see Baumann et al. 2013), indicated that human primary auditory cortex was likely located on the superior plane of the temporal lobe, a region largely hidden within the Sylvian fissure in or close to a diagonally oriented finger of tissue known as Heschl's gyrus. Additional auditory areas were also described along the superior temporal plane and portions on the lateral surface, shown in Figure 2.4, both behind and in front of Heschl's gyrus, which constitute the origins of the dorsal and ventral streams to be discussed in Chapters 3 and 4.

These gross anatomical descriptions provided the fundaments of what would become auditory neuroscience; but the nature of the processes being carried out in these regions and their functional organization remained mostly unknown. Further animal studies in the middle and later parts of the 20th century provided some hints that the auditory cortex was essential for certain aspects of behavior. In particular, one of the important conclusions to emerge was that even complete bilateral lesions of the auditory cortex did not necessarily lead to an inability to respond to sounds. Indeed, in such cases, sounds could often still be distinguished from one another if they contained different acoustical elements. For

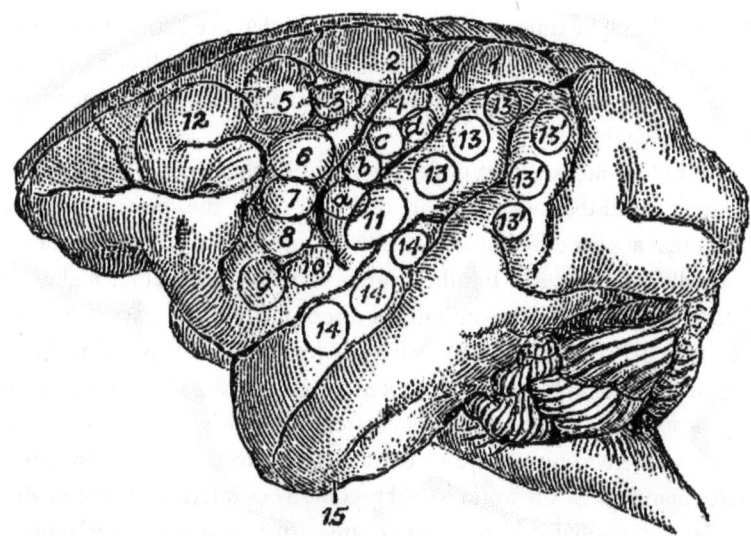

FIGURE 2.3 Early identification of the auditory cortex. Drawing by David Ferrier (1875) indicating regions stimulated in the brain of a monkey. The location of auditory cortex, number 14, along the superior temporal gyrus, was inferred by the animal's actions of pricking up its ears when the electrical stimulus was applied to this area of cortex. (Ferrier 1875); Public Domain.

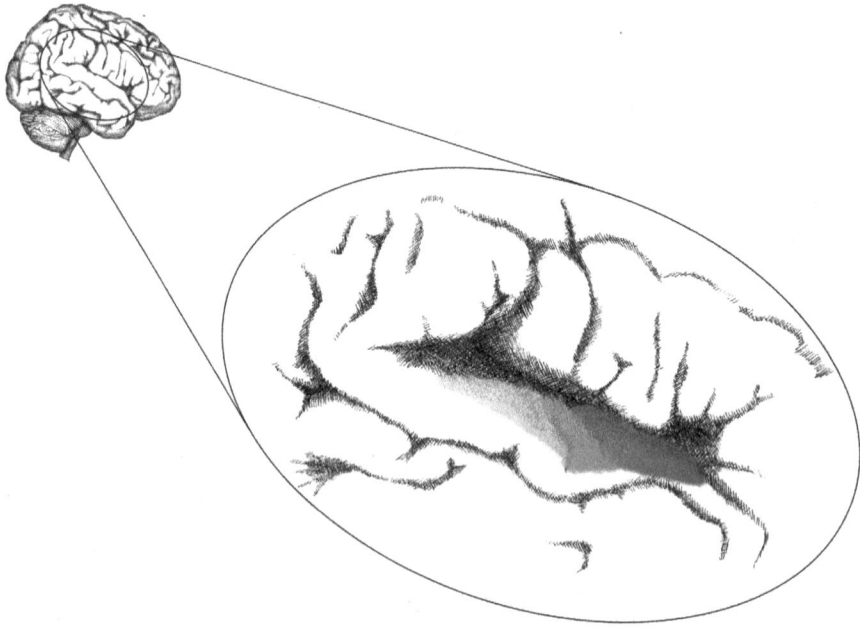

FIGURE 2.4 Heschl's gyrus and superior temporal auditory regions. Drawing illustrates superior temporal plane, normally hidden from view within the depths of the Sylvian Fissure (inset). Heschl's gyrus (orange) contains primary and adjacent regions. Other auditory cortical fields are found anteriorly (red) and posteriorly (yellow), forming the origins of the ventral and dorsal auditory pathways, respectively. Drawing by E.B.J. Coffey used with permission.

instance, if tones of three frequencies A, B, and C are presented to an animal without auditory cortex, it can still be trained to respond to the difference between any given pair of sounds; but if the task involves responding only to rising tones—AB or BC, but not BA or CB—then the discrimination fails, because there is no generalization to the more abstract concept of rising or falling.

The British physiologist Ian Whitfield put it well when he explained that one does not need a cortex to tell that two events are different; rather the cortex is necessary to tell that two events that appear different are actually the same in some way, that they represent the same event (Whitfield 1985). In other words, cortex is essential for higher-order pattern recognition and object formation; simpler discriminative processes can be carried out within subcortical structures, as we saw in the previous section. This idea is closely related to the classic invariance problem in perception, which refers to the fact that the same object may look or sound different in different environments (brighter or darker illumination, quiet or reverberant environment, etc.), yet the nervous system must represent these objects in a stable manner, else the world would become a bewildering confusion. At the time, however, the question remained—how exactly are cortical systems able to represent more abstract patterns? For that, a better understanding of the system's organization was needed.

The animal lesion research was complemented, around the same time, by a remarkable series of observations on the function of human auditory regions carried out by Wilder

Penfield. One of Penfield's principal clinical concerns was the identification and removal of epileptogenic tissue from patients so that their epilepsy would be better controlled. However, he took a conservative approach to surgical resection and did not want to produce deficits in his patients. (Indeed, this is why he hired Brenda Milner, my postdoc mentor, in the 1960s: in order to better understand the nature of the mental processes in the tissue he was planning to operate upon.) To avoid such problems, Penfield developed a direct method of exploring the function of different cortical regions: electrical stimulation of the exposed cortical surface during the surgical procedure itself, while keeping the patients awake and reporting what they experienced (Penfield 1958). This method can literally be considered the first true human brain mapping since it made possible the identification of specific regions associated with certain functions. The most well-known of these maps has become known as the homunculus, which is a representation of the human body in the motor and sensory domains, found on the precentral and postcentral gyrus, respectively. But even more important in our context is Penfield's work on "experiential phenomena"— reports from patients that they perceived or remembered something upon being stimulated in a particular region. The experiences could range from recollections of past events to feelings, faces or voices, as well as music.

These musical hallucinations, although not so commonly reported, were well-documented by Penfield and his collaborators (Penfield and Perot 1963). In some cases, the patient experienced music quite vividly: one patient, for example, thought that somebody had brought a radio into the operating theater! Others were able to name or sing the tune they heard, and in one instance, the patient could not remember the name but sang it back well enough so that one of the nurses was able to identify it as a popular tune at the time (*Rolling Along Together*). These phenomena have been repeatedly observed in other studies as well (for a thorough recent review, see Curot et al. 2017). Most critical to our understanding of auditory cortex, Penfield mapped the locations that elicited these effects, as shown in Figure 2.5 and found them to be strictly confined to the superior temporal gyrus.

Importantly, he also distinguished the sensations elicited by stimulation of Heschl's gyrus, which were almost always elementary sounds, such as buzzing or hissing, from the much more complex ones (including music) elicited outside of that region, thus implying that the primary cortex represented basic auditory information and not complex learned sequences. It is very likely that these effects were not solely related to activity in the stimulated zone, but to interactions between that region and limbic or other areas involved in memory retrieval (Gloor et al. 1982). Nonetheless, these descriptions give us tantalizing clues about how musical information may be represented in the brain and some idea of the neural substrates involved.

Physiology and Connectivity

Physiological and anatomical research on macaque monkeys has provided a wealth of evidence that is relevant to our understanding of human auditory cortex organization, despite the cross-species differences that were discussed in Chapter 1. From this literature we know that the auditory cortex can be subdivided into three broad categories, referred to as core, belt, and parabelt areas. These regions are organized in a hierarchical and parallel manner,

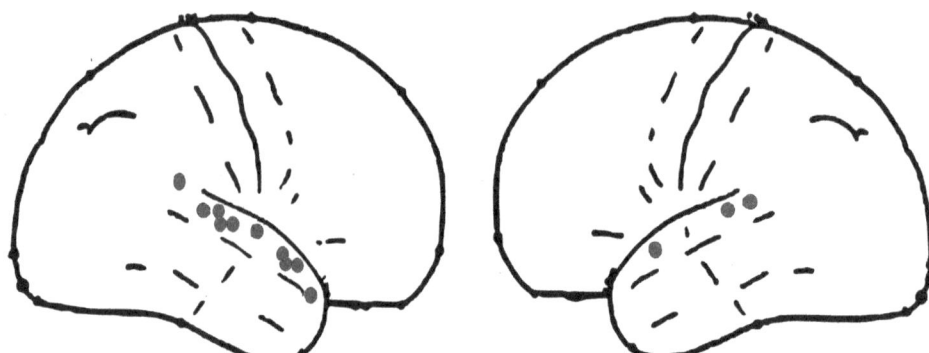

FIGURE 2.5 Brain stimulation resulting in percept of music. Summary brain map of experiential responses from epilepsy patients upon electrical stimulation. Lateral view of right hemisphere (left side) and left hemisphere (right side). The red dots indicate sites along the superior temporal gyrus which resulted in the patient reporting that they perceived music when the stimulation was applied. No sites outside this region evoked such responses. Reproduced with permission from (Penfield and Perot 1963).

based on their different connectivity patterns with the thalamus and with each other and also according to their anatomical characteristics (Kaas and Hackett 2000). Thus, the central core—which itself contains at least three separate subregions—represents the first level of cortical processing and is surrounded concentrically by belt and parabelt areas.

Each subsequent level of the hierarchy is interconnected with neighboring levels, from core to belt to parabelt but also receives some direct projections from thalamic nuclei. However, the principal input from the auditory thalamus, the medial geniculate body, goes to the core areas via lemniscal fibers, while more distal belt/parabelt areas receive more sparse thalamic connections, from the largely separate non-lemniscal thalamic pathway as discussed above (Figure 2.2B). These belt and parabelt regions also receive multisensory inputs, which may contribute to their more complex, integrative functions (Scott et al. 2017). Tonotopic organization is most clear in the early regions that receive frequency-organized input from thalamus, whereas non-core auditory fields have weaker or absent tonotopy and instead are thought to integrate information across frequencies, as well as across time windows (Recanzone et al. 2000).

The core auditory regions in the human brain can now be identified by both functional (Schönwiesner et al. 2014) and structural (Dick et al. 2012) neuroimaging methods and are typically located near the medial or central portion of Heschl's gyrus, as shown in Figure 2.6B, albeit with considerable variation across individuals. Based on homologies with the monkey, one would expect that each of the subfields within human core areas would show tonotopic organization. Indeed this does seem to be the case: for example, Elia Formisano and his group in Maastricht were among the first to use high-resolution functional imaging to show that there was an orderly map of preferred frequencies within several subfields of auditory cortex (Formisano et al. 2003), and similar results have been reported by many others as well. The precise orientation of the tonotopic fields in relation to anatomical features and how this pattern corresponds to the monkey brain is a matter of continued controversy (Baumann et al. 2013). However, it is clear that neural populations in these regions respond relatively selectively to specific frequencies.

(A)   (B)

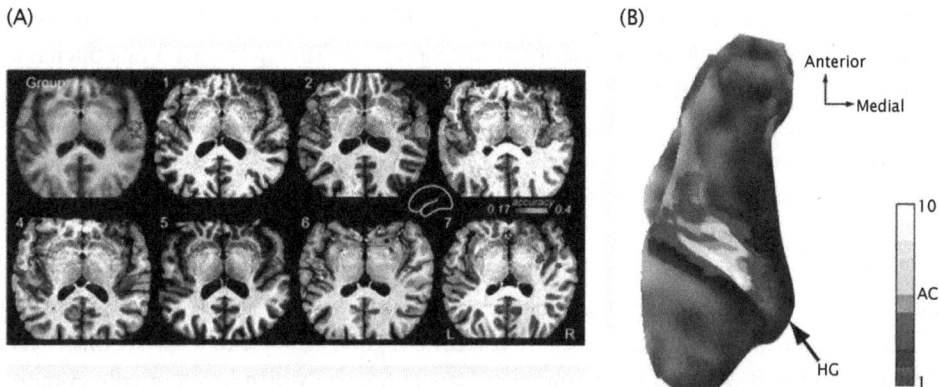

FIGURE 2.6 Location of human primary auditory area based on functional characteristics.
A: Functional MRI response in group average and each of seven individuals showing regions that most accurately classify different tonal frequencies (yellow). B: Probability map showing estimated primary auditory cortical region in the medial portion of Heschl's gyrus (HG). Color code indicates amount of overlap across individuals. From (Schönwiesner et al. 2014). Reproduced under the terms of the Creative Commons CC BY license.

More importantly for our purposes, the core regions contain neurons that are most highly sensitive to fine frequency differences—the type of differences that might be the essential building blocks for music—as consistently demonstrated across different studies using either artificial or natural sounds, both with functional imaging (Moerel et al. 2012, Schönwiesner et al. 2014) as well as with direct electrical recordings from human cortex (Bitterman et al. 2008). Functional connectivity between voxels with similar frequency tuning is highest in the core areas compared to noncore regions, especially in the right hemisphere, providing a possible mechanism for the very sharp frequency tuning in those areas (Cha et al. 2016). These findings are supported by physiological data in rats that show that recurrent intracortical connections are associated with sharper frequency tuning (Liu et al. 2007).

Lesion studies in humans show that bilateral destruction of primary auditory cortex (Tramo et al. 2002), or right unilateral encroachment onto this area (Johnsrude et al. 2000), leads to elevated pitch discrimination thresholds, although the basic ability to distinguish larger frequency differences is not abolished, in line with the animal lesion studies reviewed earlier. Thus, all these sources of evidence converge toward the conclusion that fine-grained information about stimulus frequency that might be very important for musical pitch processing is encoded within early cortical regions. We will have more to say about hemispheric differences in fine frequency processing in Chapter 5.

The Pitch-Sensitive Region

When we study physiological responses, whether measured with an electrode or with a brain imaging device, what we really care about is how those responses explain perceptual and cognitive capacities. Cognitive neuroscience is always trying to bridge the gap between physiology and cognition. In the case of frequency selectivity of neural populations in the auditory cortex, one thing we want to understand is how that property relates to the perception of pitch because of its importance to music perception. The tricky part is that pitch is a psychological construct: it does not bear a one-to-one relationship to frequency, which is a physical feature of periodic sounds. In fact, pitch may be considered a kind of prototype of more complex percepts that will be explored in the next chapters. That's because the same pitch may be perceived despite different acoustical inputs; in other words, there is a kind of perceptual constancy to pitch. In a musical context, this effect is fairly obvious since musical instruments that generate very different acoustical features (varied number and amplitude of harmonics, wide range of amplitude and frequency modulations, etc.) nonetheless readily produce the same pitch percept in listeners. A good example to illustrate this effect might be Ravel's *Bolero* (his most famous piece and the one which he reportedly eventually regretted writing as it overshadowed much of his other work); in that piece the same melody is obsessively repeated by various instruments or combinations of instruments for the entire duration of the piece; yet, despite the constantly changing timbres, the melody remains perfectly recognizable throughout.

To better understand this phenomenon, we need to consider that objects that vibrate in a periodic manner most commonly produce several different frequencies simultaneously. For example, if you pluck a violin string, it will vibrate in its full length, but also in halves, thirds, fourths, and so on. As Pythagoras already knew in antiquity, vibrations of different length objects generate different frequencies; in the case of harmonics, these frequencies are integer multiples of the fundamental. Thus, for each successively shorter vibrating segment, there would be successively higher frequencies: so, for a fundamental of, say 440 Hz, the additional vibrations would be double the fundamental (880 Hz), three times the fundamental (1760 Hz), and so forth. Musicians are well acquainted with these harmonic overtones, which arise from the physics of vibrating objects and were well described by Hermann von Helmholtz in the 1860s (Warren and Warren 1968).

For our purpose, the most relevant phenomenon in all this is that the perceived pitch generally corresponds to the fundamental frequency (i.e., the frequency of vibration of the entire length of the vibrating object) despite the presence or absence of different combinations of harmonics (which, nonetheless, contribute to timbral aspects of sound). What's more, pitch perception remains pretty stable even if the fundamental frequency itself is physically absent—if, for instance, the lowest frequency is masked by noise or eliminated by filtering, we still perceive it as a "phantom" frequency. A lot of psychoacoustical research has gone into this missing fundamental phenomenon (for review see Plack et al. 2006), but we will skip the details here; what's of interest to us is that it represents an excellent example of the brain's ability to "fill in" missing information or produce an invariant percept in the face of variable input from the environment (McDermott and Oxenham 2008).

So, what is the mechanism by which this constant percept of pitch arises? There are several strands of evidence supporting the existence of a pitch-sensitive region in the auditory cortex which integrates frequency information in an organized way and generates consistent responses corresponding to the fundamental frequency, despite variation in the harmonics present in the stimulus. This phenomenon was discovered by Daniel Bendor and Xiaoqing Wang at Johns Hopkins University, who recorded neural responses from marmoset monkeys to tones containing different combinations of harmonics (Bendor and Wang 2005). As shown in Figure 2.7, they specifically identified individual neurons in a lateral portion of the core auditory cortex that responded to a pure tone of a given frequency and also responded almost identically to combinations of harmonics corresponding to the same fundamental. For example, the neurons would respond similarly to a mixture of 300, 400, and 500 Hz, or to a mixture of 400, 500, and 600 Hz, and so forth because they all correspond to a fundamental of 100; but there would be no response to other combinations that were not integer multiples of 100. Importantly, these frequencies

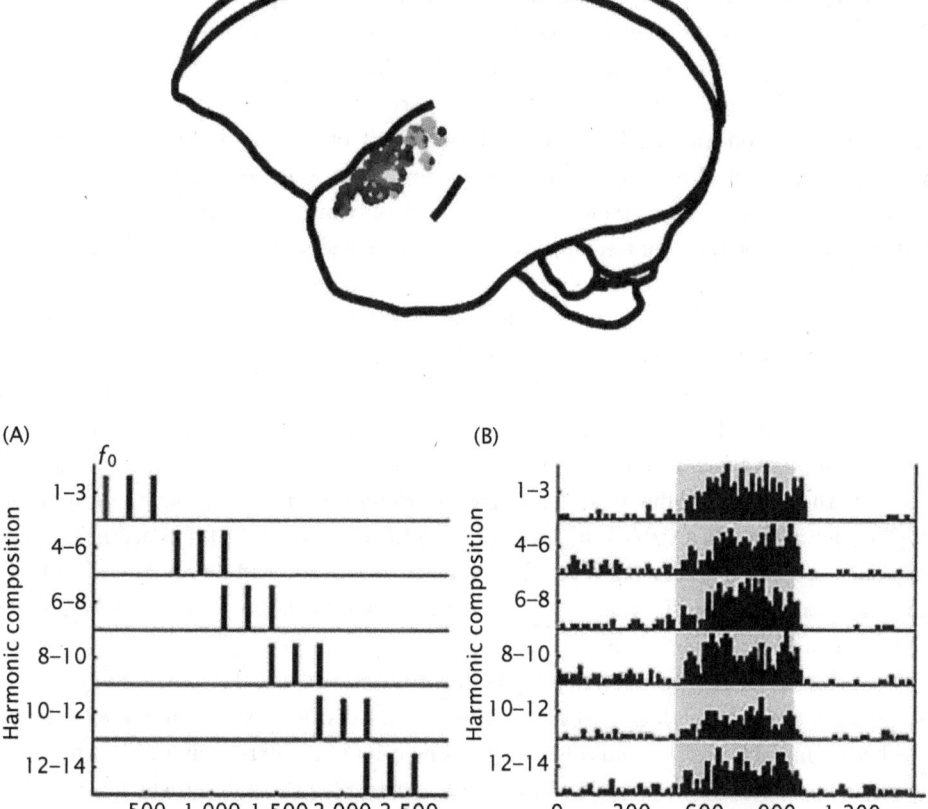

FIGURE 2.7 Pitch-sensitive neurons in the marmoset brain. Top: Characteristic frequency map of neurons in auditory cortex. Pitch-sensitive neurons were found clustered in the low-frequency portion of this region (yellow color) at the lateral end of the primary auditory cortex. Bottom: a: schematic spectrum of stimuli used for stimulation, all of which consisted of three consecutive harmonics of the same fundamental frequency (f_o), ranging from low harmonics (1–3) to high ones (12–14). b: neuronal spike count during the time each stimulus was sounded (gray shading), indicating that this neuron responded almost identically to all stimuli that shared the same underlying fundamental frequency despite containing different harmonic frequencies. Adapted by permission from Springer Nature: Springer, *Nature* (Bendor and Wang 2005), copyright 2005.

were outside of the neuron's so-called receptive fields. In other words, presentation of any of those higher tones alone would not evoke any response; it's the combination of them and their frequency relationship which make the neuron respond. In a subsequent study, these authors extended the findings by showing that activity in these neurons was related to how marmosets perceived the stimuli, thus demonstrating a direct brain–behavior link (Bendor et al. 2012).

All of this is remarkable, but, given the issues with cross-species comparisons, one needs to consider if a similar mechanism might exist in humans or if this specialized pitch region of the brain is particular to marmosets, who are a highly vocal species and likely use this ability as part of their communication system. In fact, a good deal of converging evidence now indicates that homologous pitch-sensitive areas exist across species (Bendor and Wang 2006). One piece of evidence for a human counterpart can be seen in Figure 2.8, which comes from one of the first studies I carried out ages ago as a postdoc (Zatorre 1988). I tested patients with surgical excisions of the anterior temporal lobe that either encroached onto the lateral portion of Heschl's gyrus or stopped short of this region. The task was simple: on each trial, the person heard a pair of tones and indicated which one was higher in pitch; the principal manipulation was that some tones contained both the fundamental and upper harmonics, while others were synthesized without a fundamental; but in all cases, the range of frequencies used was matched so that the response could not be based on the brightness or spectral energy. Nobody had much difficulty performing the task with the fundamental present, which is an important control, as it indicates that there was no difficulty with understanding task instructions nor with the basic concept of pitch. However, the patients with excisions encroaching onto the lateral part of right Heschl's gyrus were the most impaired when the fundamental frequency was absent, whereas patients with more anterior lesions of the temporal cortex were not different from matched controls, thus indicating that this lateral aspect of Heschl's gyrus specifically was important for the ability to form a stable pitch percept from the harmonics. At the time, the concept of a pitch-sensitive region did not yet exist. But retrospectively, the findings fit well with the idea that somewhere in the vicinity of the lateral portion of Heschl's gyrus may lie the site of these neural processes in humans.

A technical issue in studying pitch is that it is difficult to know whether the brain is responding to the individual harmonics or only to the distribution of sound energy along the basilar membrane. To avoid this problem, one can use a different approach, whereby a segment of random noise is repeatedly added to itself with a slight delay, which produces a pitch percept despite stimulating large portions of the basilar membrane rather than only one spot (Bilsen and Ritsma 1969). It is of historical interest to know that this phenomenon was first described by the 17th century Dutch physicist Christiaan Huygens, shown in Figure 2.9, who upon visiting the fountains and gardens of the Château de Chantilly outside of Paris, noticed that if he stood in a particular spot he could perceive an odd pitched sound (Bilsen 2006). He was clever enough to notice that this only happened when he stood between the noise of the water in the fountain and a staircase leading upward, and he correctly inferred that each of the steps reflected the noise back to his ear with successive delays because they were at slightly different distances from him.

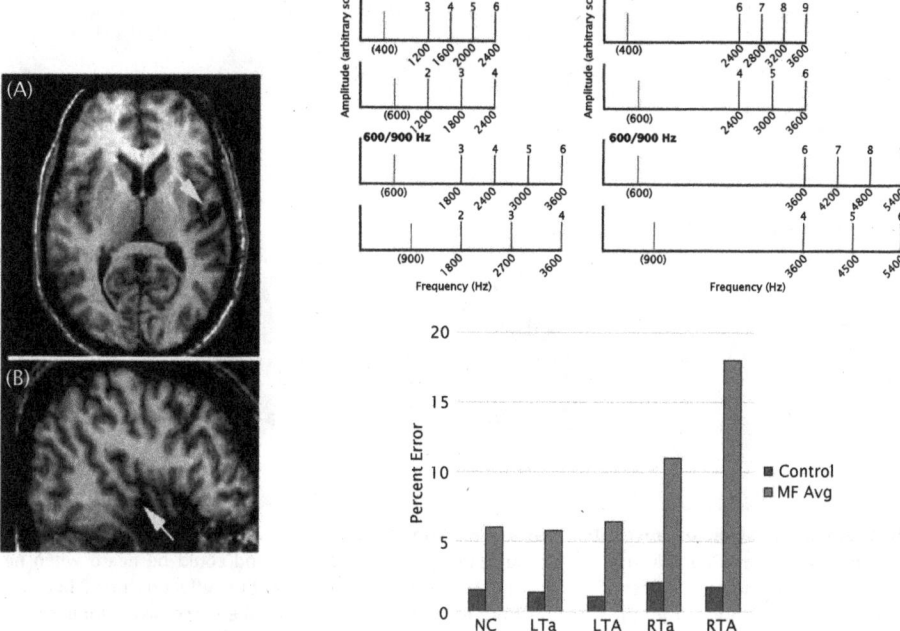

FIGURE 2.8 Deficit in pitch perception after auditory cortex excision. Left: Horizontal (A) and sagittal (B) slices of MRI scans showing incursion of lesion onto right Heschl's gyrus (yellow arrows). Right: top: schematic spectrum of stimuli used; half of the trials included the fundamental frequency and upper harmonics; the other half included only the harmonics without the fundamental. Bottom: bar graph showing mean error rates for patients with different excisions. Only the patients with right Heschl's gyrus lesion (RTA) showed elevated error rates for the missing fundamental condition (MF—orange bars); patients with more anterior temporal-lobe lesions sparing Heschl's gyrus (RTa) showed a milder deficit, while those with left-sided lesions (LTa, LTA) showed no significant difference from controls (NC). All patient groups performed well when the fundamental frequency was present in the stimulus (blue bars). Adapted with permission from (Zatorre 1988).

French chateaux are lovely places, but it's easier to reconstruct such pitch-inducing noise in the lab, which is what several studies have done. For instance, Tim Griffiths and his team at Newcastle University in the UK, were among the first to report that when noise is manipulated so as to give a perception of pitch, there is an increase in cerebral blood flow in the lateral region of Heschl's gyrus (Griffiths et al. 1998) as compared to noise of similar characteristics but without any pitch percept. The same approach was also used in magnetoencephalography (MEG) studies, which confirmed that a pitch-specific response emerged from either the central portion (Krumbholz et al. 2003) or the lateral portion (Gutschalk et al. 2002) of Heschl's gyrus. Marc Schönwiesner and I used this same type of stimulus in a case study of an epilepsy patient with an implanted electrode along the length of Heschl's gyrus (Schönwiesner and Zatorre 2008). We observed that the more medial recording sites picked up a response whenever the noise was turned on but not when the pitch began; whereas the opposite was true at the most lateral recording site, where we observed a clear response only to the onset of the pitch. More extensive direct cortical recordings with similar stimuli suggest that pitch responses can be found in variable locations in different

FIGURE 2.9 Illustration of how pitch can arise from iterated noise. In 1693, Christian Huygens visited the Château de Chantilly de la Cour. He noticed that a pitched sound could be heard when he stood between a fountain and a set of stone stairs. He correctly deduced that the effect occurred because the reflections of the noise from the fountain from each step (red arrows) take successively longer time intervals to reach the listener's ear. The pitch corresponds to the inverse of the time delay of these reflections. Drawing by F. Ahmed used with permission.

people but confirm their existence in every individual in locations roughly corresponding to the lateral Heschl's gyrus or adjacent cortex (Gander et al. 2019). These studies demonstrate the reality of a human pitch-sensitive region and suggest that it really represents pitch itself—and not other features of the stimulus.

One might be concerned that while this unusual noise stimulus is good to use in the lab, it is not representative of what is normally used in real music. It's true that such sounds are contrived. But luckily, the conclusions regarding pitch-sensitive regions are not confined to the use of this curious pitched-noise stimulus. Indeed, several functional imaging studies have reported reasonably convergent results in terms of the location of the pitch-sensitive region in the human brain using a variety of different stimuli, including harmonic complex tones (Norman-Haignere et al. 2013) and pitches that can only be perceived when different harmonics are presented separately to each ear (Puschmann et al. 2010). Besides, the neurophysiological and lesion studies that we discussed earlier using missing fundamental stimuli, also provide convergent data without using iterated noise stimuli. Therefore, we can be fairly confident that the phenomenon is not tied to any one stimulus class or experimental approach.

As we discussed in Chapter 1, one of the key concepts in contemporary psychology is that perception is not a passive, stimulus-driven, bottom-up process, but that it also involves top-down influences. The pitch-sensitive area has been proposed to participate in exactly this kind of interaction, specifically with a predictive component (Kumar and Schönwiesner 2012). According to this idea, there is a hierarchy within auditory cortex such that the pitch-sensitive region is at a higher level than earlier core areas. As such, it receives inputs from these earlier areas that are directly linked to the stimulus, allowing it to integrate information across different frequencies, while conversely it sends a predictive signal back to the earlier areas. This concept is supported by mathematical modeling of the strength of connections between earlier and later regions (Kumar et al. 2011). In other words, the role of the pitch-sensitive region is to form an abstract representation of the fundamental frequency (even if that frequency is unclear or absent) and send that information back to earlier areas, which in turn provide error signals to the higher region indicating how closely the representation, or internal model, corresponds to the actual input from the environment. Thus, at the level of basic auditory cortical function, this interplay between earlier, more stimulus-related neural representations and later, more abstract representations serves as a microcosm of what happens throughout the brain with more complex cognitive processes, which we will explore in subsequent chapters.

Representation of Harmonic Information and Octave Relationships

In the previous section, we focused on how the brain represents pitch information, which generally corresponds to the fundamental frequency in tones and usually also contain harmonics. But the harmonics themselves are important in many ways, including for grouping of different sounds, as we shall discuss below. Furthermore, harmonics are of course also represented neurally. The Johns Hopkins group once again provided compelling neurophysiological data for neural sensitivity to harmonic structure in marmoset monkeys (Feng and Wang 2017). They showed that neurons located in core auditory cortex respond best to combinations of pure tones that stand in a harmonic relationship to one another. In other words, these neurons encode the presence of tones in the environment that are related to each other as integer multiples (e.g., 150 Hz, 300 Hz, 450 Hz, and 600 Hz, which are all simple integer multiples of 150); but they do not respond to inharmonic tones that have a similar frequency difference and range, but are not related by simple integers (e.g., 173 Hz, 323 Hz, 473 Hz, and 623 Hz, which have the same distribution as the prior example—a difference of 150 for each pair—but are not multiples of one another because they don't have a common divisor, except 1 of course).

One reason these so-called harmonic template neurons do not respond to such inharmonic tone combinations may be because such sounds don't occur much in the natural environment, since only a few objects produce inharmonic tones (certain bells for instance); whereas most naturally occurring periodic sounds, including those made by an animal's own vocal cords, generate harmonic sound complexes. This finding leaves open the interesting question of whether harmonic sensitivity is an innate property of these neurons because of their communicative significance or whether they respond as they do via learning because of extensive exposure to harmonic sounds from the first moment of birth—and possibly even before.

Two functional imaging studies by Michelle Moerel and her colleagues in the Netherlands provide support for the existence of a similar phenomenon in the human brain: neural populations that respond to combinations of tones that stand in a harmonic relationship to one another (Moerel et al. 2013, Moerel et al. 2015). They measured responses to a wide variety of natural sounds and then used an algorithm to characterize the frequency profiles of each brain voxel. As shown in Figure 2.10, in addition to the more "classical" responses that favored only a single frequency, they also found spots that showed maximal activity to frequencies at integer multiples of one another (i.e., harmonics), and a specific class of voxels that were sensitive to frequencies in a relation of 2:1, that is, an octave. Each type of frequency profile tended to cluster together in particular regions of the temporal lobe, both within the primary cortex and extending beyond it, but without a very clear topography.

The exact relationship between these human fMRI results and individual neural responses in monkeys remains to be determined, but both sets of results point to the representation of more complex relationships than might be expected from a simple tonotopic organization, as described in earlier literature. These neurons, as well as the pitch neurons that are sensitive to fundamental frequency, are all responsive not just to the frequency content of periodic sounds, but also to higher-order features, and require integration of separate components of the sound. The findings of responses linked to octaves is particularly interesting in the context of music, of course, since octave equivalence represents a foundational concept of Western music theory and is also found in many other musical systems, including traditional scales from India, China, and Persia (see Savage et al. 2015), even though not all cultures necessarily make use of it (Jacoby et al. 2019).

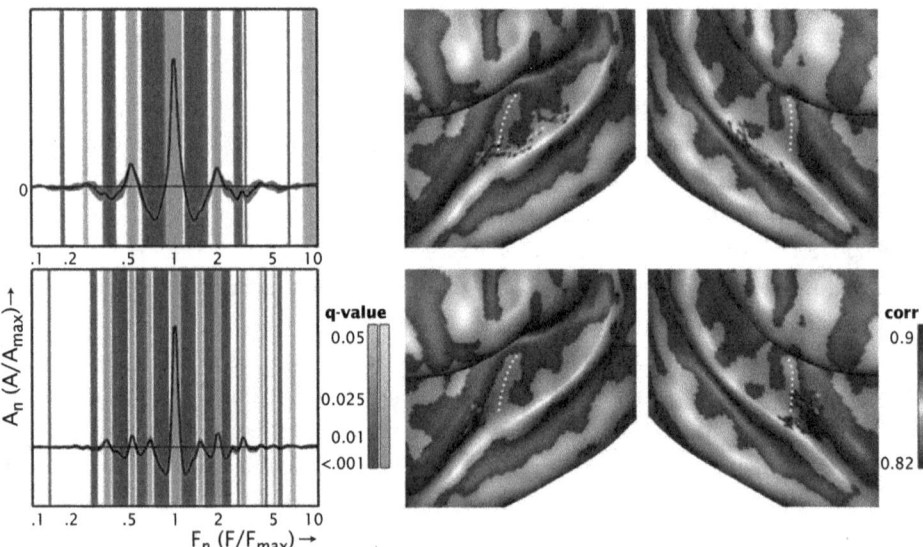

FIGURE 2.10 Neural responses to multiple spectral peaks. Left: Functional MRI plots show sensitivity to the structure of complex tones; the main peak in each graph represents the strongest response across voxels, and additional peaks (green) show the presence of sensitivity to additional frequency bands at consistent spectral intervals, including octaves, while responses away from these intervals elicit reduced responses (blue). Right: location of voxels showing multi-frequency sensitivity in locations lateral to Heschl's gyrus (white dotted line). Adapted from (Moerel et al. 2013). Reproduced under the terms of a Creative Commons Attribution-Noncommercial-Share Alike 3.0 Unported license.

Remarkably, the specialness of the octave is even observed in monkeys, who show evidence of octave generalization for melodic perception (Wright et al. 2000). Therefore, it seems likely to be a fairly basic property of auditory neural processing rather than an arbitrary construct.

The reason the octave is important is that cognitive theories of pitch perception have, for a long time (Bachem 1950), proposed that pitch is not represented merely linearly (as one might assume if frequency is just represented in the brain along a linear tonotopic dimension) but rather that its cognitive representation has a more complex structure such that tones that are an octave apart are more closely related to one another than tones spaced at any other distance. These models, which accord well with Western music theory and practice at least, generate a kind of a three-dimensional helix (Shepard 1982) or somewhat more complex geometric configurations (Krumhansl 1990). These ideas have led to the distinction between pitch chroma, which corresponds to the scale degree of each tone (C, D, E, etc.) and is represented in a circular manner versus pitch height, which corresponds to the overall brightness (C4, C5, C6, etc.) and is represented linearly.

When these mathematical pitch models are explicitly fit to brain data, as done by Moerel and colleagues (2015), they correspond well to the activity of separate neural populations for chroma and height. Consistent with this observation, Jason Warren and colleagues (Warren et al. 2003) found that pitch chroma was represented more in anterior auditory cortical regions than pitch height, which activated more posterior areas, perhaps corresponding to pattern processing versus sound-source segregation, topics we will return to in subsequent chapters. Irrespective of their anatomical location, the results from these different studies show that psychological and music-informed concepts, such as pitch chroma and octave equivalence, capture significant aspects of neural responses—or put another way, that our perception is the outcome of specific neural computations that cause us to perceive pitch as we do.

The Frequency-Following Response

The neural representation of frequency has so far been discussed in the context of the magnitude of neural responses related to the presence of periodicity in the stimulus. In other words, the research we have considered has measured the brain activity that is associated with the perception of pitch, but the activity itself is not periodic. However, sounds giving rise to pitch are periodic, and neural responses starting at the earliest levels of the auditory system do respond in a periodic manner to the undulations of the basilar membrane (Worden and Marsh 1968). This type of response is referred to as phase locking because neurons respond in phase with one another and with the input, as it systematically vibrates over time. These phase-locked responses can be measured, and in fact are ubiquitous at every level of the auditory system, from the cochlear nucleus all the way to the cortex. Although there are many phenomena tied to phase locking, here I will focus specifically on the signal known as the frequency-following response because of its importance in representing sound information relevant for pitch and hence for music as well as speech (for review see Coffey et al. 2019b).

The frequency-following response is aptly named, as it describes exactly what it does: follows the frequency. As shown in Figure 2.11, using just a single electroencephalography (EEG) channel it's possible to record the electrical fluctuations that are synchronized with both the fundamental frequency and at least some of the harmonics of a complex periodic stimulus, as demonstrated through extensive research by my scientific friend Nina Kraus and her team at Northwestern University in Chicago (Kraus et al. 2017, Krizman and Kraus 2019) and other labs as well. The frequency-following response can be elicited by a wide range of periodic stimuli, including spoken vowel sounds, musical tones, and even the iterated noise stimulus discussed earlier (Krishnan et al. 2012). One reason it has become important in neuroscience is that, unlike traditional EEG evoked responses or fMRI measures, some of the acoustical features of the sound itself are preserved in the response and can be analyzed. Therefore, it can serve as a direct index of the fidelity and precision of the auditory nervous system's encoding of periodic components of sounds (as well as some nonperiodic components, such as onsets, which I shall not discuss here).

The relevance of the frequency-following response for understanding perception, particularly in a musical context, is well illustrated by its sensitivity to musical training. Several studies from Nina Kraus's group have shown that musicians typically show higher quality of neural frequency encoding compared to people lacking such training. For example, Gabriela Musacchia and colleagues found that musically trained individuals have larger amplitudes

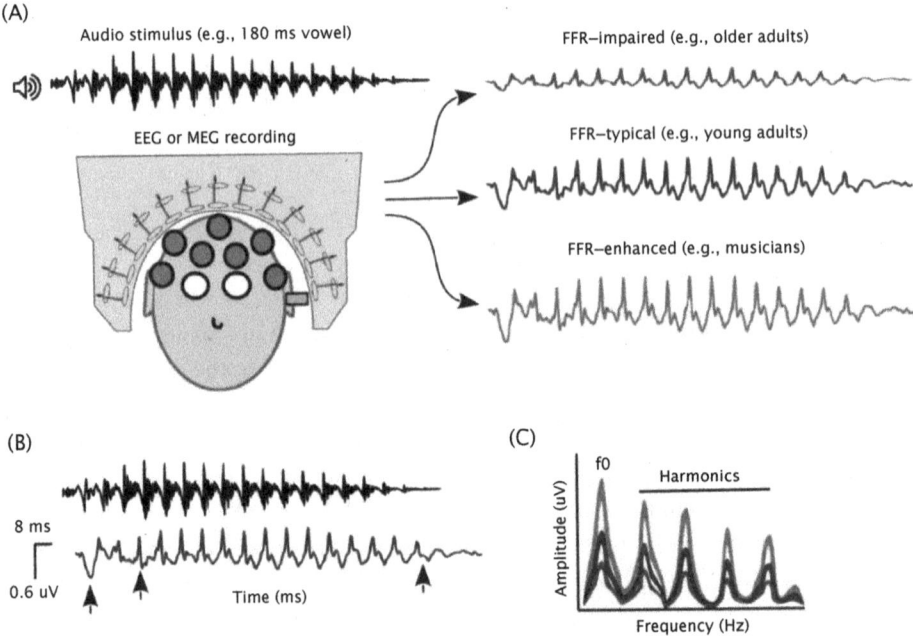

FIGURE 2.11 Frequency-following response. A: Schematic of the EEG or MEG acquisition, together with depictions of canonical responses from healthy young adults (blue), decreased amplitude in older persons (red), and increased amplitude in musically trained persons (green). B: Measures available in the time domain, including latency and amplitude. C: Measures available in the spectral domain, including amplitude of fundamental frequency and harmonics. Adapted from (Coffey et al. 2019b). Reproduced under the terms of the Creative Commons CC BY license.

and shorter latencies of frequency-following responses to both speech and musical sounds, presented in auditory-only and in auditory-visual conditions (Musacchia et al. 2007a); similarly, Patrick Wong and colleagues found that the frequency-following response of musicians followed the pitch contour of Mandarin speech tones better than in nonmusicians even though they did not speak any Mandarin (Wong et al. 2007). Both studies also showed that these effects were stronger as a function of the number of years of musical training.

These findings indicate that already at a sensory level, the musician's brain is better able to represent pitch information. Indeed neural synchrony, as indexed by the frequency-following response, predicts individual differences in the ability to discriminate musical tones (Bidelman et al. 2011; Coffey et al. 2016b), thus showing its relevance to musical perception. At the same time, enhanced neural synchrony is also related to better perception of speech-in-noise among musicians (Parbery-Clark et al. 2009, Coffey et al. 2017a). A clever demonstration of how real the enhancement can be in the musician's brain was provided by a study in which frequency-following responses to a series of vowel sounds were recorded in both musicians and untrained persons, and then subsequently sonified—that is, the electrical brain recordings themselves were played back through a speaker (Weiss and Bidelman 2015). A separate group of people, who did not have musical training, then made judgments about these sounds, and they were better able to discriminate the vowels when the sounds had been generated from the brains of musicians than from nonmusicians. So, the benefit of musical training can be picked up by listening in, as it were, to what the musicians' brains are doing.

Other related studies show that it's not only musical training which is related to better encoding but also that linguistic training can do the trick; for instance, people who speak a tonal language show better frequency-following response than those who don't (Krishnan et al. 2005). It is also possible to find these enhancements even after a limited amount of laboratory training, indicating that the cross-sectional findings with regard to musical training are not entirely due to preexisting factors (Song et al. 2008). A recent study from Bharath Chandrasekaran's lab (Reetzke et al. 2018) showed this learning-related enhancement but also showed that the change in frequency-following response occurred later in time than the behavioral improvement, suggesting that it represents a kind of sharpening of early sensory encoding most likely via top-down (perhaps predictive) mechanisms.

Where does the frequency-following response originate? For a long time most studies measuring this response attributed it to either the inferior colliculus or to a combination of subcortical nuclei, based on physiological studies showing that phase locking prominently occurs in those locations (Chandrasekaran and Kraus 2010). Figuring out the source of the response is important in attributing phenomena like plasticity and accuracy of sensory encoding to the correct level of neural processing. But because the frequency-following response is usually measured with EEG, which does not have excellent localization capacity, the question remained open until recently.

In 2016, Emily Coffey (at the time, a PhD student in my lab and now a professor at Concordia University) set out to address this question using MEG, which has much better capacity to locate sources than EEG. Emily's analyses, shown in Figure 2.12, revealed that not only could we detect frequency-following responses from each of the subcortical nuclei

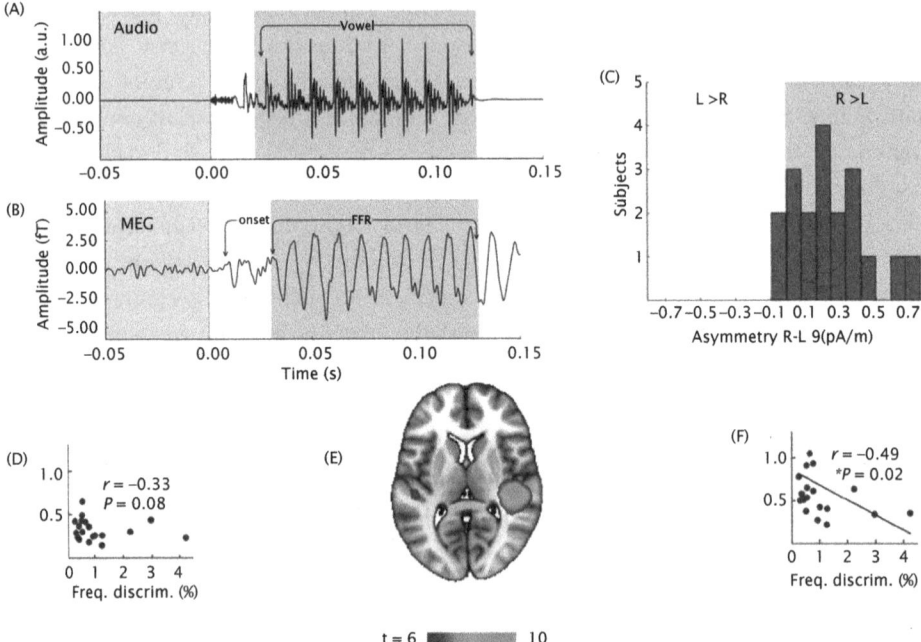

FIGURE 2.12 Asymmetry of cortical frequency-following response. A and B: Audio waveform and MEG response to the same stimulus (consonant–vowel syllable with fundamental frequency of 98 Hz), showing correspondence between the input and the measured response. C: Distribution of response amplitude differences between left and right auditory cortex, showing a stronger response over the right than the left in most individuals. E: projection of MEG data onto a horizontal MRI slice, showing location of maximal response in auditory cortex and rightward asymmetry. D and F: correlations of individual pitch discrimination ability and response magnitude measured over the left and right auditory cortex, respectively. Only the right auditory cortical response is related to individual differences in pitch processing (higher response = lower thresholds and better ability). From (Coffey et al. 2016b). Reproduced under the terms of the Creative Commons CC BY license.

(cochlear nucleus, inferior colliculus, and medial geniculate) as expected, but unexpectedly there was also a clear signal coming from the auditory cortex (Coffey et al. 2016b). The amplitude of this response was significantly larger on the right side than on the left, in keeping with a lot of data on cortical pitch representation that will be reviewed in Chapter 5, while the maximal response fell close to or slightly more posterior to the pitch-sensitive cortical region described above. This finding expanded our understanding of how this cortical region encodes pitch not only in terms of magnitude of brain activity but even at the level of phase-locked responses.

Although finding cortical responses associated with the frequency-following response was unexpected, Emily verified it in a follow-up study in which she documented that the fMRI signal from the right auditory cortex was correlated with frequency-following response amplitude as measured with EEG and, as expected, was once more found to be greater in musically trained individuals (Coffey et al. 2017c). The cortical response in this experiment could be detected even when using a piano tone with almost no energy at the fundamental, in keeping with the evidence reviewed earlier for the role of auditory cortex in computation of missing fundamental pitch. In fact, another study showed that people can be taught to attend either to the missing fundamental pitch or to the harmonics of a melody and that this manipulation is related to the strength of their frequency-following response to the fundamental frequency, thus linking the neural measure more directly with pitch perception (Coffey et al. 2016a).

A number of subsequent studies from other labs have confirmed the cortical source (Bidelman 2018, Hartmann and Weisz 2019) and confirmed the asymmetry (Gorina-Careta et al. 2021). The cortical response can also be seen with direct intracranial recordings from within human auditory cortex (Behroozmand et al. 2016, Pesnot Lerousseau et al. 2021). Many of the phenomena previously ascribed to subcortical responses have now been observed with the data arising from the cortex (see Figure 2.12); thus, the amplitude of the response from the right auditory cortex correlates with ability to discriminate pure tones, number of years of musical training, and age at which musical training started (Coffey et al. 2016b). These findings, together with other findings of possible modulation of the frequency-following response with attention (Lehmann and Schönwiesner 2014, Hartmann and Weisz 2019), all suggest that these higher-order aspects of pitch processing may depend upon the integrated responses of the entire auditory nervous system, cortical and subcortical. This conclusion emphasizes the dynamic interplay between bottom-up and top-down signals, in keeping with the auditory system's highly interconnected nature (Coffey et al. 2019b).

The implication of this line of reasoning is that the frequency-following response represents more than just a slavish reproduction of sounds in the environment; rather it may serve as an internal model of sounds and hence may have a predictive function as well. This idea is still speculative but is supported by recent MEG data showing that the neural oscillations generated by a tone outlast its offset by several cycles and that, conversely, it takes several cycles for the frequency-following response to reach the right frequency (Coffey et al. 2020, Pesnot Lerousseau et al. 2021). These results, observed at subcortical and cortical levels, suggest that there is an endogenous oscillatory component to frequency encoding, which could serve as a template, or a fine-grained predictor, of incoming frequency information, providing stability and enhancing perception of noise. This idea fits with broader concepts of how oscillations can serve predictive functions (Arnal and Giraud 2012). Even more relevant for our purposes, the concept seems quite compatible with the pitch model proposed by Sukhbinder Kumar (and referred to earlier), according to which prediction signals are generated at later levels of the cortical hierarchy and propagated back to earlier cortical levels (Kumar et al. 2011) but which we can also generalize to the entire cortical-subcortical neuraxis. As such, this idea is also reminiscent of the proposed cortical-subcortical cascade of prediction signals discussed in the context of sensory-specific adaptation (Carbajal and Malmierca 2018).

Beyond Frequency Coding: Spectrotemporal Modulations

Most of our discussion to this point has focused on how frequency information is represented at different levels of auditory function, reflecting the importance of this feature for sound perception, in general, and for music in particular. However music and many other

(A)

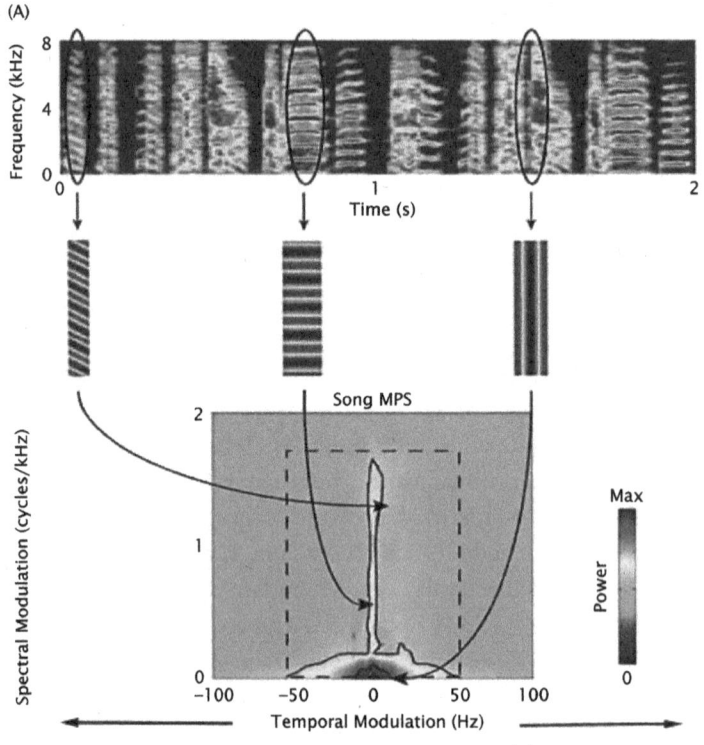

(B)

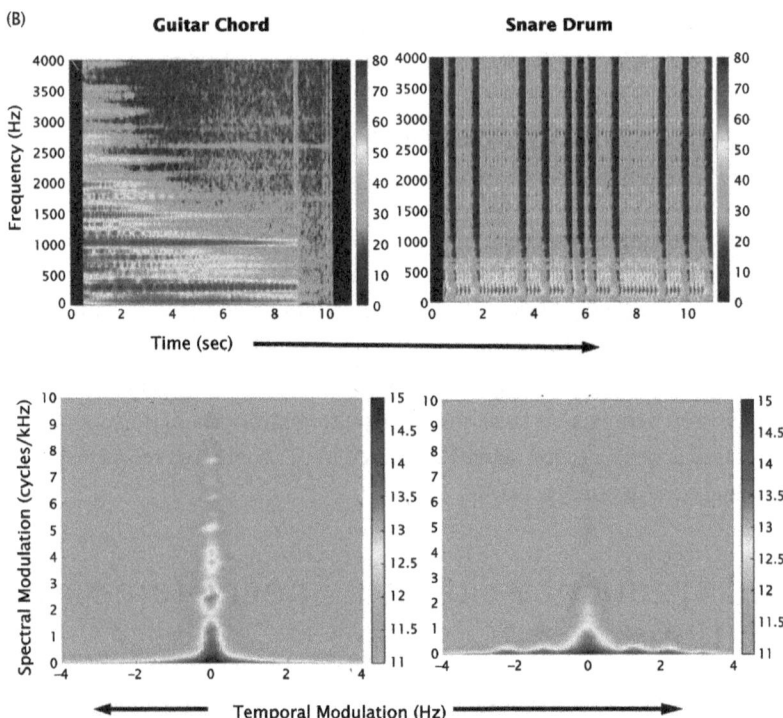

natural sounds, including especially speech, contain complex acoustical information beyond just their frequency composition or spectrum. The other principal dimension to account for, of course, is time because sounds exist in time, and hence temporal phenomena, like sound onsets, offsets, changes in intensity over time, and so forth, are all critical. To visualize the acoustical properties of sound, engineers in the 1940s devised a machine that produced a representation of the amount of energy in the frequency spectrum on the y axis, plotted as a function of time on the x axis, called a spectrogram (Koenig et al. 1946), as illustrated in Figure 2.13A.

The spectrogram may be a convenient way to depict sound on a computer screen or piece of paper, but does it have any bearing on how the brain does it? Neurophysiological studies in monkey auditory cortex suggested joint neuronal sensitivity to both temporal and spectral features (deCharms et al. 1998). These spectrotemporal receptive fields (reminiscent of feature detectors in vision, like lines or edges, that were described earlier by Hubel and Wiesel), could be nicely represented on a time–frequency graph that bears some similarity to a spectrogram.

More recently, however, further steps have been taken to understand the functional organization of neurons tuned to complex combinations of features, using the spectrotemporal modulation framework. This mathematical approach allows the decomposition of any complex stimulus into a series of simpler spectral and temporal modulation functions (Chi et al. 2005). These can be thought of as the equivalent of visual gratings, which are used in vision research to model complex scenes (and was indeed the way that the idea initially emerged in auditory neuroscience, promoted by Shihab Shamma in particular [Shamma 2001]). So, rather than representing sounds in terms of frequency as a function of time, they are represented in a spectrotemporal modulation space. These two representations are directly related to one another by the mathematical relationship of Fourier transformation (Shamma 2001).

In Figure 2.13A, we can see how a complex sound (in this case a birdsong) can be displayed as a spectrogram and then in terms of the amount of energy at any given combination of spectral and temporal modulations (Woolley et al. 2005). To illustrate how this

---

FIGURE 2.13 **Spectrograms and spectrotemporal modulation plots.** A: Top: spectrogram (frequency as a function of time) of a birdsong. Spectrotemporal content in the stimulus can be captured by analyzing each segment according to the modulations it contains; the leftmost segment consists of harmonic content with a downward sweep; the middle one corresponds to a steady-state portion with prominent harmonics; and the last one corresponds to an onset with little harmonic structure. Each of these spectrotemporal modulations can be plotted in a graph capturing the structure of the entire sound, as shown at the bottom of the panel. Reprinted by permission from Springer Nature: Springer, Nature Neuroscience (Woolley et al. 2005), copyright 2005. B:

The principles illustrated in A are shown for musically relevant examples. Two contrasting musical segments, a sustained guitar chord (opening of the Beatles' *It's Been a Hard Day's Night*), and a snare drum passage (opening of Ravel's *Bolero*), are shown at the top as spectrograms. Note the rich harmonic structure without much change over time in the former, and the prominent temporal variation without much spectral structure for the latter. Spectrotemporal modulation plots for the two examples (below) show that the guitar chord contains a lot of spectral modulation but very limited temporal modulation (more energy distributed vertically on the display), while the reverse is the case for the snare drum excerpt (more energy distributed horizontally on the display). Images created by P. Albouy, used with permission.

framework applies to music, we can look at two contrasting musical sounds (Figure 2.13B): a harmonically complex but steady-state sound, like the famous sustained guitar chord at the start of the Beatles' *Hard Day's Night*, contains high spectral modulation but low temporal modulation because of its rich distribution of spectral elements that don't change much over time (energy distribution is mostly vertical on the display). Conversely, a sound with a relatively even distribution of spectral energy but with sharp onsets and offsets, like the ostinato snare drum rhythm that permeates Ravel's *Bolero*, would have low spectral but high temporal modulation (energy distribution is mostly horizontal). Real music, like all natural sounds, would typically contain both types of modulation; but as we shall see in Chapter 5, the spectral content seems to be especially critical for many aspects of music. The advantage of considering sounds within a spectrotemporal modulation framework is that it allows for efficient representation of acoustical complexity since any stimulus can be encoded via a finite set of filters (neurons) with different combinations of spectral and temporal modulation sensitivity, allowing different acoustical dimensions to be encoded simultaneously (Elliott and Theunissen 2009).

Many neurophysiological studies have now identified spectrotemporal receptive fields in auditory cortical (Linden et al. 2003, Massoudi et al. 2015) and subcortical (Rodríguez et al. 2010) neurons across species. Therefore, it seems that the nervous system has evolved to utilize this efficient coding scheme. To determine how this concept might play out in the human auditory cortex, Marc Schönwiesner and I measured fMRI responses to synthetic sounds that varied systematically in their spectral and temporal modulation content (Schönwiesner and Zatorre 2009). As shown in Figure 2.14, we found that individual voxels in primary and surrounding areas could be identified that were sensitive either to one or the other dimension, or to both jointly, paralleling the types of responses seen in individual neurons in other species. The temporal range over which sensitivity was greatest corresponded to relatively low rates of both temporal or spectral modulation, consistent with what is found in natural sounds in the environment (Singh and Theunissen 2003), including speech and musical sounds (although these two domains differ in important ways, as we shall see in detail in Chapter 5). The distribution of these responses across the cortex varied considerably from individual to individual, but there was some indication of greater sensitivity to spectral modulation toward the lateral portion of Heschl's gyrus, as might be expected from the presence of a pitch-sensitive region in its vicinity.

Subsequent neuroimaging studies have been able to replicate and extend these findings using natural instead of synthetic sounds (Santoro et al. 2014) as well as speech sounds (Venezia et al. 2019) and have additionally noted that temporal versus spectral sensitivity are partly segregated in more posterior versus anterior auditory cortical areas, respectively. A similar result, including the anterior–posterior dissociation, has also been reported from direct electrical recordings from human auditory cortex (Hullett et al. 2016). This finding of an anterior–posterior gradient is very relevant in terms of the two pathways that emerge from auditory cortex (and will be discussed in Chapters 3 and 4) because more posterior/ dorsal regions are sensitive to timing-related features, whereas more anterior/ventral regions tend to integrate spectral information across time.

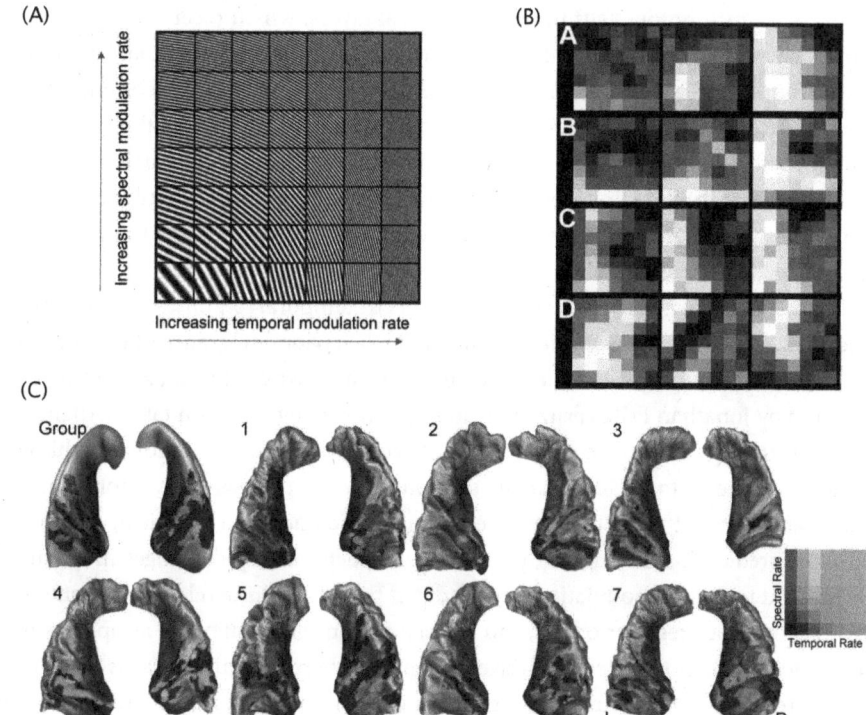

FIGURE 2.14 Evidence for sensitivity to spectrotemporal modulation in the human brain. A: Stimuli used to sample different combinations of spectral and temporal modulation. Each box represents a ripple noise that was repeated many times during functional MRI scanning. The more vertical the stripes the faster the temporal modulation; the thinner the stripes the greater the spectral modulation. B: Diagram showing the activity for individual voxels following the same arrangement of stimuli as in A. Top row: examples of three voxels that have relatively focal responses to a single combination of spectrotemporal modulation. Second row: examples of three voxels that have a preferred spectral modulation rate (horizontal on the graph) but which are not well-tuned to temporal modulation. Third row: examples of three voxels that have a preferred temporal modulation rate (vertical on the graph) but which are not well-tuned to spectral modulation. Fourth row: examples of three voxels with more complex, multi-peak tuning. C: Distribution of brain activity along the auditory cortex (surface rendering of temporal lobe viewed from the top). Colors indicate preference for one or the other dimension; most individuals show distinct patterns of tuning in different regions, but there is only weak topography across individuals (tendency for greater spectral modulation preference anteriorly and greater temporal modulation preference posteriorly). Adapted with permission from (Schönwiesner and Zatorre 2009).

All these results together lead us to conclude that auditory cortical neurons are tuned according to spectrotemporal modulations and that their properties correspond well to the features of natural sounds that might be important in a given species' environment, thus yielding an efficient way to represent these complex sounds. Communicative sounds of different species seem to be especially relevant signals whose acoustical features correspond to the tuning of these neurons (Woolley et al. 2005); and I would argue, of course, that speech and music are the principal communicative sounds of our species. The topography of these spectrotemporal neural responses appears to ride on top of a more basic tonotopic organization that is characteristic of early cortical regions. According to a recent study (Moerel et al. 2019), this simultaneous coding is explained by the presence of tonotopy within deeper cortical layers, which receive input from the thalamus, whereas spectrotemporal organization occurs in more superficial layers of the cortex. Responses to spectral and temporal modulation appear to be confined to relatively early cortical regions, as contrasted with higher-order regions that respond to meaningful categories of sound, as will be discussed in Chapter 3. As such, it is likely that spectrotemporal feature coding feeds into later stages of processing, where more abstract auditory object representations emerge that are not solely dependent on the acoustics of the stimulus.

Finally, it is important to point out that the spectrotemporal responses are not static but that they can adapt rather quickly depending on the behavioral relevance of the stimuli, as shown in elegant physiological studies involving learning tasks carried out at the University of Maryland by Jonathan Fritz (Fritz et al. 2003), or involving attention tasks (Atiani et al. 2009). These findings tie into the global theme of prediction because, according to this view, when there is a change to the local statistics of sounds in the environment, error signals are communicated upward to cortical representations, which are then modified, resulting in a new set of predictions for ongoing processing. As such, the rapid changes in sensitivity to spectral and temporal modulations, engendered by a shift in the relevance of one set of spectrotemporal features over others, can be seen within this framework as updates to an internal model, as demonstrated in these experiments where the animal learns that a given stimulus is rewarded or is trained to attend to a given stimulus. It is likely that such relatively quick modifications to neural receptive fields have important implications for musical processing, whereby expectancies are formed while listening to a piece of music based on its particular textures as they unfold in real time.

Analyzing the Auditory Scene

Stream Segregation

As alluded to in the introduction to this chapter, one of the problems the hearing system faces is that if multiple sounds are present simultaneously, their vibrations are all mixed up together at their point of arrival in the eardrum. It's as if, in a visual scene, all objects were transparent, so their outlines and colors would all blend with each other. Before further processing can occur, the different sound sources need to be disentangled, and much of this work occurs in the auditory cortex together with its inputs and outputs. There are several interesting aspects of this complex problem, which has been studied extensively with psychophysics and behavior (Bregman 1994) as well as with neurophysiology and brain imaging (Snyder and Elhilali 2017). One of the prime findings from this vast literature is that both stimulus-driven (bottom-up) as well as schema-driven (top-down) factors contribute to the segregation of sound sources. This distinction fits well with the predictive coding concept and has interesting ramifications for our understanding of the neural substrates of this ability. Here, I will consider several phenomena of particular relevance for music (as opposed to speech or other sounds).

One common stream segregation scenario is the situation where a single stream of sound splits into two percepts: for instance, when a melody in which only one tone is·

presented at any one time is nonetheless heard as two separate elements. This situation, sometimes referred to as melodic fission (Dowling 1973), is quite common in baroque music (where a violin, for instance, can play what sounds like two melodies by alternating between notes on two separate strings, as shown in the Bach partita score in Figure 2.15), but is certainly not confined to that genre, as it can also be heard in xylophone music from Uganda for instance (Bregman, 1994).

Segregation using two alternating tones differing in frequency has been extensively studied, and even with such a simple stimulus, it is already evident that both stimulus-driven and more cognitive factors are at play. Thus, increasing the frequency separation, speed (Moore and Gockel 2002), or spatial separation of the tones (Middlebrooks and Onsan 2012), all of which are clearly bottom-up factors, typically leads to greater segregation. Conversely, attention (a top-down factor) can influence whether the sequence is perceived as one or two streams (Cusack et al. 2004). Preexisting knowledge can serve as another top-down mechanism: if a target melody is known a priori, for example, it can be disambiguated more easily when it's presented interleaved with unrelated tones (Bey and McAdams 2002). In the case of real music, more abstract knowledge of musical schemas, such as harmonic or voice-leading features, could also come into play. These would constitute examples of how prediction based on past learning, could enhance the listening experience since knowledge of what's coming would enable a listener to perceive—and hence enjoy—a pattern that might otherwise be obscured.

Single-cell recordings in primate auditory cortex using this simple two-tone stimulus suggest that part of the mechanism involves frequency-specific adaptation (Fishman et al. 2001a), similar to that described earlier in this chapter. According to this idea, if the frequency separation between two tones is large enough, adaptation between them is decreased or eliminated, leading to a stronger representation of each tone of the sequence, which results in the perception of two separate streams. The degree of adaptation in cortical neurons and its buildup over time, as measured in the monkey auditory cortex, are directly related to how humans perceive these patterns, which typically require several seconds to split into separate percepts (Micheyl et al. 2005). Daniel Pressnitzer and his team in Paris have also reported effects similar to those from the cortex in the cochlear nucleus (Pressnitzer et al. 2008), the lowest structure in the auditory hierarchy; this observation indicates that this simple form of stream segregation may emerge early in the processing chain, consistent with the data reviewed above that sensory-specific adaptation is found throughout the entire auditory system.

Human neuroimaging studies with similar streaming paradigms have largely supported the interpretation emerging from more basic physiological studies, as seen in Figure 2.16. Thus, there is a correlation between the magnitude of MEG response from auditory cortex and the degree of splitting of the percept into two streams such that the response to the different tones become more differentiated when the sequence segregates, consistent with an adaptation phenomenon (Gutschalk et al. 2005). Results compatible with this model were also reported from fMRI responses measured in auditory cortex, in which greater signal is associated with larger frequency differences between adjacent tones, a situation that promotes stream segregation (E. C. Wilson et al. 2007a).

FIGURE 2.15 Stream segregation with a single instrument. Autograph manuscript of first page of the *"Preludio"* from the *Partita for Solo Violin in E Major*, BWV 1006, by J. S. Bach. Note the use of notes in different ranges, and of repeated tones, to create the percept of two simultaneous melodies, even though only one tone is played at any one time (especially visible in lines 5 and 6). Public Domain.

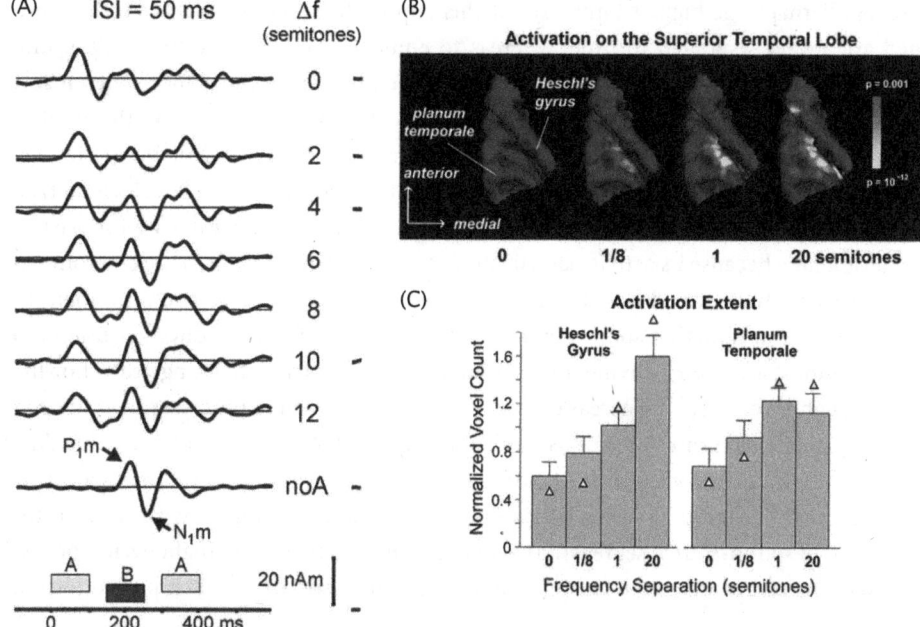

FIGURE 2.16 Physiological responses associated with stream segregation. A: MEG responses from auditory cortex to an A-B-A pattern (bottom) show greater differentiation to the B tone as a function of the pitch separation between tones. Used with permission of the Society for Neuroscience, from (Gutschalk et al. 2005); permission conveyed through Copyright Clearance Center, Inc.

B and C: functional MRI signals from Heschl's gyrus and planum temporale show greater magnitude of response as a function of increasing pitch distance between adjacent tones.

Used with permission of the American Physiological Society, from (E. C. Wilson et al. 2007); permission conveyed through Copyright Clearance Center, Inc.

Stream segregation also involves interactions between cortical and subcortical structures. Streaming of tone patterns is correlated with changes to the frequency-following response (Yamagishi et al. 2016), which as we saw above has generators in both cortical and subcortical structures. More specifically, fMRI responses to an ambiguous tone sequence occurred earlier in auditory cortex when the percept switched from a dominant to a less dominant pattern; whereas responses from the medial geniculate in the thalamus occurred earlier when the listener reverted back to the more dominant response (Kondo and Kashino 2009). This pattern of results can be explained in the context of bottom-up and top-down interactions such that when the segregation is predominantly driven by stimulus features (more dominant percept), it is represented by bottom-up influences from thalamus to cortex; but when a listener imposes an attentional or other top-down influence (i.e., a prediction) to perceive the pattern (the less-dominant percept), then the cortical response predominates.

Concurrent Sounds

These experiments using a simple two-tone sequence have much to teach us about basic perceptual mechanisms, but certainly do not begin to capture the complexity of real-world listening. Perhaps the biggest limitation of this approach is that the sounds to be segregated are always separated in time, whereas to emulate real life more fully, segregation should involve the situation when a target sound is embedded within other simultaneous background sounds. To handle this type of concurrent segregation problem, the auditory system makes use of some of the functions we have already reviewed. Notably, periodicity in target or background sounds helps to segregate sources from one another (Steinmetzger and Rosen 2015). The computations within the pitch-sensitive region discussed above may be relevant here because its activity signals the presence of a given pitch in the environment in an invariant manner and is relatively insensitive to masking or distortion. Ecologically, it is interesting to note that sounds in a natural environment that produce a sensation of pitch are almost all made by living things, usually via a vocal tract (honking geese, howling wolves, laughing children); whereas events generated by nonliving processes typically generate only unpitched noises of various kinds (branches waving in the wind, waves breaking on the shore, thunderclaps, rain). Identifying the presence of other creatures against the environmental background seems like a very important ability for survival. In fact, this facilitation of sound-source segregation via processing of periodicity might even represent the evolutionary pressure to develop periodicity coding—and hence pitch perception—in the first place.

An important feature of periodic sounds, as we have seen, is that they most often contain harmonics at integer multiples of the fundamental. This harmonic relationship not only helps to perceive harmonic sounds from noisy backgrounds, but also to pull out one harmonic sound from among other harmonic sounds, which is obviously very relevant for any music where more than one line is present. For example, if a single component of a harmonically complex tone is mistuned so that it is no longer an integer multiple of the fundamental, it tends to pop out perceptually (Moore et al. 1986). The phenomenon is also accompanied by modulation in the frequency-following response (Bidelman and Alain 2015), thus reflecting how frequency coding in brainstem and/or cortex helps segregate periodic sounds. Similarly, when two vowels of different fundamental frequency are presented together, they can be segregated based on the composition of their harmonics (Assmann and Summerfield 1994). Interestingly, computational models of this type of concurrent segregation have suggested that part of the process may involve harmonic templates that act as filters (de Cheveigné 1997), an idea that fits remarkably well with the neural sensitivity to harmonic structure discussed earlier (Moerel et al. 2013, Feng and Wang 2017). These templates could also be construed as a kind of predictor, or schema, about the expected relationships between harmonics.

Tonotopic organization has also emerged as a mechanism by which multiple simultaneous sounds that differ in frequency can be distinguished. In particular, attending to a tone of one frequency when two are present enhances fMRI activity in auditory cortical neural populations that are selectively tuned to that frequency (Da Costa et al. 2013). Similarly, using more complex simultaneous melodic patterns across three different frequency ranges (Riecke et al. 2017), it was shown that selective attention to one of the streams modulated fMRI activity in a frequency-specific way throughout the entire superior temporal gyrus; moreover, a multivariate machine-learning method applied to the brain data was able to determine which frequency band the listener was attending to. These findings, which also agree with neurophysiological recordings (Atiani et al. 2009), suggest that tonotopy can help the segregation of multiple overlapping elements by acting as a kind of a filter to help active selection of one stimulus over another.

All these mechanisms rely on the processing of spectral aspects of a compound stimulus. But temporal aspects also play a very important role in helping to disambiguate sound sources; thus, sounds that start or end together, or have a common fate as Bregman emphasized (Bregman 1994), typically group together because they are likely to have come from the same source (Darwin et al. 1995). The neural basis for this mechanism involves temporal coherence across neurons, which plays a major role in segregation together with selectivity for frequency or other features (Snyder and Elhilali 2017).

Evidence that temporal coherence provides a way to bind features together comes from some clever experiments carried out in Maria Chait's lab in London, in which the listener is presented with a complex, rapidly varying chord sequence as a background, from which a "figure" emerges, consisting of multiple tones that are temporally correlated (Teki et al. 2016). As shown in Figure 2.17, under these conditions, despite overlap in both temporal and spectral features between the target and background, MEG signals show a response emerging from posterior areas of auditory cortex, as well as the parietal cortex, shortly after the onset of the temporally correlated elements in the target sound. Although the listeners in these experiments were distracted by an irrelevant task, suggesting a more automatic, bottom-up process, it has also been shown that attention to the target sound is relevant (O'Sullivan et al. 2015). These findings are consistent with physiological experiments which also show that measures of coherence across neurons can be altered very rapidly during task performance, as would be necessary to not only track the changing features of a stimulus, but also to actively select a given target via attentional mechanisms (Lu et al. 2017).

Listening in the Real, Noisy World

In almost every experiment described in this section so far, the stimulus sequences used consisted of various types of tones that were synthesized to allow tight control over psychoacoustical features. But it's obvious that the real world, including the musical world, does not consist of such neat and clean sounds. Therefore, we now turn to experiments that attempt to look at how the brain distinguishes sounds in something more akin to a natural context. The most common example of how scientists have approached the problem is by studying real speech embedded within background noise (Nilsson et al. 1994).

The ability to perceive speech in noise, as opposed to the more artificial stimuli already discussed, depends both on stimulus-driven factors and top-down factors such as attention, higher-order knowledge of speech structure, and motor-related mechanisms (Davis and Johnsrude 2007). These paradigms typically involve a target stream of speech (words or sentences) presented against a background of either white noise (or other unstructured masking noise), which tends to affect stimulus-based features, or complex noise (involving informational masking that mimics real-world situations, where one or more speakers are in the background) (Kidd et al. 2008), which tends to involve more top-down attentional

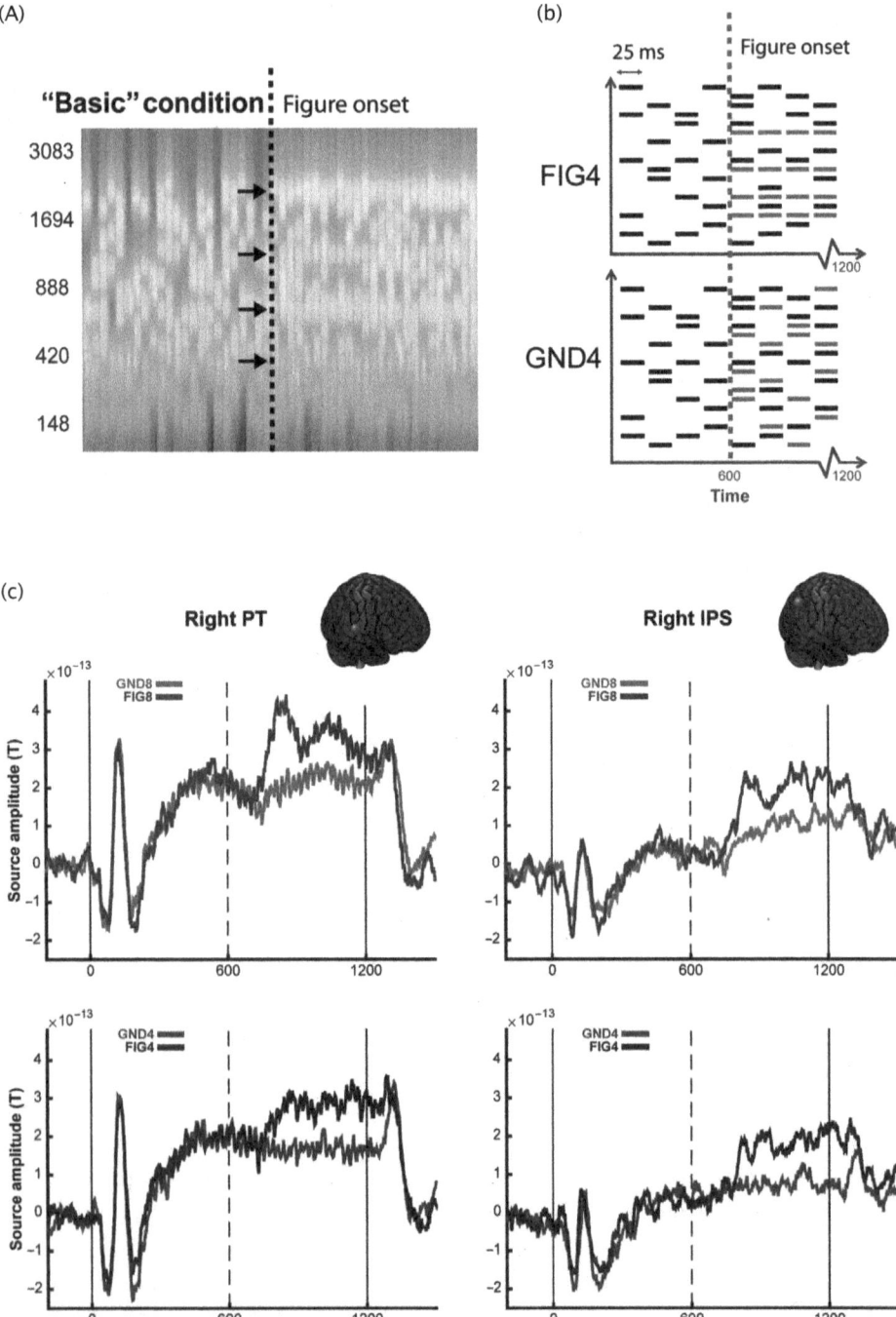

FIGURE 2.17 Neural correlates of auditory figure-ground segregation. A and B: Spectrogram and diagrammatic representation of stimulus sequence. Initially the sequence consists of randomly selected simultaneous tones, all different from one another; after 600 ms (dashed line) an additional set of fixed simultaneous tones (FIG4) is presented that repeats over time (pink lines). The resulting percept is of a chord that pops out over the background. The control condition (GND4) adds a varying set of tones that does not result in such a percept. C: MEG activity from the two conditions. A clear signal emerges from right auditory and parietal regions shortly after the repeating chord begins, corresponding to the figure-ground segregation percept. Adapted from (Teki et al. 2016). Reproduced under the terms of the Creative Commons CC BY license.

mechanisms. A recent fMRI study, shown in Figure 2.18, demonstrated that nonprimary auditory cortex responds to target sounds (speech, music, or other) in a more invariant way, whether they are embedded in background noise or not (Kell and McDermott 2019), unlike primary areas. This finding supports the hierarchical functional organization of the auditory pathway, implying that primary cortex contains representations closer to the acoustical input; whereas segregation of complex targets from background noise occurs beyond primary processing stages, in the auditory ventral stream, which represents auditory objects (to be discussed further in Chapter 3).

Many other studies have investigated the neural substrates of speech perception in the presence of distracting background noise. But since our focus is on music, we will only mention one particularly important speech finding, which is that the attended signal has a stronger neural representation in nonprimary auditory cortex when there are two competing speakers. This effect was shown by Nima Mesgarani and Edward Chang in San Francisco, who recorded activity from electrodes implanted into nonprimary auditory cortex in patients undergoing epilepsy investigation (Mesgarani and Chang 2012). They reported that the neural activity corresponding to the attended speaker contained enough information for a machine learning algorithm to reconstruct the message and, hence, determine which speaker the listener was paying attention to. A related finding, reported around the same time, showed that an attended speech stream could be tracked by looking at the correlation of its amplitude envelope with the MEG responses (Ding and Simon 2012). Thus, a stimulus that is attended shows a stronger correlation value in auditory cortex than the distractor, suggesting that neural activity in auditory cortex outside of primary regions, doesn't merely reflect the acoustical energy present (as primary regions tend to do) but can also represent the information the person is listening to, and hence incorporates top-down processes.

These findings have interesting implications for music listening for two reasons: first, because some of the same neural processes applicable for selective listening to multiple speech streams might also be relevant for music listening; and second, because it raises the

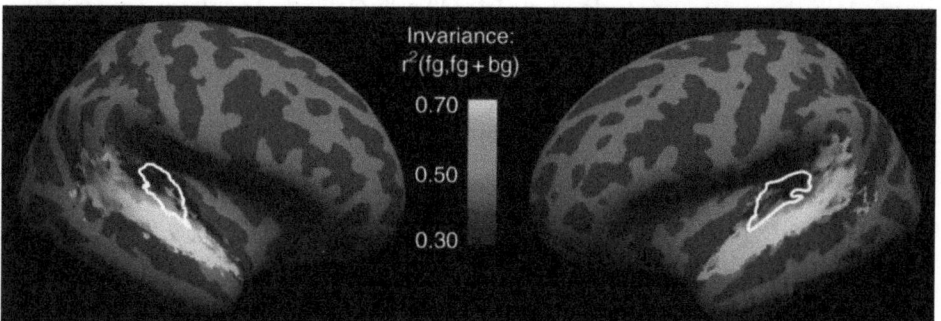

FIGURE 2.18 Neural responses invariant to background noise. Brain maps correspond to correlation between target sounds when presented in quiet or embedded in background noise. Primary cortical areas (white outline) show low correlation (blue), indicating sensitivity to presence of noise; cortical areas throughout the rest of the superior temporal gyrus show more similar responses between conditions (yellow), indicating robust representation of target sounds irrespective of background. Adapted from (Kell and McDermott 2019). Reproduced under the terms of the Creative Commons CC BY license.

question of whether experience with music with multiple parts may translate to better speech perception in noisy environments. Selective listening in a musical context might apply when trying to follow one line of music within an ensemble of multiple instruments (e.g., listening to the bass line in the dense texture of a rock band), or when listening to music with unrelated background noise (e.g., listening to the radio while traveling at high speed in a noisy old car).

To simulate the situation where one is trying to pick out a specific musical part from a mix of many parts, Niels Disbergen and I together with colleagues in Maastricht, created a polyphonic listening task (Disbergen et al. 2018). We hired a composer to write short, two-part pieces played on synthetic cello and bassoon; the listener was invited to either attend to one instrument or the other and detect the presence of occasional triplets. Thus, the task mimics real music listening but still has good experimental control. Behaviorally, we found that even musically untrained people could perform the task well with a little training. Furthermore, as expected from prior studies on acoustical cue-driven segregation, we found that the participants were aided when the two lines were played by the two instruments with distinct timbres as compared to similar timbres.

Using this task with EEG, the Maastricht team revealed that, just as had been shown previously with competing speech streams, when listeners attended to one or the other instrument, the amplitude envelope corresponding to the attended musical line could be better reconstructed from the EEG (Hausfeld et al. 2021). As shown in Figure 2.19, this effect was maximal from around 150 ms and later, suggesting that the attention-driven segregation of the two stimuli occurred after initial acoustical analysis had taken place. To understand the brain regions driving this effect, we used high-field fMRI, with the same task again, and found that the activity in auditory cortical regions and frontal cortex could successfully distinguish whether the listener was attending to the bassoon line or the cello. Taken together, these results confirm that attended streams of musical information are represented in the activity patterns of auditory regions, just as speech streams are.

A slightly different selective listening situation arises when the target sound is embedded in irrelevant noise rather than forming part of a structured piece of music. To test how well people are able to pick out a musical line under such circumstances, and to understand which cues are used to do so, my friend Emily Coffey and our team developed the music-in-noise test (Coffey et al. 2019a), which consists of short melodic excerpts presented against a background of musical cacophony (created by taking various unrelated snippets of music and mixing them together). The behavioral results showed that performance is generally enhanced by musical training, as expected, and also in the condition when the target melody is heard in advance because of the ability to use the information predictively (Bey and McAdams 2002). But, to a greater extent, this predictive ability boosted scores in musicians, indicating that they are better able to use this kind of top-down cue, presumably because they can incorporate the information into a better internal model of what to expect. Similarly, when a synchronous visual stimulus was introduced that served as a cue to help segregate the melody from the background, greater enhancement was seen among musicians, in keeping with the idea that musical training enables better integration of visual and auditory information (Musacchia et al. 2007b, Lee and Noppeney 2011). One surprise was that people who spoke more than one language were also better at this task, independent of

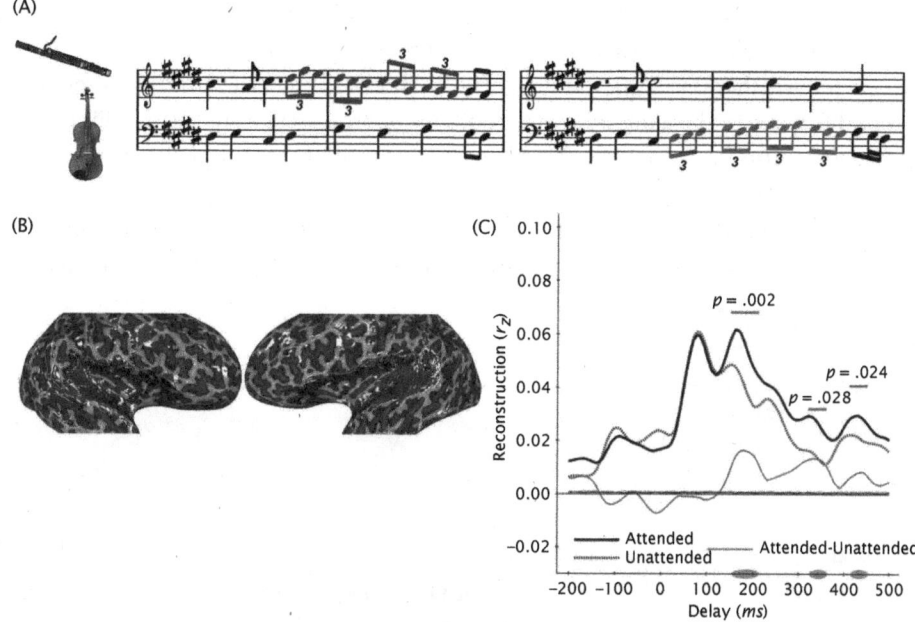

FIGURE 2.19 Selective attention in polyphonic music. A: Stimuli used for selective attention task consisting of two lines played by a bassoon and cello. Listeners were instructed to attend to one or the other and indicate the presence of triplets in the attended line. B: High-field functional MRI data during task performance showing cortical areas in temporal and frontal lobes whose activity differs depending on attentional state. C: EEG data during task performance showing reconstruction accuracy of each instrument's sound envelope based on EEG amplitude envelope; greater accuracy is observed for the attended line in the 200–500 ms range. Adapted from (Disbergen et al. 2018) and (Hausfeld et al. 2021). Reproduced under the terms of the Creative Commons CC BY license.

their musical training, which was a reversal of the usual music-to-language enhancement that has been discussed in the literature (Patel 2011).

To examine the neural basis of this ability, Emily and her team carried out an EEG study using a version of this music-in-noise task, in which multiple top-down cues were simultaneously present (Greenlaw et al. 2020). The findings showed that, as with speech, attention to the target was accompanied by better decoding of the EEG signal corresponding to the amplitude envelope of the target melody (Ding and Simon 2012). On the other hand, the frequency-following response was not enhanced, but instead corresponded to the entire mixture of sounds. These findings support the idea that top-down attentional mechanisms operate on later representations of sounds at the higher level after initial processing, as indexed by the frequency-following response.

If there are commonalities in the mechanisms used for listening to speech and musical targets in noise, then one might ask how musical experience affects speech processing. Although the evidence is not entirely consistent, a large number of studies have reported that musical training is associated with better perception of speech-in-noise (for review, see Coffey et al. 2017b), whereas not a single study has reported the reverse. Even just six months of musical training is enough to enhance perception of speech-in-noise in older adults (Zendel et al. 2019), which is an important finding since it demonstrates a direct causal effect of training, as opposed to a simple correlation (which could in theory be due to factors other than training and hence does not prove causality).

To test whether this music-training phenomenon is more related to better processing of sound features in the auditory cortex (bottom-up processes) or to better utilization of predictive cues (top-down processes), my former postdoc Yi Du, now on the faculty at the Chinese Academy of Sciences in Beijing, carried out a fMRI experiment using speech syllables embedded in different levels of noise (Du and Zatorre 2017). As shown in Figure 2.20, musicians outperformed people without musical training, as others have found, but the

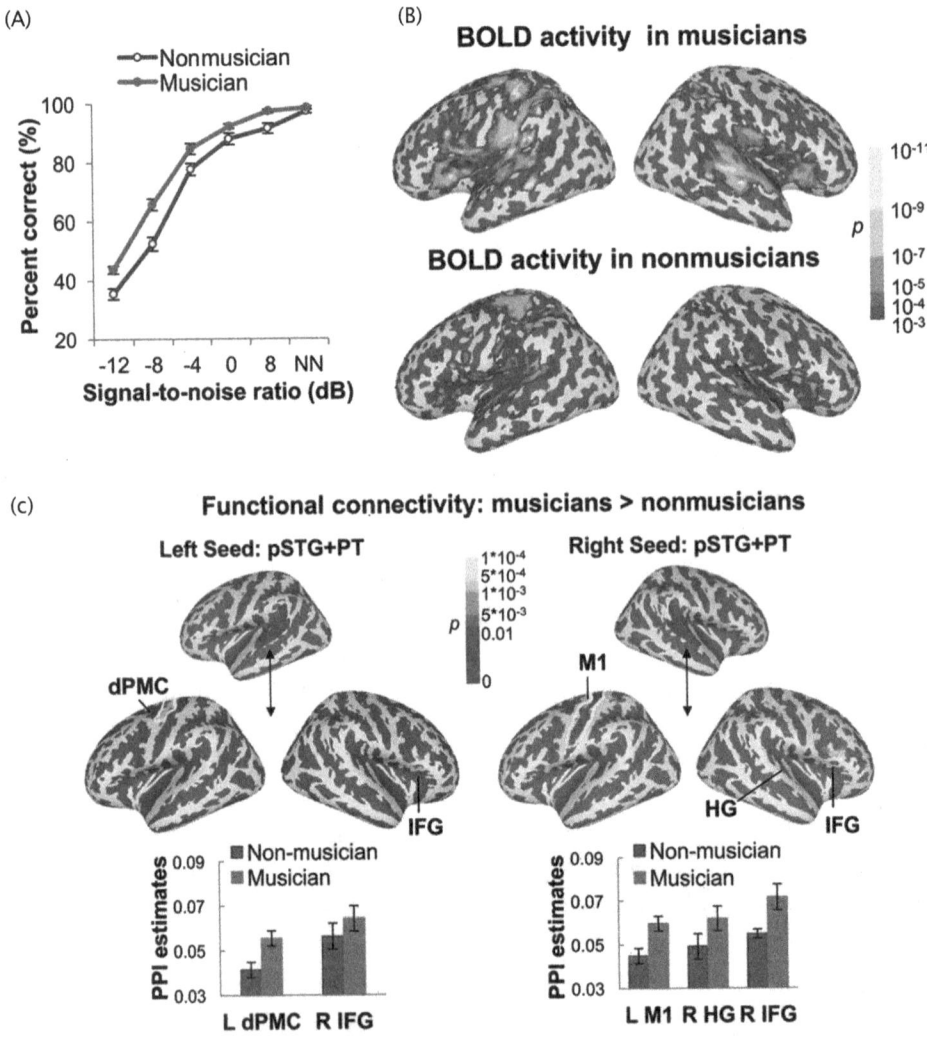

FIGURE 2.20 **Perception of speech in noise and effect of musicianship.** A: Behavioral performance showing consistent advantage for musicians at different signal-to-noise ratios (NN = no noise control condition, where there was no difference). B: Functional MRI responses during task performance shows more widespread pattern of activity among musicians in temporal and frontal lobes. C: Functional connectivity between auditory cortex and frontal/motor regions during task performance is greater in musicians. Adapted with permission from (Du and Zatorre 2017).

pattern of brain activity was more widespread in the musicians. Moreover, in the high noise conditions, the syllable identity could be decoded from frontal and motor cortical areas in the musicians. Finally, enhancements in functional interactions between auditory and frontal/motor systems were related to musicianship and better task performance.

We interpreted these results to mean that both sensory and cognitive/motor mechanisms are responsible for the advantage experienced by musicians. When the signal is relatively clear, then the acoustical cues can be handled within auditory cortex, consistent with the fact that cortical frequency-following response amplitudes are associated with better speech-in-noise perception (Coffey et al. 2017a). But once the signal is degraded, additional mechanisms involving more top-down cognitive functions, such as attention (Strait and Kraus 2011) or working memory (Kraus et al. 2012), as well as motor codes, become important in order to disambiguate the result. A complementary way of thinking about it is that predictive signals become more important when the input is degraded, whereas when the sensory information is of high quality, bottom-up signals coming from lower levels are sufficient to generate an accurate model.

In the case of selective listening of one speech stream over another competing one (as opposed to speech with irrelevant background noise), there is also evidence showing how musical training can enhance performance from a study using the envelope-tracking approach (described above Ding and Simon 2012) carried out by Sebastian Puschmann, now at the University of Oldenburg, but who was a postdoc in my lab at the time (Puschmann et al. 2019). Listeners with varying amounts of musical training attended to one of two simultaneous speech streams, while their brain activity was recorded with MEG. Once again, we found that better performance was associated with more musical experience. As expected, the amplitude envelope of the attended speech stream could be better decoded from the MEG signal than the unattended one; but to our surprise, the greater the amount of musical training, the better the *unattended* channel was represented in the brain data. In other words, rather than being better at suppressing irrelevant information, musicians have an enhanced ability to follow both the attended and unattended speech, leading to better separation of the two.

The brain topography analysis showed that only the auditory cortex was tracking the unattended speech stream, whereas a broader network, including not only auditory but also frontal and motor regions (as seen in the fMRI study of Yi Du), represented the attended part of the stimulus. This difference in the regions recruited suggests that the attended stream receives deeper processing, perhaps involving syntactic or semantic analysis. In retrospect, however, perhaps the outcome should not have been unexpected, because musical training often involves learning to attend to multiple lines simultaneously (Disbergen et al. 2018). Hence, it makes sense that this ability would come in handy during selective listening more generally. The concept of prediction is relevant here once again: keeping track of both the foreground and the background should help to anticipate the evolution of each stream over time and stabilize the segregation process (Elhilali and Shamma 2008). Since music training also involves the refinement of auditory anticipatory processes (Huron 2006), the application of this ability to nonmusical auditory stimuli as well seems plausible.

This capacity for simultaneous listening leads us to another situation relating to auditory scene analysis, which is particularly—perhaps uniquely—pertinent to music: stream integration (Uhlig et al. 2013). This ability involves the opposite of segregation; it occurs when multiple strands of sound must be integrated: as in a symphony orchestra in which many different instruments blend together to form a rich texture for example; or when vocalists in a barbershop quartet sing tight harmony in order to render a unified chord, rather than distinct voices. In many musical styles (for instance, in polyphonic music), both integration and segregation are simultaneously relevant, setting up ambiguous and enjoyable interplays between what musicians might call vertical (harmonic) and horizontal (melodic) listening (Pressnitzer et al. 2011). In the polyphonic melody study described earlier, we also included an integration condition, in which listeners had to attend simultaneously to both lines in order to detect triplets that went across the two instruments (Disbergen et al. 2018). Unlike in the segregation condition, listeners were unaffected by timbral differences between the two lines, which makes sense since attention is placed on the aggregate rather than the separate lines. During fMRI, a network of auditory, frontal, and parietal regions was consistently identified whose activity patterns could be decoded to indicate whether the listener was attending to one line only or to the combination, in agreement with other studies that also identify this extended network in the processing of melodies and their accompaniment (Ragert et al. 2014). These areas, outside the temporal lobe, suggest the involvement of more complex top-down mechanisms, some of which will be covered in Chapters 3 and 4.

Musical Imagery: Hearing in the Mind's Ear

Can you imagine *Imagine*? Most people who are familiar with John Lennon's signature tune find that they can vividly call it to mind: the distinctive, slightly world-weary voice with its hopeful message; the repetitive, broken piano chords. But can you also imagine it as sung by Lady Gaga? (She did a version of it in 2015 for the opening of the European Games.) How about as performed by a children's choir, or played on a xylophone? Many people report that they can do so, which means that they not only have an experience of the song, which is akin to the original sound, but that they can also imagine sounds they have never experienced before. This ability for musical imagery raises several interesting questions for neuroscience.

The most obvious question is, how are we able to accomplish these acts of evocation or re-creation? Embedded within this question is another: how can we investigate what appears to be an ineffable, completely internal process? Indeed, the hermetic nature of internal experiences, in general, led behaviorist psychologists to declare phenomena such as imagery to be out of bounds for years, in favor of overtly observable, and hence measurable, behavior. Yet, imagery is such a salient aspect of cognition that simply ignoring it is deeply unsatisfying. Besides, science is often confronted with measuring phenomena that are not immediately accessible or observable and hence must be inferred by their secondary effects on something that can be measured. This is certainly the case in other hard sciences: for instance, in astronomy, when the existence of an invisible planet is inferred based on how it makes an observable star wobble. So, psychology should not be any different.

The way forward in the study of mental imagery was shown in the 1970s, by cognitive psychologists like Roger Shepard, who used mental rotation tasks to demonstrate that visual imagery could be studied as systematically as any other phenomenon by measuring behavioral responses precisely correlated with imagery task parameters (Shepard 1978). Armed with this cognitive probe approach, we can proceed to investigate musical imagery in analogous ways. My friend and colleague, Andrea Halpern from Bucknell University, created one of the very first rigorous tests of musical imagery by asking people to make judgments about the pitch of certain syllables in a familiar imagined song (Halpern 1988). For example, in the first line of the song *Jingle Bells*, which of the two bolded words below is higher in pitch?

dashing through the **snow**, in a one-horse open **sleigh** . . .

Most people can easily call the song to mind and decide that the second syllable has a higher pitch. Since the pitch is not given by the text and is only available by accessing some internal stored representation of pitch information, this ability constitutes strong empirical evidence for the existence of imagery. Beyond that, in direct analogy to visual imagery experiments, this task also allowed Andrea to show that the time taken to make the judgment was a near-linear judgment of the temporal distance between the syllables probed. For instance, in the example above, it takes longer to make the judgment between "dash" and "sleigh" than between "snow" and "sleigh." This finding shows that the imagery process shares a critical feature with real listening: it unfolds over time. This is not a foregone conclusion, in the sense that one might imagine a non-imaginal internal representation consisting of something like a lookup table of the pitches in the song; under those circumstances, it would take the same amount of time to query two close syllables as it would two distant syllables. That this is not the case fits with our intuitions that imagery is akin to perception, in that one experiences the imagined event as extended in time, much as the original, real event also took place over a certain amount of time. In fact, the tempo of imagined music matches pretty well with the tempo of the original piece (Jakubowski et al. 2015).

The demonstration that imagery ability could be assessed via behavioral means prompted the study of its neural basis. The essential hypothesis in the first instance was that imagery depends, at least in part, upon the same sensory cortices that are involved in perceptual analysis of a real signal. This hypothesis follows from the behavioral data because the claim is that imagery and perception share processing features in common. If so, then it makes sense to assume that this similarity arises from similar neural computations being carried out in comparable cortical areas. Among the first direct tests of this hypothesis was an investigation that Andrea and I carried out on patients with resections of the superior temporal gyrus (Zatorre and Halpern 1993). We reasoned that if imagery and perception share an underlying neural substrate, they should suffer a common fate when that substrate is damaged, a logic previously developed in the visual domain by Martha Farah (Farah 1989). We found that patients with damage to the right auditory cortex showed deficits while performing the behavioral judgment of pitch change with familiar melodies. This finding was, as expected, based on the role of auditory cortex in representing pitch patterns and the fact that these patients were also impaired when the same task was performed with imagined sounds. In contrast, matched patients with left-sided lesions were unimpaired on both versions of the task.

In a series of functional neuroimaging studies Andrea and I examined the patterns of brain activity when people were asked to evoke previously experienced musical sounds. The most reliable finding in this series of studies was that certain regions of nonprimary auditory cortex—typically more on the right side—were recruited for imagery in the absence of overt sound. This effect was first demonstrated for a task similar to the song judgment task already described (Zatorre et al. 1996); it was later demonstrated in experiments requiring imagery of the timbre of musical instruments (Halpern et al. 2004) or the imagined continuation of a song after hearing its first few notes as a retrieval cue (Halpern and Zatorre 1999).

An interesting imagery task was developed by Sibylle Herholz and colleagues in Germany, who used MEG to measure the time-locked brain activity elicited by a (real) target tone that was preceded by an imagined melody (Herholz et al. 2008). When the target tone did not correspond with the listener's imagery of what the correct tone should have been, it elicited a neural mismatch response from the auditory cortex, when contrasted to the response associated with the correct target tone. The only difference between the two types of targets was whether they matched the listener's internally generated imagery or not. We will have more to say about the mismatch response in Chapter 3, but here it serves as evidence that neurophysiological mechanisms are producing internal auditory images, which are sufficiently real in neural terms, to modulate the neural response to the sounded target. It also shows that imagery can be temporally precise or else it would not have interfered with the time-locked response to the physically presented target. When this same task was adapted to fMRI in our lab (Herholz et al. 2012), we saw, once more, significant overlap between perceived and imagined sounds, in nonprimary auditory cortex as well as in dorsal premotor and supplementary motor areas, which implies interactions between auditory and motor systems (see further discussion in Chapter 4).

The activation of the auditory cortex in the absence of any external acoustical stimulus is, in a sense, the hallmark of imagery and can be taken to mean that sensory regions are not involved solely in the analysis of incoming information, but that they are also, in some way, the repository of perceptual information, which can be reactivated under various circumstances. This conclusion is consistent with the various pieces of evidence, discussed earlier in this chapter, indicating that auditory cortex, especially beyond the primary region, represents higher-order aspects of sound. In this context, it is useful to point out that such imagery tasks are only one way of activating auditory cortex in the absence of real sound entering the ear. For example, retrieval of a sound from memory can sometimes generate activity in auditory cortex (Wheeler et al. 2000). Similarly, anticipation of an auditory event (Voisin et al. 2006) or viewing silent lip movements (Calvert et al. 1997) can lead to auditory cortex activity, without an actual sound being present. The latter effect is likely related to a cross-modal association mechanism since moving lips (visual) and speech (auditory) have been experienced together very frequently in most people's experience.

The role of association is further supported by studies in which a sound paired with some visual input during learning can lead to recruitment of auditory cortex upon presentation of the visual cue, in the absence of the sound (Tanabe et al. 2005). The reverse process, that is, visual cortex activity in response to sound only, also occurs after brief bimodal exposure (Zangenehpour and Zatorre 2010). A musical counterpart to this association phenomenon can be observed when trained pianists (but not untrained controls) merely observe a hand playing a piano keyboard, resulting in recruitment of the auditory cortex (Haslinger et al. 2005). This topic is also related to motor-to-sensory interactions that will be discussed in greater detail in Chapter 4.

All these effects, as well as active musical imagery, may rely on similar underlying mechanisms. The similarity would be that some kind of representation of a sound is maintained in the sensory cortex that initially processed that sound, which becomes reactivated under conditions of retrieval, rehearsal, association, or directed imagery. This idea has been championed by Antonio Damasio, of the University of Southern California, whose team provided rather direct evidence that the pattern of brain activity in auditory cortices contains sufficient information to represent individual imagined items—not only music but other sounds as well (Meyer et al. 2010). In this study, the authors asked people to view silent videos of sound-producing events (a violin being played, coins dropping, a dog barking, etc.) and to imagine the corresponding sound. They then analyzed fMRI activity in auditory regions using a multivariate machine-learning classifier algorithm, which showed that the distribution of activity across these voxels was sufficient to correctly classify the specific item being imagined.

Taken together, these studies of auditory imagery show that both hearing and imagining a sound recruit similar auditory cortical areas. But beyond this topographic overlap, they do not indicate whether the underlying content of the representations is actually similar across perception and imagery, nor if the dynamic processes in imagery are the same across different individuals. To address this question, Mor Regev, a postdoc in my lab, carried out an fMRI study using intersubject correlation, a technique in which the dynamic pattern of brain activity of one individual is correlated against that of another individual, leading to the conclusion that any common result must be driven by similarity in the stimuli being processed since unrelated person-specific brain activity would not correlate (Hasson et al. 2004). This approach has been used with music before (Abrams et al. 2013), but the trick is that, with imagery, there is by definition no external stimulus—it's entirely covert. However, Mor was able to train people to memorize several musical pieces, keeping a strict tempo with a visual metronome, that subsequently had to be imagined inside the scanner (Regev et al. 2021). The results showed that the pattern of brain activity over time was indeed correlated between perception and imagery, across individuals, in several auditory cortical areas. But, more remarkably, as shown in Figure 2.21, this was also true when comparing activity across people when they were all merely imagining the same music. It was even possible to correctly classify which tune the person was imagining by analyzing the activity pattern coming from right nonprimary auditory regions.

These findings demonstrate that when we imagine music, the auditory cortex represents the information with sufficient fidelity that the contents of the conscious experience

(A)

(B)

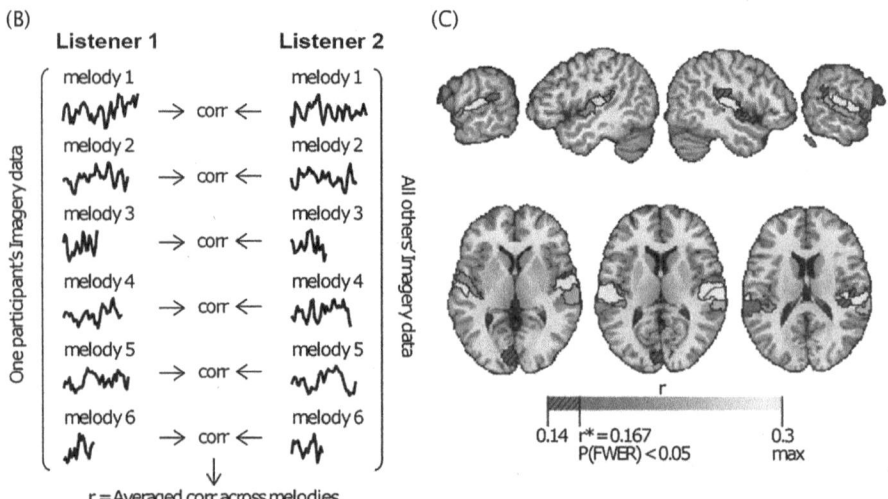

r = Averaged corr across melodies

FIGURE 2.21 Temporal pattern similarity in auditory cortex during musical imagery. A: Portion of melody that volunteers memorized for the study. B: Illustration of analysis used in which the temporal pattern of brain activity during imagery for one individual was correlated with the mean of the others, while they were imagining the same melody. C: Significant values in the inter-individual correlation are found in auditory cortical areas, indicating that the auditory cortex contains sufficient information in common across people that their internal representations of a covert process can be mapped. Adapted with permission from (Regev et al. 2021).

can be decoded, even when measured with the rather slow response of fMRI. Furthermore, the findings point to a shared representation across individuals, indicating that there must be a common dynamic code within auditory cortex that is similar in different brains. Additional, more detailed information supporting this conclusion comes from a fascinating case study of a pianist who was being evaluated for epilepsy with electrical recordings directly from the cortex (Martin et al. 2018). When he was asked to play a piece of music on a silent piano while imagining the sound, brain recordings showed similar activity to that generated by the actual sound of the piano. Even more amazing is that the spectrotemporal receptive fields generated during imagery matched those from the real sound, once again demonstrating the physiological reality of imagery.

When Internal Musical Processes Go out of Control: Earworms and Hallucinations

The results from imagery studies are reminiscent of the experiences reported by Penfield upon stimulation of auditory cortex, in that both sets of findings support the conclusion that higher-order representations of music and other sounds emerge from auditory cortex and its interactions with other structures. However, it is essential to distinguish between imagery and hallucinations, the principal difference being that in the latter case the individual believes the event to have a true, external source; whereas in the former, the phenomenon is typically under conscious control and is not confused with a sound coming from outside.

Yet, musical imagery can sometimes be triggered spontaneously, in a less controlled manner than has been typically studied in the material reviewed above. This phenomenon of involuntary, often repetitive imagery, sometimes dubbed "earworms," had received rather little scientific attention until recently (Liikkanen and Jakubowski 2020). Self-reports confirm that such episodes are common, often occur shortly after hearing music, and can last for periods of hours or more (Beaman and Williams 2010). They are also more common among people with musical backgrounds (Liikkanen 2012), which is not too surprising, since musical training is associated with enhanced imagery ability (Aleman et al. 2000) and enhanced auditory processing in general.

Although earworms get a bad rap, the majority of people actually find most involuntary imagery episodes to be relatively pleasant (Halpern and Bartlett 2011), even if unpleasant or annoying earworms certainly do occur. Although the neuroscience behind involuntary musical imagery is scant, one study did find that greater incidence of such episodes was associated with anatomical features of the right auditory and frontal cortex (Farrugia et al. 2015), which is very consistent with the functional neuroimaging data; however, it is not clear whether this relationship represents cause or effect. An intriguing idea is that when imagery becomes too repetitive and unpleasant, it could be because of a lack of control signals (Cotter 2019). These would, most likely, originate in higher-order structures that result in disinhibition of circuits that contain sound representations.

Hallucinations are a complex topic to study because they take many forms and are associated with many distinct disorders. Here, I will only discuss musical hallucinations, which are also associated with certain distinct triggering conditions. One distinction that is worth drawing concerns the attribution of the hallucinatory experience. Patients with hallucinations arising from psychosis typically believe the experience to be real. Whereas those who experience hallucinations from nonpsychotic conditions, usually understand that their phenomenological experience, though perhaps indistinguishable from reality, is nonetheless not produced by a true, external source of sound (Evers and Ellger 2004). The latter group are the focus here.

An anecdotal case report serves to illustrate the phenomenon. A number of years ago, I was called to see a gentleman who had been brought by his wife to the Montreal Neurological Hospital because he awoke her in the middle of the night to complain that he could not sleep due to the loud music coming from the neighbor's apartment. The woman was stupefied because the house was completely silent. Perhaps fearing her husband had gone crazy, she rushed him to the hospital. Notably, however, the patient had figured out by then that it could not be music from the neighbor's house after all, because it always came from the left side of space, no matter which way he turned his head, and because it persisted even after he left the house. It turned out to be caused by a small infarct in the right auditory cortex, which cleared after a day. This case is illustrative because it shows how musical hallucinations can be triggered by some abnormal stimulation of auditory cortex (in this case by the interruption of blood supply) and how this leads to a hallucinatory experience which is akin to reality, but does not lead to misattribution of the source, as it would in psychosis.

Tim Griffiths and his team in the U.K. have studied a particularly interesting type of nonpsychotic auditory hallucination, in which people experience ongoing, recognizable music in the absence of any known brain lesions (Griffiths 2000). The condition is most common in elderly people with severe hearing loss and often starts with buzzing or ringing (tinnitus), which over time progresses to hallucinatory music (reminiscent of Penfield's observation of buzzing elicited from stimulation of primary areas and music from more distal regions). These patients report hearing incessant music, sometimes with instruments and/or vocals, often from their childhood or early life (the list included rugby songs, church hymns, children's songs, light opera, and in the case of one unfortunate individual, songs by Shirley Bassey).

When Tim's group studied these hallucinations by means of functional imaging, they observed that the most active brain regions, in relation to the reported severity of the hallucination, included the nonprimary auditory cortex and inferior frontal regions. This finding was largely confirmed in a recent meta-analysis of multiple studies (Bernardini et al. 2017), and fits very well with the network of brain areas associated with normal music perception, that have been already described, and that will be discussed further in Chapter 3. The hallucinations themselves are, of course, quite reminiscent of Penfield's results with direct stimulation of the cortex, especially in that they usually involve familiar music, and hence suggest a similarity in the underlying mechanism in both cases. It is reasonable to suppose that, just as with tinnitus, there is abnormal hyperexcitability of cortical areas caused by the deafferentation of the peripheral hearing apparatus (Roberts et al. 2010). But instead of being confined to primary regions, it expands to include nonprimary auditory and frontal regions, which are normally involved with the encoding and perception of music (Vanneste et al. 2013). These networks can apparently become abnormally and spontaneously activated in the absence of real sound, resulting in the hallucinations.

Most recently, models have been developed which implicate predictive processes as the potential mechanism that is responsible for at least some hallucinatory experiences (Kumar et al. 2014). Recall that this model proposes that in a normally functioning system, ascending sensory information cascades from the periphery to progressively higher levels of the subcortical/cortical hierarchy, generating prediction error signals; while descending influences are viewed as predictions, which are modified or updated according to the sensory error signals that are received at each level from the level below. In the case of hearing loss, the ascending error signal is degraded, providing little or no corrective (prediction error) input to higher-order areas. These latter regions which, as we have seen, contain well-structured representations of musical information, are thus left to generate predictions without the usual constraints coming from lower centers. According to this concept, it is these predictions which generate the phenomenology of hearing hallucinatory music, via their influence on lower levels. This idea also fits with the observation that these complex hallucinations often emerge after a time, during which the person experiences simpler percepts of ringing or hissing, more like tinnitus. This phenomenon would be explained by the different hierarchical levels that are progressively affected over time by the disturbed input signal, as it propagates to higher levels.

Reprise

This chapter has provided an overview of the organization of the auditory cortex, emphasizing the aspects that are most relevant for music perception. The anatomy of the system—with its rich network of ascending and descending connections between subcortical structures, core auditory cortex, and belt and parabelt areas—provides the substrate for the encoding of relevant acoustical cues and enables the interplay of ascending, bottom-up sensory signals, with top-down, control or prediction signals, as shown in the sensory-specific adaptation phenomenon. The early stages of processing encode frequency information of periodic signals, coming from the outside world, according to the principle of tonotopy, while later stages represent pitch in an invariant manner, relatively divorced from specific frequency content.

The fundamental frequency of complex tones is also represented in cortical and subcortical structures via phase locking of neurons, which can be measured as the frequency-following response. This response is sensitive to musical training and is predictive of overall fidelity of sound encoding. The representation of pitch information can be seen as an example of predictive mechanisms such that earlier structures feed frequency information to later structures, which form an abstract model of the pitch and in turn send predictive signals back to earlier structures, leading to stable and invariant representations. More complex harmonic and octave relationships between frequencies are also represented in later stages. Neural units in auditory cortex also represent sounds, with or without periodicity, by jointly coding their spectral and temporal modulation content, providing an efficient way to encode natural sounds with their widely varying acoustical features.

The principles of hierarchical organization and predictive coding explain some of the remarkable functions that the auditory system is capable of. Analysis of complex auditory scenes with multiple overlapping events is accomplished via a combination of mechanisms. Sensory-specific adaptation comes into play to explain the segregation of ongoing signals containing multiple frequencies, as neural signals are disinhibited for frequencies that are further apart. Tonotopic frequency organization can act as a filter for sound events that occur in different frequency bands. Pitch-sensitive responses can be used to segregate periodic from nonperiodic sounds. Attentional modulation of one sound over others is accomplished via top-down modulation of the target sound. This ability is enhanced in musicians, both for musical and speech signals, and is related to better sensory encoding as well as better predictive top-down effects.

The ability to imagine sounds also rests upon mechanisms within auditory cortex. Belt and parabelt areas most likely contain representations of complex, previously experienced sounds, including musically relevant sounds. Perception and imagery processes overlap in these cortical areas. These internal representations are reinstated in consciousness, in a controlled way, during imagery. Brain activity in these regions contain sufficient information to determine what sound is being imagined, and there is a common imagery code across individuals when they imagine the same music. The evocation of imagined sound most likely depends upon top-down control signals. These processes can go awry when there is a disturbance of either the top-down control or the bottom-up input signal. In the case of repetitive involuntary imagery, lack of top-down control may be involved. In some cases of hearing loss, peripheral deafferentation results in a degraded ascending signal, which leads to descending predictive signals without constraint from any input, resulting in auditory and, sometimes specifically, musical hallucinations.

3

Communicating Between Auditory Regions and the Rest of the Brain

The Ventral Stream

Music extends over time. But it's also evanescent. Once a sound is made, it literally vanishes into thin air, as the sound waves dissipate. This concept is so obvious that we tend not to think about it much. But it presents a significant problem for the nervous system: how can the relationships between successive sounds be understood if each one is gone before the next one is sounded? All sounds extend over time, but in music, it is the pattern—the way the sounds are structured across time—that carries its complex and often beautiful messages. Therefore, in order for us to be able to perceive and understand music, we need a neural system that can capture and store acoustical events from our environment so that we can operate on them. A major part of the mechanism to bridge this time gap emerges from the connectivity of the auditory cortex with ventral brain regions.

In this chapter I focus on the way that auditory information is processed within the ventral stream. One of the key features of the ventral stream relates to its capacity to maintain auditory information in an online memory system, thereby enabling sound elements that have been experienced, but are no longer physically present, to bind to one another. This process is critical to perceiving the simplest musical structures, such as the scale steps that form a melody or the timing of rhythmic sounds in relation to one another. Thanks to this capacity to retain sounds, we can build longer, time-invariant representations, enabling us to overcome the natural limitation of sound and create larger structures. Based on this capacity, another key function of the ventral auditory system is to create representations of complex sounds and organize them into different categories. Such representations are not limited to music but also include various other types of sound categories, such as speech sounds, voices, or other sounds from the environment. These representations, in turn, can

From Perception to Pleasure. Robert Zatorre, Oxford University Press. © Oxford University Press 2024.
DOI: 10.1093/oso/9780197558287.003.0003

become a permanent part of our long-term memory, which is dependent on interactions of the ventral stream with medial temporal lobe memory mechanisms. Long-term memory plays an important role in interpreting perceived sound patterns in relation to stored knowledge about regularities in such patterns.

The ability to compare sounds to one another and recognize patterns supports another function that is critical to the main concept of this book: the capacity to make predictions about future musical events based on past events. Unless the brain can maintain a representation of information in some way, both in the short term (what did I just hear?) and in the long term (how does what I am hearing now compare with what I have often heard in the past?), it would be impossible to have an idea of what's coming next based on prior experience. As we shall see, expectancies derived from statistical consistencies of sound patterns critically depend on structures within the ventral stream. But, before we get to that, let us first describe the organization of the connectivity from the auditory cortical regions to the rest of the brain.

The Auditory Ventral and Dorsal Pathways

In Chapter 2, we saw how basic sound properties are handled in the early components of the auditory brainstem and cortex. Those functional characteristics reflect the features that our brains are tuned to. But information must also flow from there to the rest of the brain for more complex cognition to occur. The basic organization of corticocortical outputs from (and inputs back to) auditory areas can be summarized broadly in terms of two major pathways, the ventral stream, shown in Figure 3.1, and the dorsal stream, shown in Figure 4.1. The ventral pathway, which is the topic of this chapter, interconnects the primary auditory cortex with other auditory areas located in the temporal lobe in front of primary regions, and in the inferior part of the frontal cortex, as well as with medial-temporal structures. The dorsal pathway interconnects auditory cortex with regions located further back and in the dorsal areas of the brain, including premotor regions, and plays a role in many cognitive functions, especially auditory-motor operations, which will be covered in detail in Chapter 4.

We know that this ventral–dorsal distinction is somewhat of an oversimplification because of the complexity of the pathways connecting distinct cortical regions. In fact, it might be more appropriate to speak of multiple streams beyond just two, subdividing into several sub-streams, each with different projection targets. Moreover, these pathways display some degree of asymmetry across the two hemispheres (Takaya et al. 2015), a topic that is covered in greater depth in Chapter 5. Additional evidence for subdivisions within each pathway have been identified for certain special functions, including language (Friederici 2016). Furthermore, interactions between the two streams must exist to be able to solve many real-world perceptual problems (Cloutman 2013). For example, to play the next chord in a sequence of chords on a guitar requires representing the chord pattern (ventral stream) as well as organizing the position of the fingers on the fretboard (dorsal stream), indicating that there are additional connections present that have not yet been fully charted. So we should not think of the two streams as entirely independent from one another.

Despite all these caveats, the dorsal–ventral distinction has proven extremely useful, both in organizing our understanding of the system and in generating hypotheses about its

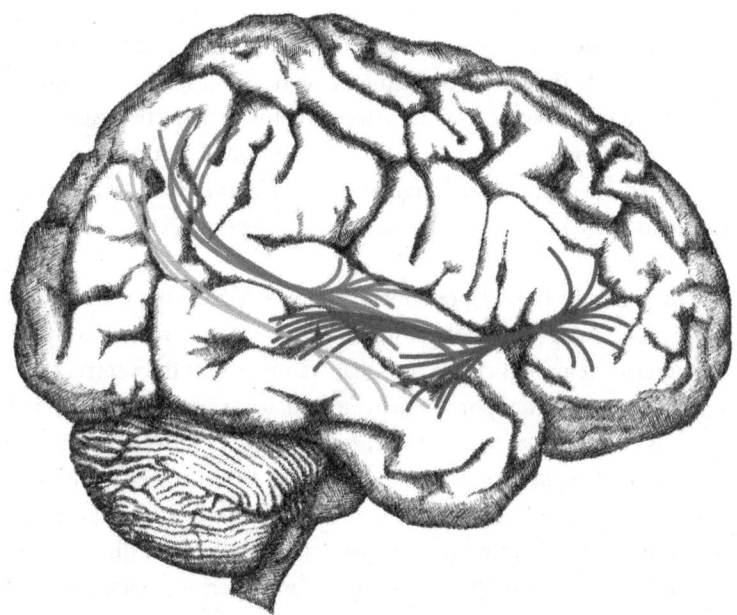

FIGURE 3.1 Auditory ventral stream. Diagram shows simplified schematic depiction of the anatomical connectivity of auditory regions along the superior temporal gyrus with regions throughout the temporal lobe and with regions of the inferior frontal lobe. Colors represent different fiber tracts. Connectivity pattern based on diffusion MRI adapted with permission from (Petrides 2014). Drawing by E.B.J. Coffey; reproduced with permission.

functionality. As a heuristic, this framework is also valuable because it parallels a similar distinction previously made in the visual system (Kravitz et al. 2013; Gallivan and Goodale 2018), which, in turn, allows for the alignment of functions that depend on auditory–visual integration. Along with other anatomists and physiologists, Josef Rauschecker and his colleagues have provided a clear-cut model of the structural and functional connectivity patterns of the auditory dual streams, based not only on assumed cross-species homologies but also on neuroimaging data in humans (Rauschecker and Scott 2009; Rauschecker 2018). Both this model and a related one, developed by Greg Hickok and David Poeppel (Hickok and Poeppel 2007), focus primarily on speech processing. Joyce Chen, Virginia Penhune, and I also applied these concepts of dual-stream organization to the processing of music (Zatorre et al. 2007). We make use of all these ideas to structure our understanding of the ventral stream in this chapter and the dorsal stream in the next.

Auditory Ventral Stream Anatomical Connectivity

Based on tracing studies of white-matter fibers in the monkey brain, the anatomical interconnections between different cortical regions within the temporal lobe are rich and complex. As a general rule, one may conclude that there are hierarchically organized connections starting in early auditory cortex, which project progressively to more anterior subdivisions of the auditory cortex, thus forming the principal axis of the ventral stream (Kaas and Hackett 2000). These anterior auditory cortical areas, in turn, send connections to the inferior frontal cortex, with an orderly topographic organization, such that the anterior portions of the temporal lobe project to the inferior frontal cortex (Petrides and Pandya 2009), in contrast to the connections originating in posterior auditory cortical areas that mostly terminate in more posterior and dorsal frontal regions (Petrides and Pandya 1988; Romanski et al. 1999a).

Connectivity between the temporal lobe and the inferior frontal cortex can be tricky to sort out, in part because portions of the inferior frontal cortex receive converging inputs from both the anterior temporal cortex, or what we might call the ventral stream proper, and also from more posterior temporal-lobe regions that are generally considered to be one of the subdivisions of the dorsal stream (Petrides 2014). In practice, when interpreting responses in the inferior frontal cortex, it is often difficult to know which pathway may have been more important in the transmission of information. However, most of the evidence we review in the context of music-relevant processes points toward the direct ventral connection between anterior portions of the temporal lobe and inferior frontal cortex.

The anterior portions of the superior temporal gyrus, corresponding to auditory parabelt areas, not only interconnect with the lateral frontal areas but also with the orbitofrontal cortex (Hackett et al. 1999), which is of particular importance because this connection most likely provides the interface to the reward system, as we discuss in more detail in Chapter 6. In addition to the cortical connections to the frontal lobe, projections flowing along the ventral stream also terminate in medial-temporal areas, such as the hippocampus and parahippocampal gyrus, which are important for memory formation (Munoz-Lopez et al. 2010), as well as in various nuclei of the amygdala (Stefanacci and Amaral 2002).

These conclusions are largely based on work done in non-human species and thus include the inherent limitations of assuming that such patterns are identical to those in humans. However, human post-mortem studies do indicate a certain degree of cross-species similarity in the architecture of auditory areas (Hackett et al. 2001; Tardif and Clarke 2001). Diffusion imaging studies also provide evidence for some similarity in organization of the connections. For example, topographically organized projections from the auditory areas in anterior temporal lobe to inferior portions of the frontal cortex have been observed with diffusion magnetic resonance imaging (MRI) (Frey et al. 2008), consistent with what is found in other primates. One of the few studies that used diffusion imaging to directly compare monkeys, chimpanzees, and humans found that all three species possessed dorsal and ventral fiber tracts to and from auditory cortex but that they differed in their relative strength and lateralization (Balezeau et al. 2020). While this finding supports the validity of cross-species comparisons, it simultaneously serves as a caution against the assumption that the organization is identical. Diffusion imaging in humans also confirms that anterior temporal regions within the ventral stream are connected to some subdivisions of the amygdala (Abivardi and Bach 2017).

All these connections are generally bidirectional, allowing auditory information coming from the environment to flow upward, from early processing areas through the entire ventral stream, while also allowing top-down influences, from frontal and other regions, to modulate processing in the earlier areas. In addition, another important feature of this system is that even though there is topographic segregation in the connectivity to frontal cortex (Romanski et al. 1999b), some frontal cortical regions serve as a kind of convergence zone for ventral and dorsal streams since information coming from both streams interacts there (Rauschecker and Scott 2009). Thus, the entire system can be thought of as a loop with reciprocal interactions that enable the functionality needed for specific aspects of auditory cognition to emerge.

Functional Roles of the Auditory Ventral Stream

Parallels to the Visual System

The original impetus for the idea that sensory processing could be understood within the framework of two functional streams came from the now-classic research of Mortimer Mishkin and Leslie Ungerlieder and their colleagues in the visual domain (Mishkin et al. 1983). These authors amassed substantial evidence that visual processing could be characterized in terms of representations of objects within a visual ventral stream, contrasting with representation of spatial information in the dorsal stream, although the latter pathway can perhaps be better conceptualized as pertaining to visuomotor and action processes (Milner and Goodale 2006), as we shall see in Chapter 4. An important concept in this literature is that of a processing hierarchy: according to this view, as one moves anteriorly, progressively more complex aspects of objects are represented within the ventral stream, culminating in abstract neural representations that can respond invariantly to certain stimuli.

Following this view that the visual ventral stream processes object information, investigators have sought a parallel functional organization in the auditory system. However, the question of what exactly might constitute an auditory object is still a matter of debate (Griffiths and Warren 2004), especially because some of the kinds of features that are relevant for visual objects (such as colors or edges, for instance) have no obvious auditory counterpart. Whereas, conversely, certain high-level auditory features that pertain to music (such as meter or tonality) have no obvious visual counterpart. For our purposes, however, it is sufficient to conceive of auditory patterns as analogous to visual objects, to the extent that objects are characterized by fixed relationships of their constituent elements. That is, a musical pattern has certain invariant properties, such that it hangs together and allows recognition. Musically relevant features that define an auditory pattern might be provided, for instance, by a minor scale, or by the dotted rhythmic sequence characteristic of a tango, or by a typical chord progression used in blues music. Each of these might be considered a stable feature which defines a pattern since they are recognizable when played on different instruments, at different speeds, or at different levels of loudness. Because we have experienced them in the past, we have a permanent record of the relationships between elements that characterize these patterns. These patterns can then be combined into larger units (a bluesy tango in minor mode, perhaps) leading to more permanent and even more abstract representations of themes, riffs, songs, movements, or even larger musical units. In this sense, the idea is not so dissimilar to vision, in that particular visual features that are characteristic of an object—say its color, shape, or texture—are combined into a whole, which allows permanent representations to emerge in the ventral stream (Reddy and Kanwisher 2006). Keeping these issues in mind, let us consider the evidence for the existence of auditory representations within the ventral stream.

Lesion Effects

Some of the earliest evidence showing the importance of auditory cortical areas in the anterior temporal lobe for the processing of musical patterns came from studies of patients with lesions to these areas. In a seminal study that strongly influenced my own career in this field, Brenda Milner studied how cortical excisions affected auditory processing in patients undergoing surgical resection for control of epilepsy (Milner 1962). She reported that patients with anterior temporal lobe removals on the right side (but not the left) were impaired on a task that required the listener to identify an altered note in a melodic pattern (See Figure 5.1). The hemispheric asymmetry is of great interest, in its own right, and is dealt with in Chapter 5. But for our current discussion, this study showed that these ventral stream regions were critical for melodic pattern analysis since the deficits appeared after surgery but were not present before.

Starting in the 1980s and continuing for two decades, my students and I carried out a series of studies with this same patient population. It should be remembered that functional imaging had not yet been developed when we began this work. As such, this kind of classic lesion-behavior analysis using carefully designed cognitive tasks was the best technique then available, following in the rigorous style established by scientists like Hans-Lukas Teuber (Gross 1994) and, of course, Brenda Milner. Examining the effects of surgery for epilepsy had the advantage of a relatively homogenous and well-controlled population, and with the collaboration of many of our neurosurgeons at the Montreal Neurological Hospital, especially Drs. William Feindel and André Olivier, we had adequate documentation of the location and extent of the surgical excisions. Furthermore, since the most frequent target of the surgery was the temporal lobe, and since at least some portion of the superior temporal cortex was always included in the resection, we had a good model for looking at auditory function.

With Séverine Samson, the very first PhD student in my lab (now a professor at the University of Lille in France), we observed that lesions within the anterior portion of the temporal lobe disrupted processing of musical patterns (Samson and Zatorre 1988), in agreement with Milner's prior findings. We were also able to show the importance of anterior temporal cortex for both short- and long-term retention of pitch information (Zatorre and Samson 1991). Adapting a task that was originally developed by Diana Deutsch, who had suggested that pitch memory was separable from verbal memory (Deutsch 1970), we found that patients with right temporal lobe excisions were less able to maintain pitch over a brief time interval filled with interference tones, even though they were unimpaired in making the judgment if the same time interval was silent, as seen in Figure 3.2. In another study we found that both right and left temporal lobe damage slowed the learning of novel melodies. However, longer term retention (24 hours) was specifically impaired after right temporal damage, whereas retention of a word list was impaired after left temporal damage (Samson and Zatorre 1992). In later work, my PhD student, Cathy Warrier, and I also found that right temporal lesions impair the ability to use melodic context to make pitch judgments (Warrier and Zatorre 2004), thereby providing further evidence that the anterior temporal lobe is important for integrating patterned pitch information across time since the context extended over several seconds. Many other studies with similar patients (Liégeois-Chauvel et al. 1998), or with more diverse lesion populations, have been carried out (summarized in Sihvonen et al. 2019). Although the results of these studies can be quite heterogenous, the broad outlines are consistent enough to be able to conclude that many aspects of both perceptual and mnemonic processing rely on the ventral auditory stream.

Although the dual-stream concept was only nascent at the time and was not yet articulated in the auditory domain, the perceptual deficits we and others observed after temporal lobe lesions are consistent with the proposed role of the ventral auditory stream in pattern analysis. However, since these tasks typically require a comparison of sequentially presented sounds, it is hard to dissociate perception from memory function, as some kind of memory capacity must also necessarily be involved. Severe difficulties in short-term pitch retention were also observed in people with damage to the right frontal cortex (as shown in Figure 3.2), consistent with the idea that the ventral stream circuitry, linking the anterior temporal cortex to the frontal cortex, is important for immediate retention. On the other hand, the deficit in longer-term encoding of melodies is more likely to be explained in terms of the communication between anterior temporal areas and medial temporal areas, which are involved in the formation of more permanent memories.

Early Neuroimaging Data

Analysis of behavioral deficits in individuals with well-documented and consistent lesions provides direct causal evidence for the importance of a given structure to a given function. However, such studies also present numerous limitations. One of the most significant issues is that it is difficult to infer the network interactions between brain regions that undoubtedly underlie complex cognitive functions from the effects of damage to a single region. Furthermore, even with relatively controlled surgical excision, there is both a disruption of cortical function as well as an interruption of connectivity due to cutting of white-matter fibers, making interpretation more uncertain. Functional neuroimaging has its own limitations as well, notably its inability to demonstrate causal relationships. Nonetheless, its advent in the 1990s allowed cognitive neuroscientists unprecedented opportunities to study the neural correlates of behavior at the whole-brain level in healthy volunteers. In our specific case, it allowed us to build upon the understanding we had developed from the lesion studies of music-related functions with much greater detail and sophistication than had been previously possible.

The very first functional imaging study that I carried out focused on speech perception. I had expected that a study on speech would most likely have a greater impact than one about music, which at the time might still have been a bit risky (plus, the scans were expensive, and I didn't have that much funding). However, I couldn't help but sneak in an additional condition in my experiment, in which people had to make judgments about

(A)

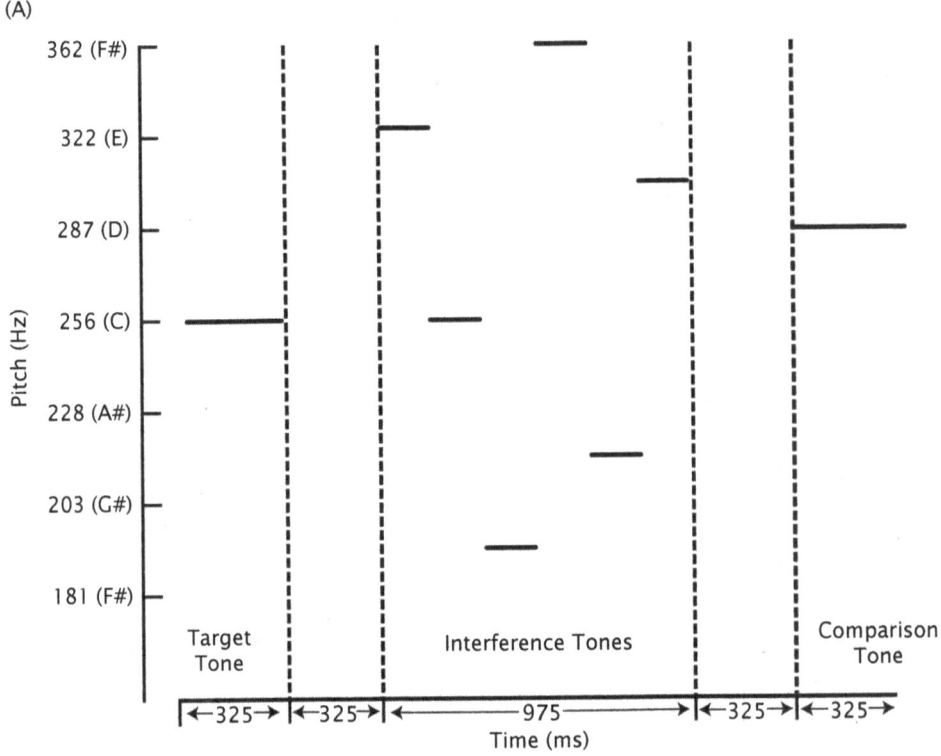

(B)

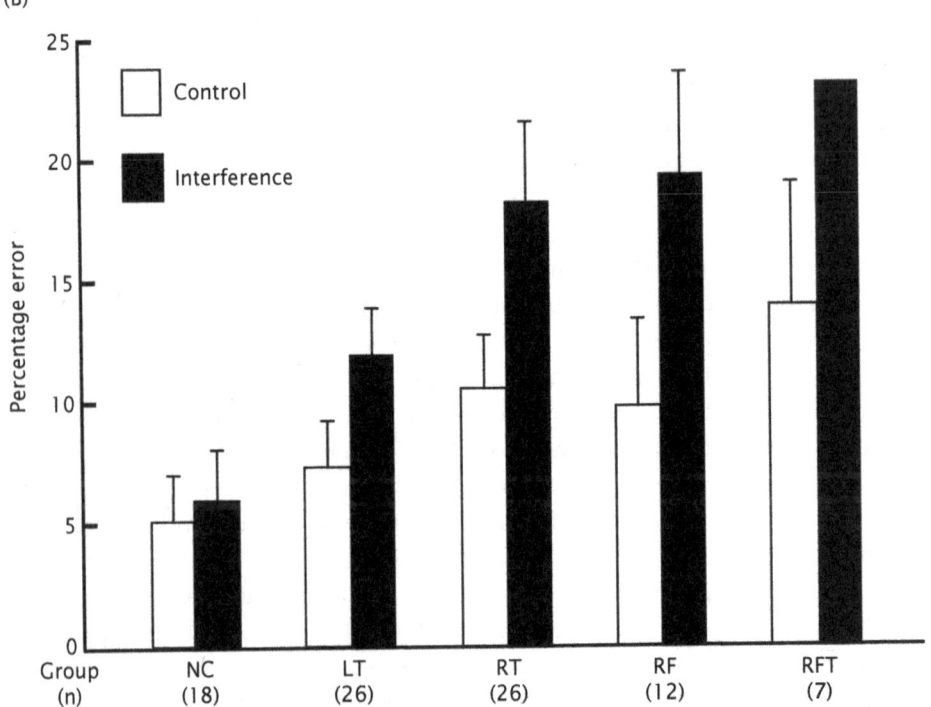

whether the pitch of one syllable was higher or lower than the pitch of the preceding one (Zatorre et al. 1992). The results included activation of the right inferior frontal cortex in this pitch comparison task, which did not occur with judgments of linguistic features on the same stimuli. We concluded that the right frontal cortex was involved in judging pitch changes, which was in congruence with the lesion literature.

In a second neuroimaging study, we carried out a more explicit analysis of perception and retention of pitch in a musical context. We used simple eight-note tonal melodies (the same homemade ones that had proven sensitive to right temporal-lobe lesions [Zatorre 1985]) and asked people to perform each of three tasks: to just listen; to compare the first two tones and judge the pitch as rising or falling; and to compare the first and last tones of the melody while ignoring the intermediate tones, mimicking the previously used Deutsch paradigm (Zatorre et al. 1994). The idea was to manipulate the load on auditory working memory in a systematic manner, without changing the stimulus or response, so as not to confound the outcome. The results showed first that just listening to melodies (compared to an acoustically matched control sound) generated activity in the right superior temporal region within the auditory ventral stream. Second, they confirmed the importance of the frontal cortex, both inferior and dorsolateral, for retention of pitch information, as these regions responded during the active comparison of the first two tones but did not respond during passive listening. And third, they also showed a much more complex distributed network of regions, including auditory cortex, parietal, and several frontal areas, which were involved in the more demanding working memory task of comparing the first to the last tone. Other early studies investigating working memory tasks yielded similar outcomes, implicating, in particular, the auditory cortex and inferior frontal areas, often with a right-hemisphere bias (Holcomb et al. 1998; Griffiths et al. 1999). However, these studies, as well as those employing more complex working memory conditions (Gaab et al. 2003; Hickok et al. 2003), also reported more distributed responses throughout the posterior and dorsolateral frontal areas.

A very relevant paper, focusing more on pattern processing than working memory, came out of the UK in 2002, from the labs of Roy Patterson and Tim Griffiths, who performed a systematic investigation of melody processing by varying the complexity of stimuli from sequences of noise without pitch, to sequences with a fixed pitch, to random and diatonic melodies (Patterson et al. 2002). The results showed an orderly hierarchy of neural responses along the superior temporal gyrus, such that there was a topographically organized response as one proceeded anteriorly along the auditory ventral stream. Thus,

FIGURE 3.2 **Role of right temporal and frontal regions in retention of pitch**. A: Diagram of stimulus used in pitch retention task: after a target tone was sounded, six random interference tones were presented, followed by a comparison tone which could either be the same or different as the target. The control condition (not shown) involved the same target and comparison but without interference. B: Performance of patients with excisions of the left or right temporal cortex (LT, RT), right frontal cortex (RF), right frontal and temporal cortex (RFT), and controls (NC). The largest increase in error rate was observed for patients with right temporal and/or frontal removals, and this effect was most significant for the interference condition compared to the control condition. Reproduced with permission from (Zatorre and Samson 1991).

noise activated areas around Heschl's gyrus, while fixed pitch sounds engaged the pitch-sensitive area described in Chapter 2, and melodic sequences involved the most anterior regions, especially in the right hemisphere (see Figure 3.3). A similar pattern was observed in subsequent studies as well (Warren and Griffiths 2003). These results were interpreted as evidence for the kind of cortical hierarchy proposed by dual-stream models in monkeys (reviewed above and in Chapter 2), where more anterior areas respond to increasingly more complex patterns, as one ascends from primary to belt and parabelt areas. Such an explanation also fits with the view that ventral processing entails more integration across time since more complex patterns extend over a longer period.

Although the methods were relatively primitive compared to what we can do today, this first decade of neuroimaging provided important and exciting results that were essential for laying the foundation for understanding the role of the ventral stream in auditory tonal processing. But much more detail remained to be added to achieve a fuller understanding of the functional characteristics and organization of the auditory ventral stream. Specifically, the two important functions that were uncovered by this work revolve around the twin questions of the mechanisms for maintenance of auditory information in a temporary working memory system and the nature and localization of the representation of

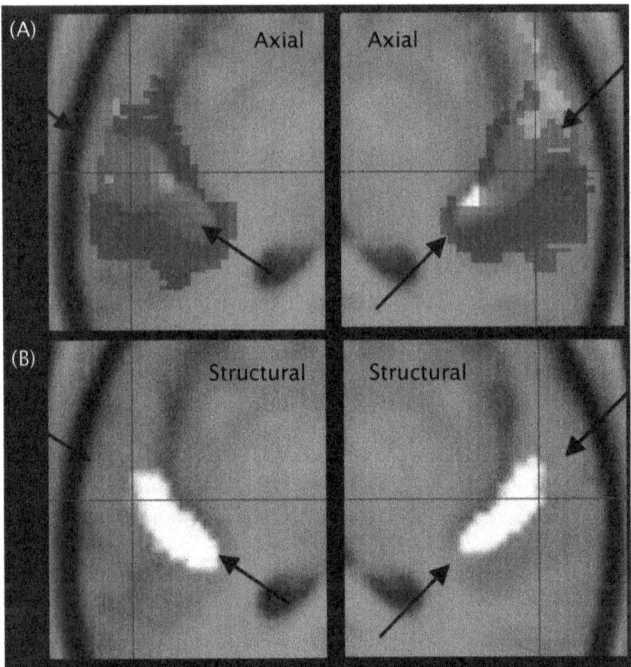

FIGURE 3.3 Hierarchical functional organization of auditory regions for melodic patterns. A: Functional MRI activity associated with four conditions. Dark blue: noise compared to silence; Red: fixed-pitch sequence compared to noise; Green: diatonic melodies compared to fixed pitch; Light blue: random pitch sequence compared to fixed pitch. Note the greater activity in and around Heschl's gyrus (shown in white in panel B) for noise stimulus, and progressively more anterior activation for more complex tone patterns, with a predominance in the right hemisphere. Reproduced from (Patterson et al. 2002) with permission from Elsevier.

stimulus features. We turn next to more recent studies on these two questions, dealing with the working memory function first.

Working Memory Retention

The lesion data and early imaging studies consistently implicated anterior auditory cortex and inferior frontal regions in tonal working memory. These basic conclusions were generally in line with a great deal of research in other domains, notably in the literature regarding visual working memory circuitry (D'Esposito and Postle 2015). But these first findings on tonal processing did not specify how the different aspects of this complex process may be carried out. Indeed, one of the difficulties in addressing that question is that even a seemingly simple task, such as deciding if a tone heard now is similar to a tone heard a few seconds ago, might involve many distinct cognitive functions (encoding, maintenance, monitoring, suppression of irrelevant information, retrieval, decision-making, and response selection). It is, in part, for this reason that I use the term "working memory" here (as opposed to "short-term memory," which is preferred by some authors) to emphasize the active, multifaceted nature of the skill, while still distinguishing between maintenance or retrieval of information versus acting upon or manipulating the information. The latter processes are covered in Chapter 4 as they are more dependent on dorsal stream structures. Although recent studies have been able to disentangle some of this complexity, we should also note that studies on working memory in a musically relevant context still represent a tiny minority of the total research, most of which is carried out in more conventional domains, such as verbal and visual processing (Owen et al. 2005). As such, much remains unknown.

Several functional imaging studies on auditory working memory focused on comparing tonal versus verbal retention; for example, work from Stefan Koelsch's lab used a clever paradigm in which tones and speech sounds were presented simultaneously for retention (Koelsch et al. 2009; Schulze et al. 2011). The findings emphasized the overlap of neural responses for both types of stimuli, again involving widespread areas within temporal, frontal, and parietal lobes along with subcortical regions. Although these findings support the role of ventral stream structures in working memory, the tasks most likely engaged a wide range of other structures too because they involved monitoring and selecting the right type of information (verbal or tonal) from complex multidimensional stimuli. On the other hand, when verbal and tonal trials are presented separately, the patterns of activity still tend to overlap but are more focused in the ventral stream (Albouy et al. 2019b).

One critique of these sorts of studies is that they focus on retention of the pitch of one or a few tones devoid of appropriate context, which is arguably irrelevant to music perception since one would normally perceive and need to maintain more complex relational patterns in mind. Another related issue is that they use artificial experimental tasks that may have little resemblance to real music processing. With regard to the first issue, it should be pointed out that the results are similar whether people are asked to retain a single pitch or an entire pattern. Studies using either simpler or more complex melodies instead of single tones, for instance, have largely shown engagement of the temporo-frontal circuit (Griffiths et al. 1999; Foster and Zatorre 2010b; Albouy et al. 2019b). So it is likely that this system is relevant for both simpler and more complex working memory performance.

Regarding the second critique, it is true that artificial tasks are commonly used because they provide the necessary experimental control and allow one to examine different aspects of working memory. However, it need not follow that such tasks do not reflect what happens with music listening in a real-world context. A nice demonstration of this conclusion comes from a study from Elvira Brattico's group in Helsinki (Burunat et al. 2014), in which listeners were scanned while they heard an entire piece of music (one of my favorites, as it turns out, Piazzolla's *Adiós Nonino*), without performing any explicit task. The authors analyzed the brain data by looking for activity that was sensitive to the repetition of the various themes used in the music. They found, among other things, a strong response in several frontal areas, including the inferior frontal cortex, which implies that this network of brain structures was involved in the recognition of the repeated motif, which of course could not happen without good working memory capacity.

As an aside, it's interesting to note that repetition plays a significant role in music. My friend, the musicologist David Huron from Ohio State University, has documented that repetition is prevalent in music from many diverse traditions (including Inuit throat singing, Caribbean Calypso, Ghanian drumming, 15th-century Chinese music, and so forth). He estimated that, on average, approximately 94% of musical passages lasting longer than a few seconds are repeated at some point within a given work (Huron 2006). This degree of repetition would be very strange in other art forms (imagine reading a novel, or watching a movie, and finding that an entire passage or scene was repeated). Repetition in music may have several different functions (Margulis 2014), but for it to be effective at all, it clearly requires a solid working memory capacity, or else one would not notice that a section had been repeated. Conversely, because sounds are so evanescent, repetition serves to consolidate more longer-term sound memories during a piece. Perhaps more interestingly for our story, repetition is also a way to increase predictability, as David Huron suggests. As such, working memory becomes part of the machinery to enable better internal models of ongoing and upcoming events and hence plays a role in predictions. We shall take up this thread in Part II of this book.

Working Memory Maintenance and Retrieval

Most of the studies reviewed thus far have examined the entire working memory process. But what happens specifically during the retention interval? This time period, when a sound has been presented and a listener must retain the information with no further input, is especially interesting because one can study the endogenous process of maintenance decoupled from the stimulus-evoked responses. This is difficult to do with functional MRI (fMRI) because of its relatively limited ability to track neural events over time. However, by acquiring functional data with fast temporal sampling, Sukhbinder Kumar and colleagues in the UK were able to plot out the responses associated with each phase of a tone retention task (Kumar et al. 2016). As seen in Figure 3.4, their results showed sustained activation in auditory cortex during retention, as well as in inferior frontal areas and hippocampus, in keeping with our ideas about the involvement of the ventral stream in maintenance.

Furthermore, functional coupling between these regions was enhanced during the maintenance phase, indicating that these structures form a functional network. Finally, they also report that the specific content of the item in memory (in this case, its pitch) could be decoded with pattern analysis techniques from activity in the auditory cortex and frontal regions during the silent interval. A complementary fMRI study was carried out by Philippe Albouy and several of our colleagues, who acquired their scans toward the end of each trial, thus corresponding mostly to the retention period (Albouy et al. 2019b). They were able to show that people who had a better memory for tones also showed increased responses in the right temporal and inferior frontal cortex (Figure 3.5), favoring the idea that the frontal cortex enhances tonal representations on an individual basis during retention.

Electroencephalography (EEG) or magnetoencephalography (MEG) offer particularly good methods for tracking brain activity in the absence of any stimulus-evoked response because it is possible to isolate the brain activity that occurs during the silent period, in between stimuli, and is thus mostly driven by cognitive processes. Several labs have made use of this approach to study the retention of tonal stimuli in working memory. One of the first studies to use this methodology showed that MEG oscillatory activity arising from auditory and frontal regions is enhanced during the retention period of a simple auditory comparison task, thus implicating these structures, once again, but specifically pointing to their role in the retention of information in working memory (Luo et al. 2005). Stéphan Grimault and colleagues from the BRAMS lab also used MEG to show that there is a neural response during a silent retention period, which increases as a function of the number of tones to be retained, thus showing a direct link to working memory capacity (Grimault et al. 2014). As expected, this response originates from auditory areas, as well as from several frontal-lobe locations, both inferior and more dorsal.

Identifying the auditory cortex and frontal regions as nodes in a working memory network is important, but it does not directly tell us what role they each may play or how information flows from one region to another. One approach to address this question is to use mathematical models of activity patterns to investigate the causal influence of bottom-up versus top-down connections between auditory and frontal areas, a method known as directed connectivity. An application of such modeling showed that top-down influences, from right inferior frontal cortex to auditory cortex, are necessary to account for patterns of EEG responses to sound (Garrido et al. 2007). This same top-down connectivity is implicated in tonal working memory and its disruption in amusia (Albouy et al. 2013), as we discuss below. More generally, these patterns of directed connectivity are consistent with the view that working memory depends on control signals arising in frontal regions, which interact with sensory processing areas, including, of course, the auditory cortex, where information about incoming sounds is initially represented.

The hypothesis that frontal cortical regions enable control over sensory processing in general has been developed into a detailed and influential model by my McGill colleague Michael Petrides (Petrides 2005). He proposes that the frontal cortex is organized according to a topographic principle, such that areas close to the motor cortex are directly involved in organizing actions, while more anterior regions are responsible for progressively more abstract representations of plans and goals. Along a dorsal–ventral axis, more ventral frontal regions interact directly with sensory regions for active judgments about stimuli, while more dorsal frontal areas are involved in a host of higher-order processes, notably including the monitoring of temporal organization of information (Petrides 2000). In the context of working memory, this latter distinction would entail the engagement of ventral frontal regions—within the auditory ventral stream, for our purposes—for retention and retrieval operations; whereas the selection of responses that do not depend as explicitly on the stimulus features would engage the more dorsal frontal cortex. This formulation can explain many of the varied results discussed above that show that the straightforward maintenance

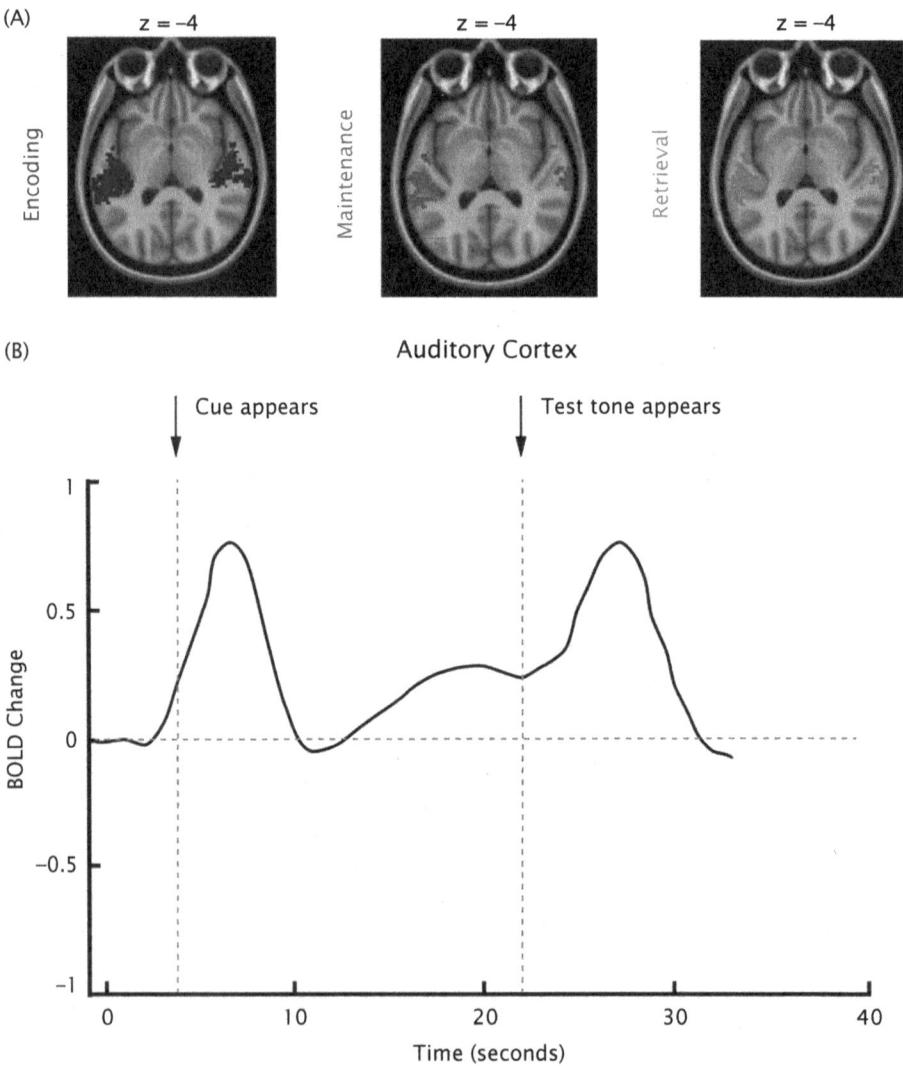

FIGURE 3.4 Auditory cortex activity during pitch retention. A: Functional MRI during a pitch memory paradigm showing activity in auditory cortex during the initial presentation of a cue tone (blue), during the silent retention phase (red), and after the comparison tone has been presented (green). B: Activation profile over time in auditory cortex showing largest responses to the presentation of tones, with sustained activity ramping up during the retention interval. Adapted with permission from (Kumar et al. 2016).

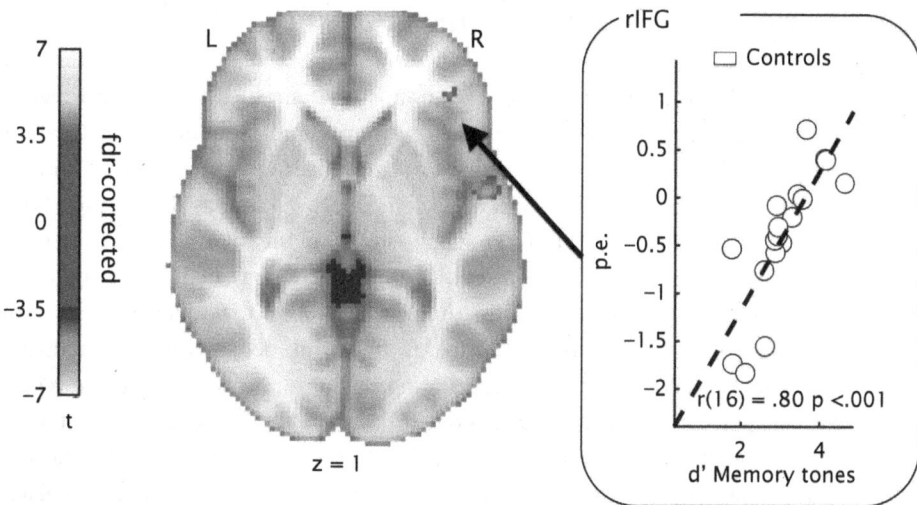

FIGURE 3.5 Role of right inferior frontal cortex in pitch retention. Left: Functional MRI activity that correlates with individual differences in pitch retention performance. Right: Scatterplot showing brain activity (p.e. = parameter estimate) as a function of performance accuracy (d-prime): individuals with good performance have greater recruitment of right frontal cortex compared to those with poorer performance. Reproduced with permission from (Albouy et al. 2019b).

and retrieval of tonal information consistently involves the auditory cortex and inferior frontal cortex, whereas more complex tasks that involve multiple cognitive demands usually involve both ventral and dorsal frontal regions.

In an auditory context, the model would specifically predict that active retrieval of information in working memory would generate interactions between ventral frontal areas and auditory cortex. This idea was directly tested in an fMRI study by Penelope Kostopoulos and Michael Petrides: several melodies, coming from different spatial positions, were presented to listeners who, after a delay, had to retrieve either the identity of the melody or its location from working memory (Kostopoulos and Petrides 2016). As predicted, recruitment of the ventrolateral frontal cortex was observed during this active retrieval task; but also the functional interaction between the frontal region and auditory cortex increased during task performance within the right hemisphere. Given the anatomical connectivity of these structures (Petrides and Pandya 2009) it makes sense to conclude that active retrieval involves top-down control signals, originating in frontal cortex, applied to information being held on-line in auditory areas.

Having established the role of ventral stream structures in these different aspects of working memory, we now turn to the second major function of this pathway, as suggested by the earlier lesion and imaging studies: the representation of stimulus features and the formation of stable sound categories. Such representations are enabled by the capacity of the ventral stream to encode and structure incoming information. As such, working memory and feature representations can be viewed as part of the same overall process. Although our focus is, of course, on music, there is much to be learned from studies of nonmusical processing, which is the topic we turn to next.

Voices and Other Strange Sounds

Outside the specific domain of music, several neuroimaging studies have lent credence to the idea of a hierarchical organization in the auditory system, such that cortical regions along the ventral stream are involved in representing more complex auditory object features in comparison to areas closer to the primary cortex. For instance, in one study, anterior areas were found to be more active to spectrally complex vowel sounds than to pure tones or band-passed noise (Chevillet et al. 2011), supporting the existence of a hierarchical organization. In a study done in our lab, we presented people with a range of difficult-to-identify environmental sounds that varied in distinctiveness and found that anterior and ventral areas of the temporal lobe were most responsive as the sounds became more differentiated from one another (Zatorre et al. 2004). This finding, again, supports a role for the ventral stream in processing complex object features rather than sound mixtures that are not heard as coherently belonging to an auditory object. Using EEG approaches, Micah Murray and his colleagues in Switzerland also found evidence for ventral stream (anterior temporal and inferior frontal) processing of sounds from the environment and also showed that processing in this pathway can be very rapid, within 70 milliseconds (Murray et al. 2006).

In another study that is slightly more relevant for music, Amber Leaver and Joseph Rauschecker from Georgetown University demonstrated that early cortical areas responded to acoustical features, but not consistently to sound categories. They showed that while a certain degree of category specificity for musical instrument sounds emerged in right anterior temporal regions (contrasting with specificity for speech in left anterior temporal areas), other stimulus categories, such as birdsongs, did not show such specificity (Leaver and Rauschecker 2010). I should note here, that not every study has supported the role of ventral stream regions in the representation of auditory objects, because in some cases, areas that are more posterior to the primary auditory cortex have been found to be sensitive to abstract auditory categories, such as sounds produced by human actions (Giordano et al. 2014) or the sounds made by tools (Lewis et al. 2005). In these cases, the more posterior location may be related to links to the motor system, which, as we shall see in Chapter 4, are more integrated with processing in the auditory dorsal stream.

The human voice constitutes a very prominent category of sounds in our everyday lives. The voice, of course, is of relevance for music, as well as for speech, but it is also a vehicle for communication and expression of emotion. How is it represented in the brain? More than two decades ago, the first neuroimaging study on this question was conducted by Pascal Belin, a postdoc in my lab back then but now a professor at Aix-Marseille University. We were inspired by the work of Nancy Kanwisher and her colleagues at MIT, who reported specialization of certain visual ventral stream areas for the processing of faces as well as other categories of visual objects (Kanwisher and Yovel 2006). We reasoned that the voice could be considered as an "auditory face" because it is linked to personal identity and provides the vehicle for transmitting emotion, just as the face. As such, we expected portions of the ventral auditory stream to represent voices, which is indeed what we discovered: regions of the superior temporal sulcus, just below the superior temporal cortex itself, responded more strongly to vocal sounds, whether they were speech, song, laughter, or grunting, than to other nonvocal categories of sounds (Belin et al. 2000).

Subsequent work established the reproducibility of the result, including a parallel result in the monkey brain (Petkov et al. 2008)—a nice reversal of the usual monkey-to-human discovery chain. New fMRI studies, shown in Figure 3.6, also detailed the existence of at least three separate "voice patches" along the superior temporal sulcus (Pernet et al. 2015), which may play different functional roles; for instance, the most anterior of these regions, corresponding to the higher-order processing stage of the ventral stream, may be involved in identifying voices of individual persons (von Kriegstein et al. 2003). Furthermore, there are interesting interactions between the visual face regions and voice areas that are important for multimodal person identification (Blank et al. 2014). All this evidence is consistent with the hierarchical organization of the ventral stream, including the idea that the most abstract object representations are found at the most anterior sites and include information from multiple modalities.

Of particular relevance to music is the obvious fact that the voice is hugely important in carrying musical information in songs, a fundamental aspect of all music in our species (Mehr et al. 2018). Is there some particular specialization of the ventral stream with regard to vocal music? According to one recent study examining cortical responses from electrodes on the surface of the brain, the answer is yes: Sam Norman-Haignere and colleagues from MIT found that some of the recording electrodes in anterior parts of the temporal lobe responded very selectively to sung music, more so than could be explained by merely the sum of speech and nonvocal music responses (Norman-Haignere et al. 2022). This finding may signify a category selectivity for perceptual analysis of vocal pitch patterns. However, it is not clear how it relates to the voice areas themselves (is it a subspecialization within the specialized region?). The relationship between how vocal timbre is analyzed, compared to other musical instrument timbres (Leaver and Rauschecker 2010) also remains to be worked out.

Feature Sensitivity or Processing Domains? Focal or Distributed?

In order to achieve a better understanding of how the ventral stream represents sounds patterns, there are several issues to consider. One unresolved question is whether the nature of the representation is best thought of in terms of specific domains of processing, such as music or speech, or whether it is best characterized in terms of a hierarchical conjunction of successively more complex features that feed forward from earlier areas to more anterior regions. A related, but not identical, question is whether the organization of the ventral stream is more focal, with specific cortical modules responding to specific types of sounds, or whether it is more distributed, with neural responses from widespread regions contributing to the way information is represented. These kinds of issues have also played out in other domains, like vision, and are in no way unique to the auditoryauditory/music literature. Indeed, the question of distributed versus localized processing reflects very general concepts about brain organization that have been discussed since the beginning of neuroscience (see, e.g., the ideas of Karl Lashley in the early 20th century [Pribram 1982]).

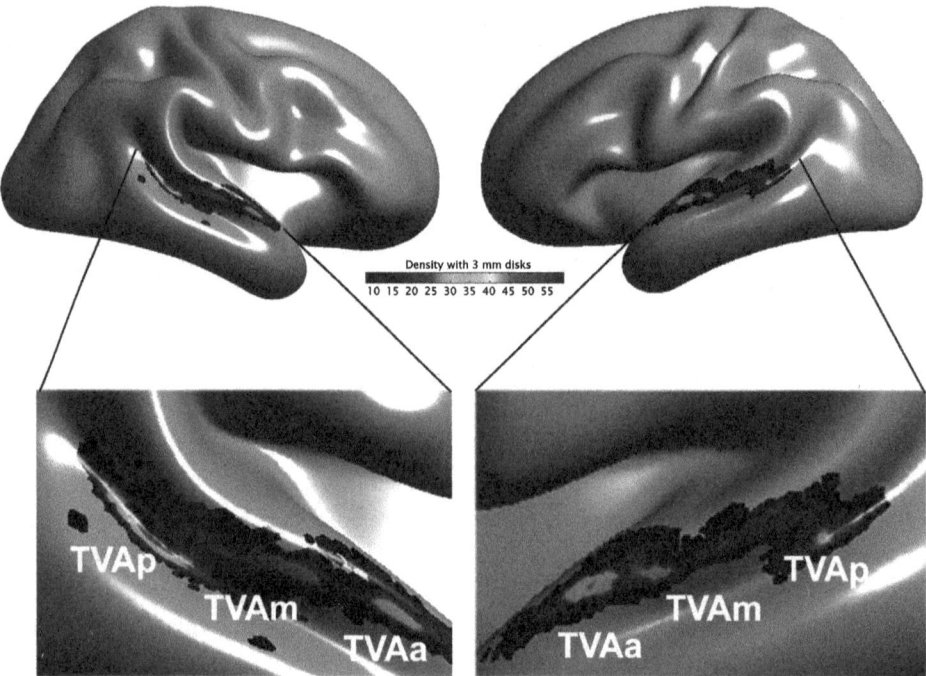

FIGURE 3.6 Temporal voice areas. Density map showing the location of functional MRI responses to vocal compared to nonvocal sounds throughout the superior temporal sulcus, superimposed upon a dilated surface representation of the two hemispheres. Activations are concentrated in temporal voice areas (TVA) in anterior, middle, and posterior clusters, with a relative right-hemisphere predominance (left side of figure). Reproduced under the terms of a CC BY 4.0 license from (Pernet et al. 2015).

One practical problem in answering these questions is that evidence for specialized responses is often difficult to interpret. This is because any two categories of sound will most likely differ quasi-systematically in acoustical characteristics as well as in the category membership (for example, musical sounds tend to be rich in harmonic structure, whereas sounds from nature, such as wind or rain, are not). This problem has been addressed in some studies by attempting to match the acoustical features across categories (Leaver and Rauschecker 2010; Angulo-Perkins et al. 2014; Agus et al. 2017), or by controlling for them with statistical approaches (Giordano et al. 2014). These approaches show that some of what might be considered specificity to categories is actually driven by simple acoustical differences; but they also reveal that low-level physical features are insufficient to account for all aspects of the neural responses, which must therefore represent more abstract aspects of auditory objects.

A thornier problem is whether to interpret these sorts of findings as indicative of the brain being organized in terms of specialized modules for handling certain categories of sound, or whether such categories emerge from the distributed activity of many different neural populations. The studies reviewed above can't easily resolve this theoretical problem. However, at least one recent fMRI study argues for a temporal cortical region that is specifically selective to music *qua* music. Sam Norman-Haignere and colleagues presented listeners with many brief sound clips, including music, speech, and various sounds from the environment (Norman-Haignere et al. 2015). However, instead of comparing the categories to one another, they used a voxel decomposition method that clusters together the sounds based on the neural responses to them (see Figure 3.7). The advantage of this method is that the resulting patterns reflect the stimulus-driven organization of the brain data and are hence unconstrained by any experimenter-imposed hypothesis. The results supported the existence of music-specific responses in bilateral auditory ventral stream regions, consistent with prior studies (Leaver and Rauschecker 2010; Angulo-Perkins et al. 2014), but go further in proposing a specific selectivity for music in this region.

So, does this mean that the human brain contains a music-specialized module, as has been argued on the basis of other evidence (Peretz et al. 2015)? Maybe. But studies using other techniques suggest a more distributed representation of sound categories. Many fMRI studies that use the multivariate pattern analysis approach to investigate this question across modalities and cognitive domains tend to tell a somewhat different story. For instance, one of the first papers to look at representations of sound categories using this approach examined the processing of sounds made by guitars, singers, and cats (Staeren et al. 2009). The results showed that information contained in wide swaths of the auditory cortex, including peri-primary regions, contributed to distinguishing the various categories, arguing against either a focal or a hierarchical organization. Other studies using the multivariate approach, for example, involving learning of novel sound categories, have come to a similar conclusion, that representation is distributed (Ley et al. 2012).

However, a distinction should be drawn between the way a computer algorithm is able to use information drawn from the brain activity versus the way that the brain itself uses that activity (Bouton et al. 2018). On the one hand, it is possible for information to be present across many brain areas, such that it can be decoded by an algorithm, while the neural processes doing the decoding are nevertheless focally and hierarchically organized. On the other hand, while the evidence for focal music selectivity (Norman-Haignere et al. 2015), as well as for voice selectivity (Frühholz and Belin 2018), is compelling, this domain-specific concept does not answer how such a higher-order representation is built up. What are the inputs to the specialized region? Are the inputs themselves organized focally or in a distributed manner? How do elementary features combine such that a response to a complex and heterogenous category, such as "music," emerges?

The issue of focal versus distributed representation of higher-order categories or of lower-level features is not solely an artifact of different analysis methods, however. Evidence for focal organization can also be observed even with multivariate pattern analysis techniques. In a musical context, two studies have looked at the way in which fundamental aspects of melodies are represented. We know from music cognition that contour and interval information are the basic building blocks of melody perception (Dowling 1978). To investigate how melodic contour might be represented, Yune-Sang Lee and collaborators presented listeners with a variety of ascending or descending scales or arpeggios and asked them to detect a change in contour (Lee et al. 2011). Multivariate pattern analysis of the fMRI data generated focal responses in the right superior temporal sulcus, within the auditory ventral stream and in the parietal lobe, where the contour information could be decoded by the classifier, as shown in Figure 3.8A. To examine how musical interval

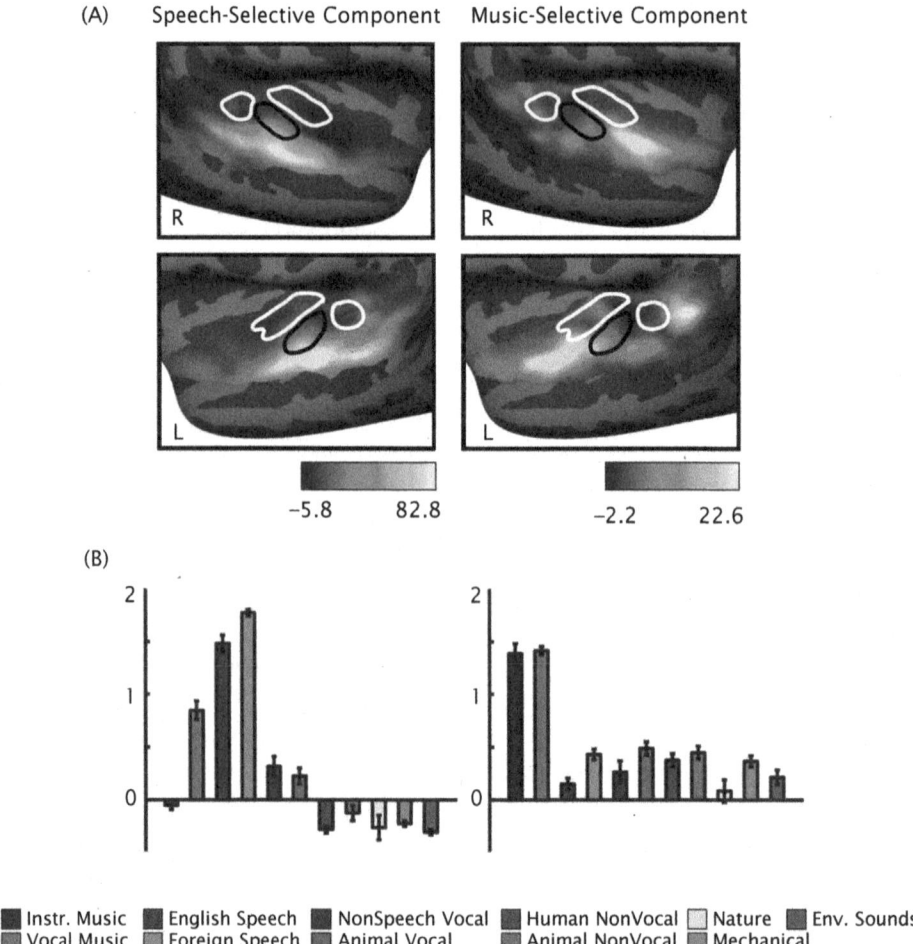

(A) Speech-Selective Component   Music-Selective Component

(B)

Instr. Music English Speech NonSpeech Vocal Human NonVocal Nature Env. Sounds
Vocal Music Foreign Speech Animal Vocal Animal NonVocal Mechanical

FIGURE 3.7 Selectivity for music and speech categories in temporal cortex. A: Maps of voxel weights for speech- and music-selective components in functional MRI data. Black and white outlines indicate low- and high-frequency primary areas, respectively. Strongest weights are found in anteroventral portions of the temporal lobe (yellow). B: Response magnitude of many different sound categories for the speech- and music-selective components shown in (A). Strongest responses are found for speech and vocal music in the speech component, and for instrumental and vocal music in the music component. Adapted from (Norman-Haignere et al. 2015) with permission from Elsevier.

information is processed, my former PhD student Mike Klein also used a multivariate approach (Klein and Zatorre 2015): he created a matrix of three ascending melodic intervals (minor third, major third, and perfect fourth) and presented them at each of three pitch heights (C, C#, and D). These two dimensions are orthogonal to one another (because any interval can be presented starting at any pitch). As such, it is impossible to know what interval was presented by simply knowing the pitch of the starting tone, thus ensuring that we are looking at the encoding of the pitch relationships rather than absolute values. When we programmed the algorithm to identify brain areas that could correctly classify the musical intervals, irrespective of the pitches they were made up from, we observed very focal responses in the right superior temporal sulcus (ventral stream) and in the left intraparietal sulcus (dorsal stream), as shown in Figure 3.8B. These two studies show that both interval and contour information appear to be represented abstractly in portions of the auditory ventral stream.

Thus, these multivariate studies would tend to support not only the concept of focal representation (despite using a technique that is able to uncover distributed activity), but they also argue in favor of a hierarchical organization since these aspects of melodies, contour, and intervals are basic elements (Dowling 1978) that would presumably be combined into more complex patterns of the type detected in previous studies (Patterson et al. 2002; Warren and Griffiths 2003), within temporal-lobe areas closer to the primary auditory cortex. This idea, that more complex responses emerge from combinations of simpler elements, is nicely supported from a different neuroimaging approach. A musical piece (a Brahms piano concerto) was presented, either normally or scrambled, at different timescales (at the level of single measures, phrases, or entire sections), while listeners were scanned with fMRI (Farbood et al. 2015). A topography of sensitivity to the different processing timescales emerged, such that the early auditory areas were sensitive to the shortest time windows, with progressively later auditory stages showing sensitivity to longer and more well-organized structural elements, culminating with the right frontal cortex which responded most reliably only to the entire, unscrambled piece (see Figure 3.9). Thus, deciphering relationships in music that spans many minutes (e.g., comparing a theme that appears once to a related theme that occurs much later in the same music) would be handled via the most highly developed endpoints of the ventral stream hierarchy, including frontal areas where the representations would be the most abstract. These processes would also require working memory, to hold on to increasingly long patterns as they arrive, and long-term memory, to compare incoming sounds to stored templates, both of which are instantiated via the ventral stream, as we have already seen. A similar processing hierarchy across multiple time windows also seems to hold for the analysis of speech signals (Lerner et al. 2011), showing that this principle is a general one that the nervous system implements to analyze complex, time-varying signals.

To circle back to the principal point of this chapter, we can admit that many questions remain about exactly how sound objects are represented: it is not yet entirely clear what the neural coding principles are, what features are represented, how they are combined, how abstract categories are formed, and so forth. But for our purposes, *how* the information is represented is a bit less critical than *that* it is represented, and especially that it happens within the circuitry of the ventral pathway. The reason this is important for our story goes back to the concept of prediction, which can only happen if a sufficiently accurate record of past events has been created, so that new events can be judged based on probabilistic knowledge generated by that store of information. If this account is correct, then we should be able to see direct evidence that the ventral stream system responds differentially, depending on the predictability of the pattern of sounds. We turn our attention to that topic next.

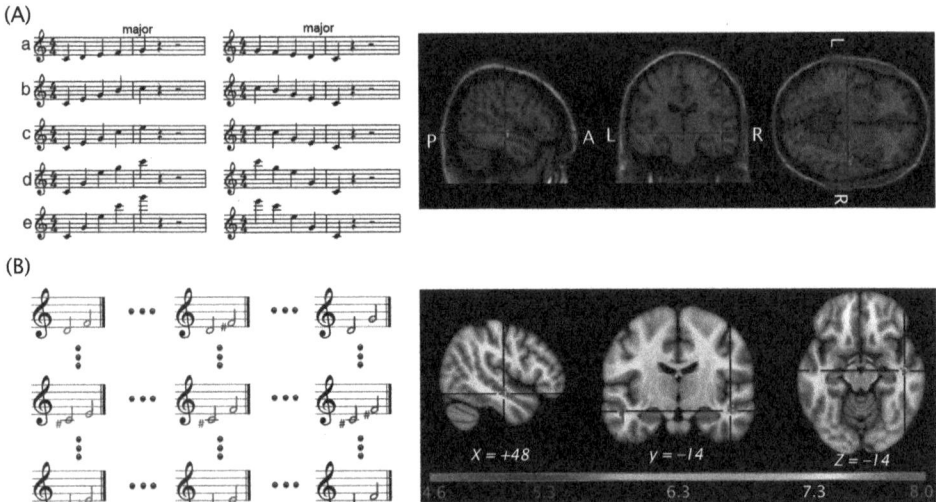

FIGURE 3.8 Representation of musical scales and intervals in auditory ventral stream. A: Region in right superior temporal sulcus where multivariate pattern classifier distinguishes contours of ascending and descending scales and arpeggios. Reproduced from (Lee et al. 2011) with permission from Elsevier.

B: Region in right superior temporal sulcus where significant multivariate pattern classification accuracy was observed for musical intervals (minor thirds, major thirds, and perfect fourths) irrespective of the individual pitch values. Reproduced with permission from (Klein and Zatorre 2015).

Predictive Processes in the Ventral Stream: Mismatch Responses

Research begun in the 1970s, notably by Risto Näätänen and his colleagues at the University of Helsinki, uncovered the phenomenon that any discriminable change in an ongoing stream of auditory information can lead to a neural response that is measurable via EEG (for a thorough recent review see Näätänen et al. 2019). This response, which consisted of a negative deflection in the EEG, is often referred to as the mismatch negativity and was first demonstrated by comparing the response to a particular "deviant" sound to the response of a repeated "standard" sound. For example, a tone sequence of one frequency that is interrupted at random, infrequent intervals by a different frequency would elicit a mismatch response. Notably, the frequency of the standard or deviant sound can be exchanged, which indicates that it is not a response to the sound features per se but rather a difference in the response as a function of how expected it is: the deviant sound is less expected because it occurs infrequently and unpredictably. It is also important to note that the mismatch response appears whether the listener is actively attending to the stimulus stream or is engaged in a totally different attention-demanding task, such as watching a movie. As such, mismatch responses were initially thought to represent the operation of an automatic or "pre-attentive" sensory memory system by which stimulus features were encoded for a brief period of time, thus setting up a basic expectancy which is ultimately violated by the change in physical feature content of the deviant stimulus.

(A)

(B)

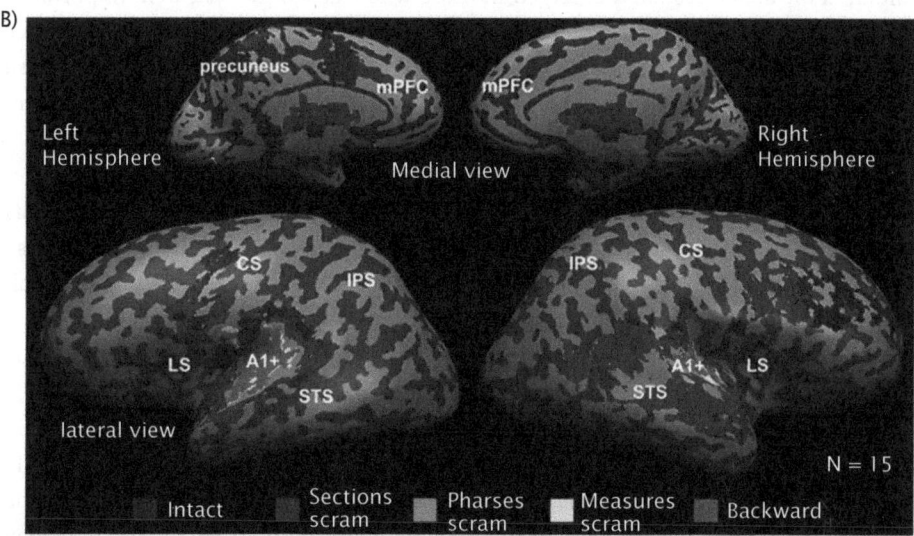

FIGURE 3.9 Hierarchical processing of musical segments. A: Musical score of Brahms Piano Concerto in D Minor, third movement, used to investigate effect of scrambling at different time scales: measures (yellow lines), phrases (green lines), and sections (blue lines). B: Functional MRI results showing topography of sensitivity to different levels of scrambling. Hierarchically earlier levels of the auditory cortex respond equally to all manipulations, while progressively more distal areas, especially along the STS and anteriorly, respond more as the musical segments become longer and more well-organized. Frontal regions only respond to the intact music. Reproduced under the terms of a CC BY 4.0 license from (Farbood et al. 2015).

After more research was conducted, it became apparent that this explanation of the mismatch response was insufficient because many studies showed that similar responses could be elicited in the absence of any consistent difference in physical stimulus features. That is, sensitivity to more abstract nonphysical differences was also observed. For example, if pairs of tones of rising pitch are presented at various pitch levels, and then a descending pitch pair appears every so often, a mismatch response is generated, even though the descending pitches were also present in earlier ascending pairs (Saarinen et al. 1992), as shown in Figure 3.10. Furthermore, this phenomenon occurs whether the stimulus is attended or ignored (Schröger et al. 2007). So, in this situation, the violation is not due to a novel sound being presented but rather is due to a switch in the order of sounds, which is a kind of a rule, or regularity, that the auditory system has abstracted from the stimulus flow.

Even more interesting is that such rules, or more complex ones as well, can emerge over the course of the experiment, indicating that on-line learning can influence the responses. Mari Tervaniemi and her colleagues in Helsinki demonstrated this phenomenon with a musically relevant feature: melodic contour (Tervaniemi et al. 2001). In this design, the standards consisted of a sequence of tones with a given contour (up-up-up-down), while the deviant had a different contour (up-up-down-up); but the pitches used for each pattern were different on each trial. Initially, this type of violation did not elicit a mismatch response. But, after a certain amount of experience making judgments about contour, a mismatch response emerged in those who learned the task well. Importantly, once learned, the mismatch happened even when the stimuli were ignored, when listeners watched an unrelated silent movie. This kind of result argues in favor of a mechanism that integrates information over time and can modify its representations of events with experience.

A mismatch response to certain musical sounds can also depend on long-term musical training, as shown by numerous experiments in which the responses of musicians display greater sensitivity to subtle deviations in pitch, harmony, or rhythm than the responses of nonmusicians (Koelsch et al. 1999; Vuust et al. 2005; Brattico et al. 2009). Remarkably, one such study even showed greater mismatch responses among musicians to modifications in each of two polyphonic melodies, indicating that learning can lead to the encoding of regularities of separate events even if they occur simultaneously (Fujioka et al. 2005). Studies of children undergoing musical training have also reported that the modulation of the mismatch responses emerges as musical expertise is acquired (Chobert et al. 2014; Putkinen et al. 2014), indicating that enhancements are not necessarily pre-existing but are driven by experience. There are dozens more studies in this mismatch literature (for review, see Paavilainen 2013), which all point in the same direction: they illustrate the auditory system's capacity to generate predictions about upcoming stimuli, based both on the immediate context (within an experimental trial or as a result of recent experience) as well as on longer-term knowledge (such as occurs after training or long-term exposure). But, where in the brain do these mismatch responses emerge from? The answer would appear to be the auditory ventral stream.

Since research in this area began, these kinds of auditory mismatch responses were shown to originate in the vicinity of the auditory cortex; although, as seen in Chapter 2, certain kinds of very basic change detection can even be observed in subcortical structures. The precise location and distribution of the auditory cortical generators vary, depending on various features of the stimuli and test protocol. But generally they involve areas within the superior temporal gyrus and sulcus. For example, some authors have noted that different deviant features lead to responses in different subregions, such as pitch-related features which tend to engage more responses on the right side (Molholm et al. 2004; Tervaniemi et al. 2006); whereas temporal deviants tend to elicit responses on the left (Vuust et al. 2005), a topic to be taken up in detail in Chapter 5.

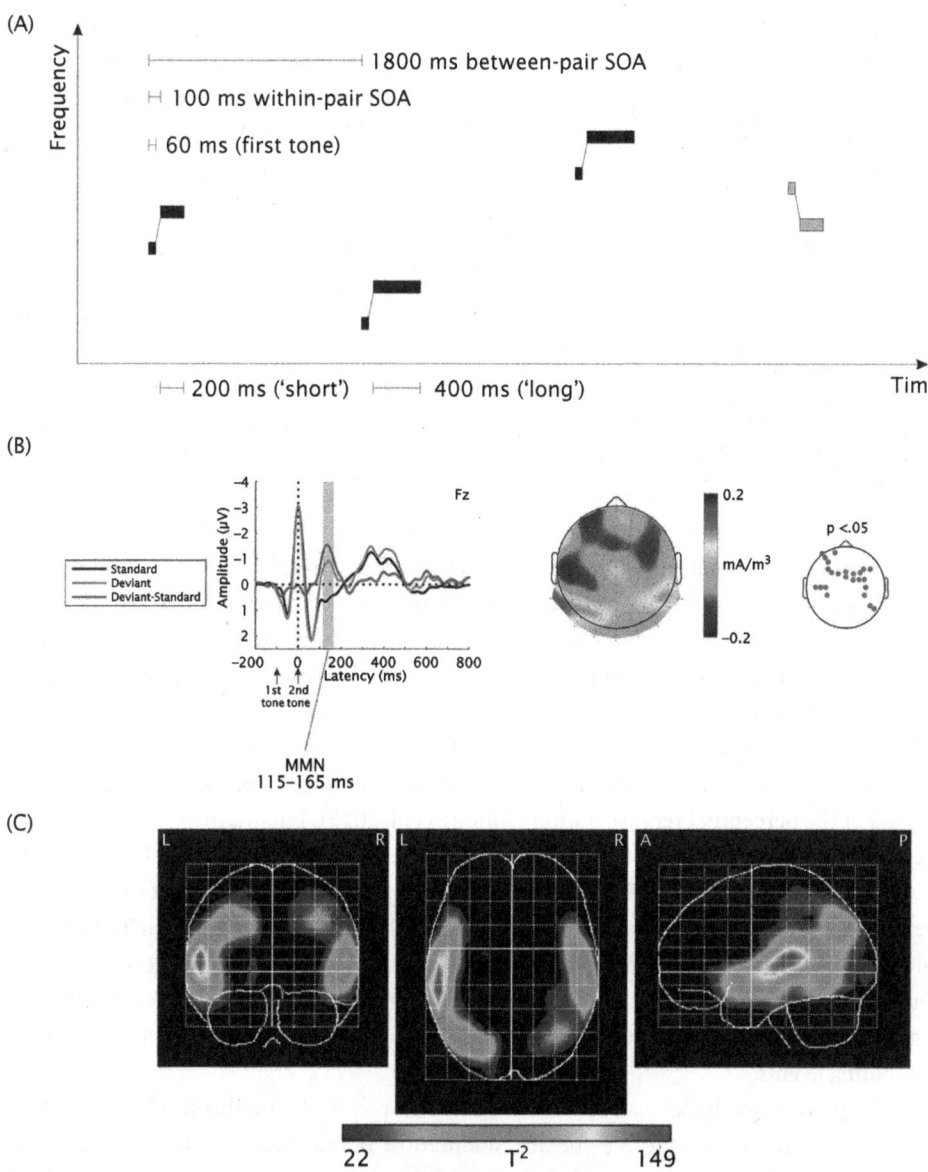

FIGURE 3.10 Mismatch response to violation of abstract rule. A: Stimulus sequence in which pairs of ascending tones (black) are presented frequently (standards), while pairs of descending tones (grey) occur unexpectedly and only occasionally (deviants). B: Electroencephalography event-related recording (left) and scalp distribution (right) of mismatch response to deviant tone pairs. C: Source reconstruction of the mismatch response showing origins in bilateral auditory cortex. Adapted from (Schröger et al. 2007). Reproduced under the terms of a Creative Commons Attribution License.

Several studies have also implicated a separate mismatch response, originating in the inferior frontal cortex (Giard et al. 1990; Opitz et al. 2002), which is one important target region for auditory ventral stream cortical projections. This response occurs significantly later than the one originating in the auditory cortex and has a different oscillatory signature (Fuentemilla et al. 2008), suggesting they play different roles. Marc Schönwiesner and colleagues collected EEG and fMRI data, using a paradigm involving duration deviants, and confirmed mismatch responses coming from both auditory cortex and right inferior frontal areas, each with different latencies, as shown in Figure 3.11 (Schönwiesner et al. 2007). But whereas the auditory cortex responses showed progressively greater magnitude as the deviant stimuli differed more from the standards, that was not the case in the frontal cortex, which responded similarly to all changes. That dissociation suggests that the auditory cortex mismatch is more related to registering a change at the sensory level, whereas the frontal response is less tied to the stimulus features and is instead more abstract in nature. These frontal responses may also be related to shifting attention to a change in ongoing stimulation (Escera et al. 1998).

The detection of novelty in such experiments typically involves a timescale that is much longer than sensory memory (Mäntysalo and Näätänen 1987). Therefore, it is more compatible with the operation of a working memory mechanism, which, as we have seen, is related to the interactions between auditory and frontal regions within the ventral pathway. In fact, formal modeling of directed connectivity between these regions in the context of these oddball paradigms has shown that the later responses are dependent upon feedback connections from frontal to auditory regions (Garrido et al. 2007), as shown in Figure 3.12. This idea also holds in the context of a same/different melody comparison test, in which functional interactions between right frontal and temporal regions are enhanced specifically for trials in which a change is present. This finding suggests that bottom-up error signals from auditory to frontal cortex indicate that an unexpected event has occurred, while the top-down influence back to auditory cortex represents an updating of the perceptual representation (Albouy et al. 2015). Finally, the role of the frontal region in response to unpredicted change is also suggested by studies with frontal-lobe lesioned patients (Alho et al. 1994; Alain et al. 1998) who show alterations of the mismatch response. Such patients are well-known to have both working memory deficits and more global difficulties in using information about changes in environmental events to plan future actions. Such deficits can be accounted for by the failure to update representations based on incoming input with a concomitant inability to make good predictions about upcoming events.

Sensitivity to change can either be elicited by physical stimulus features (in which case, it resembles the sensory-specific adaptation phenomenon described earlier in Chapter 2) or by more abstract properties of the deviant compared to the standard, as we have just discussed. However, there is also another class of mismatch responses, which are not elicited by a change in an ongoing sound stream, per se, but rather are related to higher-order processes. In the music domain, this phenomenon is best illustrated by the perception of harmonically unexpected chords. Many cognitive studies have shown that listeners who are familiar with a musical harmonic system, such as Western tonality—even if they lack formal musical training—have internalized the regularities underlying the order of harmonic progressions (Tillmann et al. 2000), or what may be termed musical syntax, by analogy to linguistic syntax (Patel et al. 1998). Thus, the processing of an expected chord in a harmonic sequence (e.g., a tonic following a dominant chord) is facilitated, as opposed to an unexpected one (e.g., a tonic

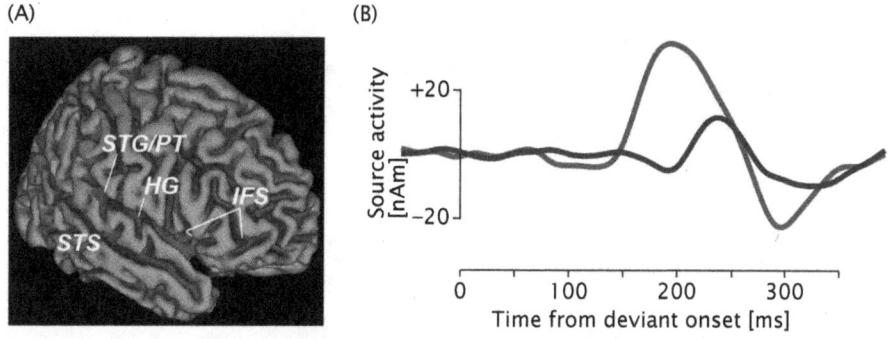

FIGURE 3.11 Mismatch responses from auditory and frontal regions. A: Functional MRI responses to duration deviants originate in auditory ventral stream regions (red) and in inferior frontal cortex (blue). B: EEG mismatch responses to deviants are elicited earlier from auditory areas (red) and later from frontal areas (blue). Adapted with permission from (Schönwiesner et al. 2007).

following a subdominant chord or even more by a chord drawn from a different key) (Bigand et al. 1999). The neural correlates of these expectancies can be seen in neural activity, particularly in the inferior frontal cortex, as measured both with MEG (Maess et al. 2001) and fMRI (Koelsch et al. 2002a).

A particularly clear demonstration of this effect was provided by Barbara Tillmann and her colleagues from the University of Lyon, who used two harmonic sequences ending with the same chord, in a fMRI study (Tillmann et al. 2006) shown in Figure 3.13. In one case, this final chord constituted a harmonically expected sound because it constituted the tonic and hence conformed to expectations of tonal music. Whereas, in another condition, the final chord acted as a subdominant and hence was more unexpected. The manipulation guarantees that the unexpectedness arises from knowledge of musical syntactic rules and not from some more superficial physical difference, of the kind used in simpler mismatch protocols, because the mismatch response is due to the identical chord; the difference only arises due to the context in which it appears. The deviant chord in this experiment elicited brain activity in auditory regions but also in the inferior frontal cortex, thus showing that the ventral stream circuitry is sensitive to knowledge of harmonic relatedness. Interestingly, this type of mismatch response does not habituate, even after repeated presentations of the same deviant (Guo and Koelsch 2016), indicating that the underlying schematic knowledge of the musical rule is quite stable. This makes sense to any musician: no matter how many times you play the same passage incorrectly (which happens all too often to those of us less gifted), it does not sound any better.

Therefore, in addition to its role in mismatch responses (to physical or abstract deviations from expectations derived from the immediate acoustical context), the ventral stream also appears to be responsible for a higher-order kind of prediction. This faculty arises not directly from properties of the stimulus stream itself but rather from how the stimulus conforms (or not) to expectations based upon long-term memory knowledge built up from experience, most likely via statistical learning (Rohrmeier and Rebuschat 2012). As we shall explore in later chapters, this mechanism has very important implications for how musical sounds that are unexpected, and hence surprising, are interpreted and assigned

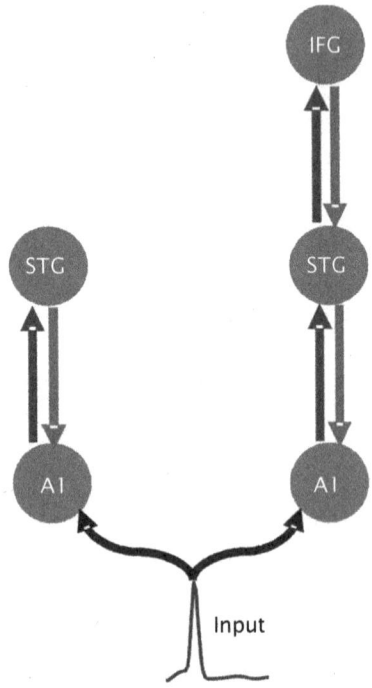

FIGURE 3.12 Computational model of auditory-frontal connectivity. Diagram illustrates simplified dynamic causal model of forward (dark gray) and backward (light gray) connections between primary auditory cortex (A1), surrounding auditory areas on the superior temporal gyrus (STG), and right inferior frontal gyrus (IFG). Inclusion of backward (top-down) connections generates the best fit to neural responses generated by a tonal deviant protocol. Reproduced with permission from (Garrido et al. 2007); Copyright (2007) National Academy of Sciences, U.S.A.

affective values. Right now, however, we turn to one final question in this chapter: what happens when there is a disruption of functioning in the ventral stream?

Congenital Amusia: A Disorder of the Auditory Ventral Stream

Sometime around the turn of the millennium, my friend and colleague Isabelle Peretz from the BRAMS lab and the University of Montreal, told me the story of a curious case she had encountered. "Monica" was an otherwise unremarkable 40-something-year-old woman, pursuing a master's degree and living a normal life, except that, by her own report, she could not perceive music at all; in fact, she indicated that music sounded like noise. Her musical disability had lasted her whole life, but it became more obvious after her marriage to a college music teacher (proving that love overcomes all). Yet, she had higher-than-average intelligence and memory, no hearing deficit, no language problems, and had not suffered brain damage nor any kind of deprivation as a child that might have led to the musical disorder.

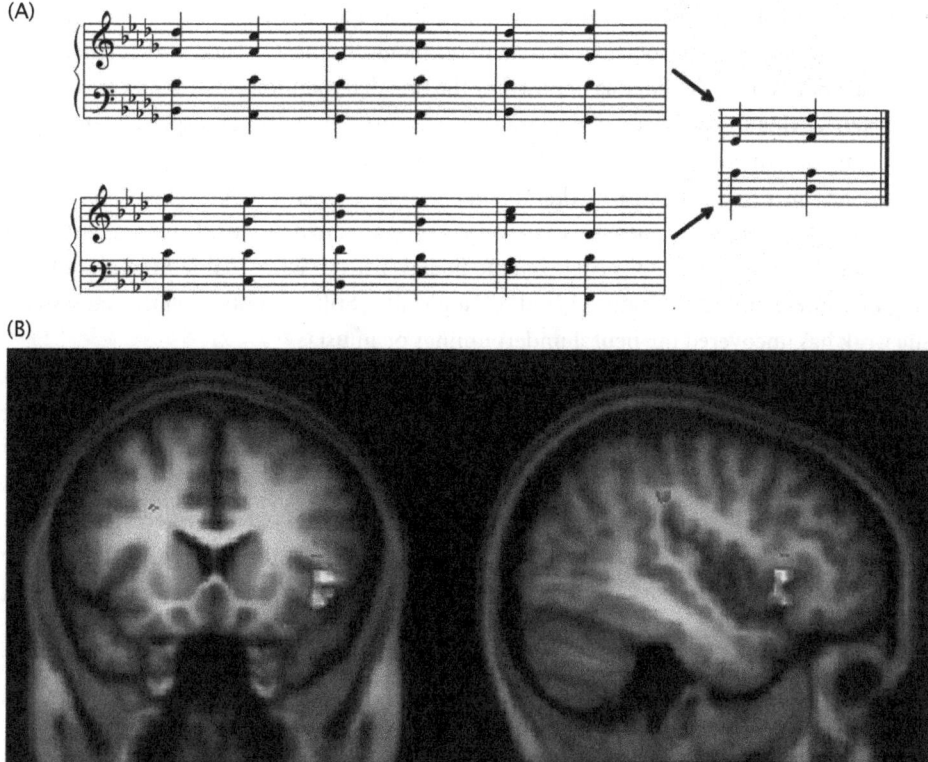

FIGURE 3.13 Neural response to harmonic information depending on context. A: Two different chord sequences are used as context, followed by the same two chords at the end of each sequence (cadence). These two chords are interpreted differently (more-expected tonic chord in the first sequence, less-expected subdominant chord in the second sequence). B: Functional MRI response in inferior frontal cortex in response to the less expected chord. Adapted from (Tillmann et al. 2006) with permission from Elsevier.

My initial response was skepticism that there could be such a specific phenomenon of complete inability to perceive music. So, Isabelle suggested that I check it out myself in my own lab. And of course, she was absolutely correct—Monica was, in fact, completely amusic. A variety of tests confirmed her musical inability: she could not distinguish melodies from one another nor judge if two rhythms were the same or different; she could not recognize musical excerpts that had been presented to her earlier; she was unable to detect glaringly wrong notes inserted into melodies (Peretz et al. 2002). Monica's deficit seemed to be linked to a very basic inability to detect fine-grained pitch differences, but the full nature and source of the problem remained to be understood. This case report turned out to be just the tip of the iceberg, as Isabelle and her students soon uncovered many more cases like this one and reported on them in a series of remarkable studies with rigorous attention to detail (Ayotte et al. 2002; Hyde and Peretz 2004; Peretz et al. 2009; Nan et al. 2010; Vuvan et al. 2018a).

Amusia, or a severe inability to perceive musical sounds, can be observed in two forms. One is an acquired condition caused by brain damage, most often to right-hemisphere frontotemporal areas and their connections (Sihvonen et al. 2019). But, of most direct interest here is the congenital form, characterized by its appearance at an early age in the absence of any other psychosocial, sensory, or neurological problems, and which tends to be more of a pure music disorder as opposed to the one that manifests after stroke or trauma, for instance. This so-called congenital amusia—the form that Monica displayed—has been informally described for a long time. In fact, several famous individuals have been known to have had it (e.g., in the movie *The Motorcycle Diaries*, based on Che Guevara's eponymous memoir, there is a scene where Guevara is supposed to dance the Cuban mambo; but due to his amusia, he cannot recognize it at all and instead dances something more like a tango, earning him the nickname "Mambo-Tango kid"). Still, it is only recently that systematic work has uncovered the neural underpinnings of amusia.

There is a great deal of behavioral literature detailing many interesting aspects of congenital amusia (Tillmann et al. 2015), such as: how it interacts with speech processing (it mostly doesn't, except in situations involving prosodic pitch intonation [Thompson et al. 2012] or lexical tone [Nan et al. 2010]); how pitch and rhythm are affected (pitch is always impaired, while rhythm is often, but not always [Hyde and Peretz 2004], affected); and its heritability (it seems to run in families [Peretz et al. 2007]), to name just a few. One of the more important behavioral findings is that statistical learning of pitch patterns is impaired in people with amusia, even if they demonstrate intact learning of the statistical contingencies of speech sounds (Peretz et al. 2012). This result is important because, as we discussed in Chapter 1, statistical learning is central to the formation of internal models, without which it is not possible to make predictions about upcoming events, leading to perceptual and cognitive difficulties in processing patterns of stimuli.

Thorough behavioral research has confirmed that the main deficits in congenital amusia include impaired perception of pitch patterns, and also impaired learning and reduced working memory, specifically for pitch information but not for verbal materials (Williamson and Stewart 2010; Tillmann et al. 2016). These functions pertain to the ventral stream, as we have seen throughout this chapter. As such, it makes sense to expect that the behavioral dysfunctions would be linked to both functional and anatomical disruption within this circuitry, as indeed they are.

The anatomical substrate for amusia was first discovered by my late friend and colleague Krista Hyde, who used structural MRI to show that cortical thickness was abnormally increased in amusia in an auditory region of the right superior temporal lobe and in the right inferior frontal cortex (Hyde et al. 2007), as shown in Figure 3.14A. Although one might have expected a decrease in this measure, increases in cortical thickness can be considered a marker of cortical malformation in certain developmental disorders (Thompson et al. 2005); hence, the increased thickness is compatible with a likely genetically determined congenital malformation.

The next step was then to characterize the anatomical connectivity between these brain regions in amusia. Krista partnered with Tim Griffiths's lab, to test two independent samples of amusics, one in Montreal, the other in the UK. The analysis of structural MRI data from these two groups showed a consistent reduction of white matter in an area underlying the right inferior frontal cortex, in line with the hypothesis that the right fronto-temporal

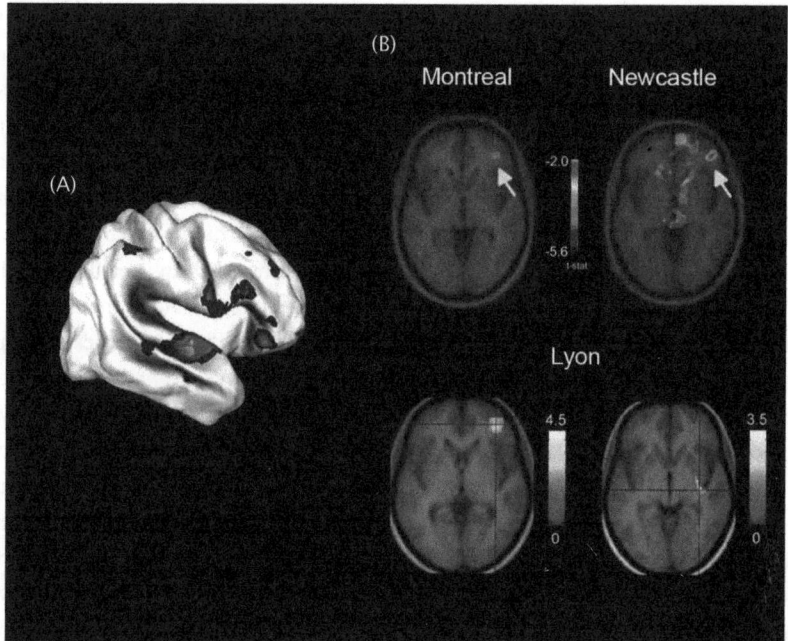

FIGURE 3.14 Anatomical features in congenital amusia measured with structural MRI. A: Surface rendering showing areas of abnormally increased gray matter (cortical thickness) in right auditory cortex and inferior frontal lobe in amusia compared to controls. Reproduced with permission from (Hyde et al. 2007). B: Horizontal slices showing areas of reduced white matter in inferior frontal lobe in separate samples of amusic individuals from (top) Montreal, Newcastle, and (bottom) Lyon. Reproduced with permission from (Hyde et al. 2006, Albouy et al. 2013).

regions are altered not only in their morphology (as seen in the cortical thickness analysis) but also in their connection pattern (Hyde et al. 2006). Reduction in white matter structure was also subsequently observed in yet another amusia sample from Lyon, in almost the identical spot as in the two prior samples (Albouy et al. 2013), as shown in Figure 3.14B. Finally, using a diffusion imaging technique to track white-matter pathways, Psyche Loui and Gottfried Schlaug at Harvard confirmed that people with amusia showed reduced fibers in right-hemisphere tracts that interconnect temporal and frontal regions (Loui et al. 2009), making this set of findings very robust.

Given this evidence for anatomical disruption of the ventral stream, one would also expect altered functional responses in the same network. Evidence from different measures of brain activity supports this conclusion, but exactly how the responses in auditory cortex may change in amusia has been debated. Several studies seem to suggest that processing of auditory input is actually intact in the auditory cortex of amusics. One of Krista's studies found that fMRI activity patterns to pitch variation were unaltered in amusics, compared to controls (Hyde et al. 2011). Similarly, the pitch-sensitive region of auditory cortex, described in Chapter 2, was found to have normal characteristics when measured with fMRI in amusia (Norman-Haignere et al. 2016).

Looking at electrophysiological responses, Isabelle Peretz together with colleagues at the University of Helsinki found that evoked responses to mistuned tones in melodies also showed a normal pattern in amusic individuals (Peretz et al. 2009). This latter result was all the more remarkable because the degree of mistuning was very small (a quarter tone), and many of the people with amusia who were tested could not reliably discriminate this small pitch difference. Interestingly, this dissociation between preserved neural sensitivity to pitch differences, yet absence of conscious awareness of them, fits well with existing behavioral data showing that people with amusia are poor at explicitly rating how well a given pitch fits into a tonal melodic context, yet nonetheless show intact priming of response times, thus demonstrating a behavioral influence of melodic expectedness (Omigie et al. 2012). It's as if the answer is inside their brains, but the information is not accessible to them.

At first, this pattern of results seems difficult to explain. But it becomes clearer when we consider evidence of disruption at a network level in the ventral pathway rather than considering only isolated responses from the auditory cortex. Thus, in the study mentioned above (Hyde et al. 2011), even though fMRI responses in auditory cortex were normal in amusics, the functional connectivity between auditory cortex and inferior frontal areas in the right hemisphere was reduced during perception of pitch patterns. This reduction in connectivity between the right inferior frontal and auditory regions was also subsequently observed by the Lyon team, in a resting-state condition without any stimulation (Lévêque et al. 2016), thus showing a disturbance of intrinsic interactions within the ventral auditory stream, in accord with the anatomical evidence reviewed above. The apparently normal responses previously documented in auditory cortex do suggest that the deficit is not to be found in low-level processes; yet, when tested under circumstances that require more interaction within the network, evidence of abnormality in the auditory cortex can emerge.

This conclusion is supported by an MEG experiment conducted by Philippe Albouy and colleagues in Lyon, who tested the ability of amusics to discriminate between a pair of six-tone melodies differing in pitch by one tone (Albouy et al. 2013). The physiological response to the first tone of the melody was largely the same in amusics compared to control individuals, in agreement with the earlier data discussed above, but the responses to subsequent tones were significantly delayed and decreased in amplitude. It is likely that the processing of melodic tones after the first one reflects ongoing integration of information in working memory; thus, a deficit in the capacity to encode these tones in amusia is most likely a consequence of dysfunction in the working memory network. Critically, as shown in Figure 3.15, these authors applied the same modeling technique mentioned earlier in this chapter (Garrido et al. 2007), which tests for the causal influences of auditory and inferior frontal regions on one another. They found that in amusics compared to controls, there was a reduced directed connectivity during the encoding phase of the task, in the feedback part of the loop between right inferior frontal and auditory areas, indicating that predictive coding was disrupted (Albouy et al. 2013). In contrast, in the retrieval phase, there was a greater-than-normal increase in connectivity from auditory to frontal areas, suggesting that more prediction error signals were being generated since no good model was established that would inhibit such bottom-up inputs (Albouy et al. 2015).

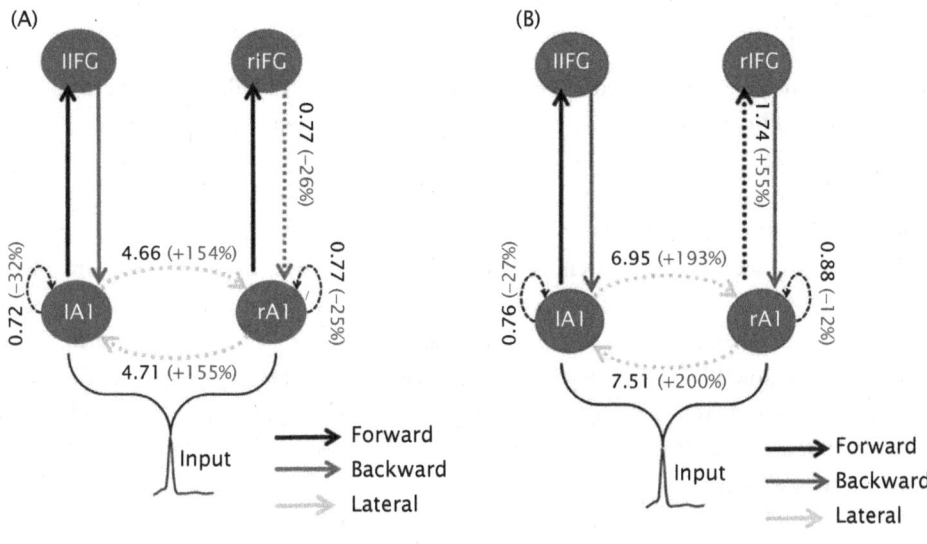

FIGURE 3.15 Dynamic causal models of interactions between auditory and frontal regions in amusia. A: Reduced top-down influence of right frontal area on auditory area (gray dotted line) during the encoding phase of the task in amusics compared to controls. B: Increased influence of auditory area on right frontal area (black dotted line) during the retrieval phase of the task. Reproduced under the terms of a Creative Commons Attribution License from (Albouy et al. 2015).

These observations are very important because they specifically point to the role of feedforward and feedback connections: this mechanism is disrupted anatomically in amusia, as shown by the white-matter analyses, and the functional consequence is that neither top-down influences on early auditory cortex function nor bottom-up inputs to frontal regions occur normally. The critical role played by behavioral feedback is vividly demonstrated by the fact that even normal listeners can show amusia-like behavior, when they are given random feedback about the correctness of their answers during learning (Vuvan et al. 2018b). Furthermore, disruption of frontal-to-auditory cortex modulation can explain the phenomenon whereby people with amusia are unaware of certain acoustical events, despite showing neural or behavioral traces of sensitivity to them. Conscious awareness and perception requires network interactions beyond early responses in sensory cortex, as proposed, for example, by Stanislas Dehaene (Dehaene and Changeux 2011). Therefore, in the absence of top-down neural signals from higher-order areas, low-level processing is not necessarily accessible to consciousness. Conversely, one can infer that in normal processing, these interactions provide the substrate for the ability to perceive musical patterns, in perfect agreement with all the other evidence reviewed in this chapter.

Reprise

In this chapter, we have traced the organization and functionality of the ventral portion of the auditory pathway. Broadly speaking, the evidence reviewed supports the conclusion that this circuitry allows for three critical functions in auditory cognition. First, the temporo-frontal ventral circuitry is important in the maintenance and retrieval of auditory information in working memory, as shown by neuroimaging and lesion studies. Second, this system is also responsible for the ability to form and store representations of sound patterns, both in the short term (as might occur during continuous listening or performance of an online task) and also on a longer-term basis, including statistical learning (when more abstract categories may be encoded). Third, the capacity to register deviations from expectations during perception depends upon this system, as shown by a variety of mismatch responses coming from temporal and inferior frontal regions.

All these functions are closely related to one another. Specifically, in order to form any kind of higher-order representation, it is necessary to maintain and concatenate incoming sounds to one another in real time, to allow computation of their temporal or spectral relationships; hence, the formation of perceptual categories depends upon working memory maintenance and retrieval mechanisms. Similarly, without some kind of stored representation, it would not be possible to detect a deviation from expectations; indeed, expectations could not form at all, if there were not some record of regularities experienced over time. Given their interrelationships, it might be best to consider these functions as manifestations of a single underlying mechanism—perceptual analysis is carried out in superior temporal auditory cortical areas, where derived representations are stored, while various kinds of top-down control mechanisms, originating in frontal cortex, act upon the stored perceptual information. Once this system is in place (i.e., after sufficient exposure to environmental patterns), new sound patterns can be evaluated actively in relation to known information. In this context, deviance responses are important, not so much as markers of deviance, per se, but as an index that ongoing perception is guided by expectancies, generated automatically, and based upon each individual's history in the world of sound.

The evidence from amusia is important in two respects. First, it allows an understanding of the disorder itself, which, in turn, may illuminate related disorders and/or provide avenues for remediation (Whiteford and Oxenham 2018). Second, on a more theoretical level, this research has important implications because it provides the necessary causal evidence for the proposed role of the ventral auditory pathway in all of the functions described in the preceding paragraph. If those functions truly depend on this system, then its disruption should lead to problems in those functions. Given the consistent anatomical and functional evidence that amusia is associated with the disconnection of temporal from inferior frontal regions, primarily in the right hemisphere, and given the behavioral deficits in memory, learning, and perception of music that are the defining features of amusia, it follows that those cognitive functions depend on the integrity of the ventral auditory pathway.

The detailed findings in amusia also support the specific role of predictive processes in perception; since functional interactions between anterior temporal and ventral frontal regions are impaired in amusia, it is difficult to build up predictive models of sound patterns, preventing their encoding and deeper processing. The inability of people with amusia to perceive music can thus be ascribed to a breakdown in the ability to generate predictions, which conversely suggests that such mechanisms must be critical for music perception more generally.

4

Communicating Between Auditory Regions and the Rest of the Brain

The Dorsal Stream

In order for anyone to perceive and enjoy music, someone has to produce it. Generating music requires a complex interplay between motor representations that allow exquisite temporal and spatial control over the muscles, and sensory representations that guide those actions and allow the performer to determine whether the sound produced matched the intention. Conversely, the perceiver must also have a sufficiently abstract and robust mental representation of the musical pattern, such that subtle variations and manipulations introduced by the performer can be appropriately interpreted. These mental representations build upon the perceptual analysis steps described in previous chapters but go beyond them to become more dynamic and flexible, thus allowing both listener and performer to transform sound patterns in various ways. Often, perceivers are also producers of musical sound, like in group music-making, adding an additional layer of feedback interactions to the process.

The interface between the auditory cortex and the dorsal regions of the brain—the parietal lobe, premotor cortex, and other dorsal regions of the frontal lobe, shown in Figure 4.1—provides the substrate for many such remarkable functions that are essential to music. The characteristics of this system have been extensively studied in the context of neurophysiological and anatomical research work in monkeys. But how that knowledge relates in any detail to the organization of the human brain remains somewhat of an open question. There is a great deal of research on dorsal stream functionality in humans, but a lot of it focuses on visuospatial processes or on speech and language. Therefore, we must infer a lot of what may be pertinent to music somewhat indirectly from both animal research and from human research on visuospatial or language tasks. Nonetheless, recent progress in

From Perception to Pleasure. Robert Zatorre, Oxford University Press. © Oxford University Press 2024.
DOI: 10.1093/oso/9780197558287.003.0004

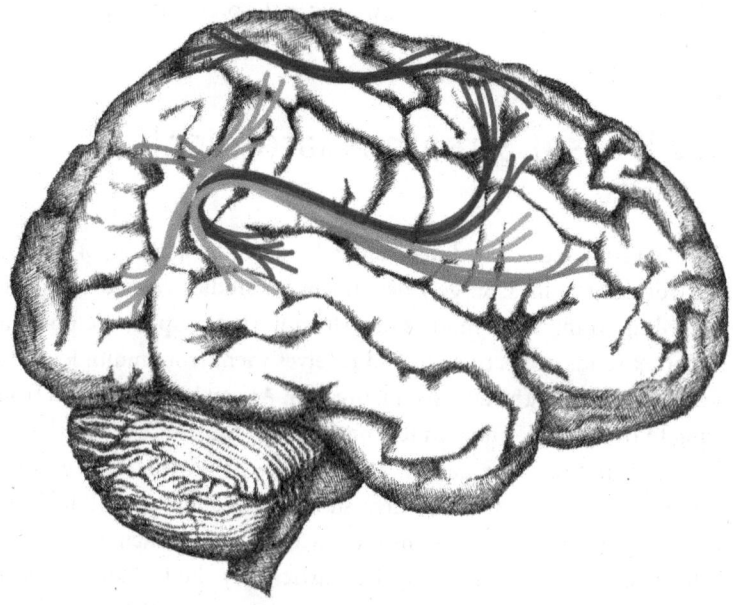

FIGURE 4.1 Auditory dorsal stream. Diagram shows simplified schematic depiction of the anatomical connectivity of auditory regions within the superior temporal gyrus with dorsal brain regions, including the parietal cortex, dorsal and ventral premotor cortex, and several regions of the frontal lobe. Colors represent different fiber tracts. Connectivity pattern based on diffusion MRI adapted with permission from (Petrides 2014). Drawing by E.B.J. Coffey; reproduced with permission.

auditory and music-related studies has given us good insights into how the functions of the dorsal stream, gleaned from studies in other domains, turn out to be important for musical processing. Some of the most important of these functions include the ability to interpret sound patterns in terms of abstract relationships between elements, to actively manipulate these elements in working memory, to represent temporal relationships between sounds, and to generate and monitor actions to produce sounds.

In this chapter, I will first focus on the ways in which dorsal circuitry contributes to generating and operating on perceptual representations. I will then turn to the sensory-motor aspects of this system. In a general way, this set of reciprocal interactions between higher-order perceptual and motor mechanisms can be thought of as providing the means to generate predictions about upcoming events based on past or current events and to update those predictions in real time. The events can be perceptual: anticipating when a chord change will come next in a song or knowing when to expect a new instrumental voice to enter, for instance. Or they can also be actions—adjusting the tension on the vocal cords, or speeding up the striking of a drum—in response to how the music evolves. As we have already seen in the previous chapter, these capacities are central to the processing of music in the ventral stream. But a similar principle also applies to the dorsal stream, just as much or more, albeit with some important differences, particularly in terms of the temporal dimension. Most of the operations carried out within the dorsal stream are highly time-locked, as required for a sensory-motor interface that must operate in real time; this feature contrasts with the ventral stream operations, which as we saw in the previous chapter tend to be more time-independent.

Auditory Dorsal Stream Anatomical Connectivity

As with the ventral stream discussed in Chapter 3, our understanding of the connectivity of the auditory dorsal stream is largely dependent on studies carried out in nonhuman primates. According to the same models described in that chapter, the posterior part of the auditory cortex sends projections to and receives them from multiple dorsal cortical targets via several distinct pathways (see Figure 4.1). As can be appreciated from the diagram, referring to this complex anatomical network as "the dorsal stream" does not capture the full complexity of the situation. Nevertheless, it provides a useful heuristic to understand many of the phenomena of interest. In this chapter, we will focus on the two most relevant components of this system for our objectives, which are the connections from the auditory cortex to the parietal lobe, particularly the intraparietal sulcus (IPS), and to the dorsal premotor cortex (Rauschecker and Scott 2009, Rauschecker 2018). Portions of the frontal cortex that receive inputs from the auditory dorsal stream also send projections to the dorsal part of the striatum (Haber and Knutson 2010), providing an interface with the reward system; this interaction will be discussed in more detail in Chapters 6 and 7.

Although the dorsal premotor cortex receives both a direct and indirect input from the auditory cortex, the latter via the parietal cortex (Petrides 2014), it is not yet clear whether same or different populations of neurons within the dorsal premotor cortex are involved. It is also mostly unknown how these different pathways may be engaged when interactions between auditory and premotor cortex are observed. Connections from the posterior auditory cortex also terminate in the ventral premotor cortex (encompassing Broca's area) and in more anterior parts of the inferior frontal cortex. All these projections are largely reciprocal and hence form part of a continuous corticocortical loop, in which information may be thought of as proceeding from auditory cortex through to the dorsal regions and also back toward the auditory cortex in the reverse direction.

Functional Role of the Dorsal Stream

Space, Action, or Other Things?

By analogy to the well-established concept that the visual dorsal stream is involved in visuospatial processes (Mishkin et al. 1983), it was initially proposed that the pathway involving posterior auditory regions and parietal areas was dedicated to the analysis of auditory space (Tian et al. 2001a). However, a competing idea, also drawn from the visual domain, stipulated that the dorsal stream is best conceptualized as pertaining to movement and action (Milner and Goodale 2006) and the transformations required to effect actions from initial intentions (Cui 2014).

A good deal of evidence from neurophysiological and human studies supports a role for this pathway in auditory spatial analysis (Alain et al. 2001), as well as in the integration of auditory and visual information into a supramodal spatial representation (Sestieri et al. 2006). However, no cortical topography of auditory space has ever been found within the auditory cortex (unlike the visual system, whose cortical organization is inherently spatial, from the retina onward), which would seem essential if auditory space is directly represented in this pathway. Furthermore, neurophysiologists have found that spatial information is encoded by ensemble activity of widely distributed neurons—not only in posterior areas but throughout the entire auditory region (Furukawa et al. 2000). Conversely, many neurons in the posterior auditory cortex simultaneously code for both spatial as well as spectro-temporal characteristics of sounds (Tian et al. 2001b). Although a number of neuroimaging studies indicate that posterior auditory cortical regions are recruited in auditory spatial tasks (Maeder et al. 2001), recent work suggests that this pattern may be more related to the coding of each spatial hemifield (Ortiz-Rios et al. 2017). Recruitment of the parietal portion of the auditory dorsal stream is most likely related to actions applied to sounds (Zatorre et al. 2002b), which clearly engage the IPS since movements must occur within real three-dimensional space, thereby requiring a sensory-motor-spatial mapping.

The findings from the literature are thus more compatible with an action-related interpretation, which is also supported by data showing that the parietal cortex contains regions that are specific to performing the actions of reaching, pointing and grasping (Grefkes and Fink 2005). Another important feature of the posterior auditory cortex and its projections to/from parietal areas, which supports the importance of action representations in the dorsal stream, is that neural responses in these regions have a higher temporal resolution than those in anterior/ventral regions (Camalier et al. 2012, Jasmin et al. 2019); the latter instead tend to integrate across time, as discussed in Chapter 3. This ability to encode time information accurately would be highly relevant to the organization of actions, which must not only take place in the real spatial world but also must be organized in time. Precise timing of movements is critical for many actions, including speech; but it's particularly important for playing music, of course, not only to achieve timing and coordination of one's own fingers, limbs, or vocal tract but also for coordinating actions while playing with others. But as we shall see shortly, the neural representation of time in relation to music also has interesting abstract properties beyond the timing of individual actions, involving higher-order music-specific features, such as rhythm and metrical hierarchies.

So, does this all mean that the dorsal pathway is only relevant for motor/action operations? This conclusion would also appear to be too narrow because it would be incompatible with its role in many complex cognitive skills that do not explicitly involve action, including musical ones. An arguably more powerful approach to describe the functional role of the dorsal stream is to shift the emphasis from stimulus features, or motor actions engaged by them, to the computations necessary to generate the relevant behavior. Research from the visuospatial domain (Andersen 1995, Gallivan and Goodale 2018) has suggested that in order to execute a seemingly simple action, such as reaching for a target, the brain must compute a transformation of information from one reference frame (such as the position of the item on the retinal surface) to another reference frame (the position of the target in the real world in relation to the agent). This type of transformational operation can be quite powerful, as it can be applied in a more abstract way on any input (not only visual) or output (not only motor). Hence, it may explain why this dorsal system is involved in many cognitive tasks, including, for instance, mental arithmetic, which shares some features with visuospatial tasks (Dehaene 2011).

A conceptually related account of dorsal-stream functionality is that it is involved in the creation of internal models for sensory-motor integration (Rauschecker 2018). The concept of transformation across reference frames can be incorporated into this broader scheme since a model refers to a representation of the environment and also of the body, such that one can interpret events and act upon them. And to do so, it is necessary to take into account the constantly changing nature of outside events, especially how their processing changes as a function of one's actions upon them, which requires these transformations from one reference frame to another.

Music Perception and Transposition

One of the first realizations that the dorsal stream was involved in musically relevant functions beyond sensory-motor ones came from work on the perception of transposition. Transposition refers to the ability to recognize a tonal pattern, such as a melody or a chord sequence, irrespective of the absolute values of its individual pitches, and as such, is a cornerstone of Western musical practice. In fact, transposition may be the most frequently used musical transformation (van Egmond and Povel 1996). Transposition works only if the pattern is encoded as a series of interval relationships rather than in terms of individual units. From a psychological perspective, it can be considered as a prime example of perceptual constancy (a topic we also encountered in Chapter 3) because the pattern is invariant across different pitch heights or musical keys (you can recognize *Happy Birthday* whether it is sung by your squeaky four-year-old or by your gravelly-voiced uncle). The Gestalt psychologists of the early 20th century recognized this principle and specifically pointed to melodies as evidence that perception depended on shapes, or wholes ("gestalts") and not just the sum of the parts (Von Ehrenfels 1937). Many behavioral studies have shown that even musically untrained people can easily recognize a transposed melody (Attneave and Olson 1971, Dowling and Harwood 1986), although musical training can enhance this ability, especially for unfamiliar patterns (Dowling and Harwood 1986). This ability can also be seen in infants (Plantinga and Trainor 2005), indicating that it is either inborn or that learning it requires only minimal exposure.

To understand the neural basis of transposition, we carried out a series of neuroimaging studies with my former PhD student Nick Foster, in which listeners judged whether two unfamiliar short melodies were the same or different, as shown in Figure 4.2. In the control condition, the two items were in the same key (i.e., all the pitch values were the same except the one tone that differed on half the trials); while in the transposition condition, the two melodies were in different keys (i.e., all the pitch values differed, but the musical intervals were held constant). We observed clear recruitment of the IPS in the transposition condition relative to the control task; moreover, the magnitude of activity in this region was directly proportional to the individual success rate of the individual, thus directly linking brain response to behavior (Foster and Zatorre 2010b). This finding was replicated in a

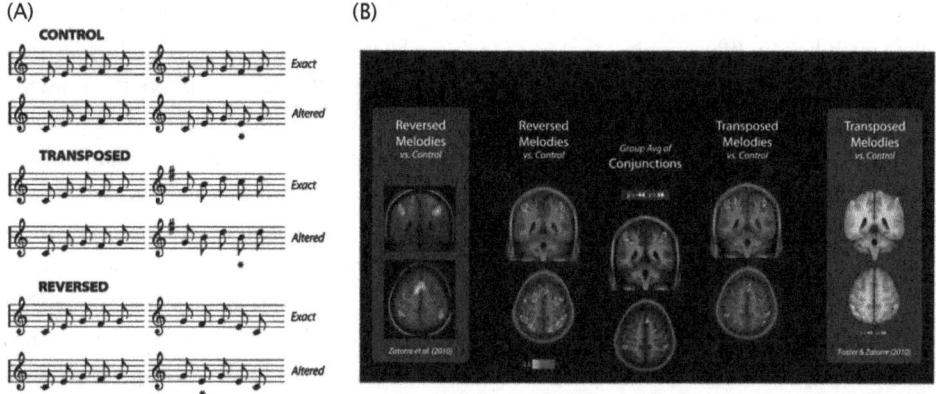

FIGURE 4.2 Contribution of intraparietal sulcus to musical transformations. A: Stimuli used to test musical transformations using same/different tasks. Top: control condition in which a short melody is repeated with or without a changed note (indicated by asterisk). Middle: Transposition condition in which the comparison melody is shifted in pitch relative to the target melody. Bottom: Reversed condition in which the comparison melody is reversed in time relative to the target melody. B: Middle images: Functional MRI activity in intraparietal sulcus and other dorsal-stream structures to the tasks shown in A, with the conjunction across tasks shown in the middle. Left and right images: Functional MRI activity in intraparietal sulcus from previous studies examining reversal and transposition separately. Note similarity across all three studies. Reproduced from (Foster et al. 2013) with permission from Elsevier.

subsequent study (Foster et al. 2013) and was also linked to anatomical features of the IPS (Foster and Zatorre 2010a).

Could this pattern of results be explained by visual processing? There is a good deal of behavioral evidence suggesting the existence of a relatively automatic link between visual spatial position and pitch height (Rusconi et al. 2006). There is also behavioral evidence for a link between visual mental rotation and auditory transformations: Cupchik, Phillips, and Hill (2001) showed that individual differences in mental rotation ability correlated with a melody reversal task. Furthermore, Douglas and Bilkey (2007) showed an association between impaired musical ability in amusia and impaired visual mental rotation. Therefore, the activity in dorsal areas seen during transposition might reflect engagement of some sort of visual representation of the melody. Indeed, activity in visual cortices was observed in our experiments (Foster and Zatorre 2010b), perhaps implicating some parallel implicit visual process going on during task performance. However, recruitment of visual areas was similar across transposition and non-transposition control tasks, and the degree of activity in visual areas did not correlate with performance on any of these tasks, indicating no specific role for visual processing in transposition.

Moreover, if the recruitment of IPS were directly related to some kind of one-to-one correspondence between pitch and visuospatial representations, one would expect that activity in the region would scale with pitch height. We tested this prediction by comparing activity during transpositions at different pitch heights—thus an octave constitutes a large difference in pitch height, a fifth is intermediate, and a semitone is small. Brain activity in the IPS did not covary with pitch height; instead, we found a correlation with key distance. Melodies separated by small pitch height differences, but in distant keys (e.g., one semitone, such as C versus C#), elicited a large IPS response; whereas those separated by an interme-diate pitch height difference but that were in related keys (e.g., C versus G) elicited less IPS activity. Although the octave was the largest pitch height difference we tested, it essentially elicited no IPS activity, since a melody transposed by an octave stays in the same key as the original. The degree of relationship between keys is based on the number of shared tones; thus, the keys of C and G are related because they share all but one scale tone, whereas C and C# are far from each other because they share only two scale tones. This finding indi-cates that IPS response is not related in any straightforward way to a visuospatial mapping nor to a tonotopic map. Instead, and more interestingly, it suggests an abstract relationship according to key distance, a concept in keeping with music theory, where key distance has been developed and codified in terms of the well-known "circle of fifths." The psychological reality of the circle of fifths is also supported by behavioral data and cognitive models ex-ploring key relationships in music perception (Krumhansl 2001). I always enjoy this type of finding, where a neural response corresponds to a musical relationship, because it seems to link neuroscience with concepts derived by music theorists based on their formal under-standing of tonal relationships. The musical concepts are real enough, and we can find their instantiation in the brain.

Recruitment of this dorsal system is not limited to pitch transformations, as it can also be observed with temporal transformation. A few years ago, with my friend and colleague Andrea Halpern from Bucknell University we devised a difficult musical test as a way to probe active musical imagery ability: we asked people to listen to a tonal pattern and deter-mine if it was the same as another pattern which was reversed in time relative to the first. In functional magnetic resonance imaging (fMRI) data, we observed recruitment of IPS and other dorsal-stream structures, including frontal cortex and supplementary motor area (SMA) (Zatorre et al. 2010). In a follow-up study, we showed that transposition and reversal activations within the IPS overlap in voxels at an individual level (Foster et al. 2013) (see Figure 4.2), suggesting that both operations call upon similar neural circuitry. The dorsal stream can also be engaged for verbal tasks that require reordering words in time (Rudner et al. 2005), suggesting that the function is not limited to tonal information.

What are the common operations between transposition and reversal? I would argue that both require transforming a pattern from one reference frame to another, in keeping with the proposed role of the dorsal stream articulated above. Although the dorsal stream is known to be involved in many types of cognitive tasks (e.g., multiple-demand system, see [Woolgar et al. 2018]), increasing evidence points to its specific role in sensory-motor and cognitive transformations, including, for example, visual mental rotation (Zacks 2008, Gogos et al. 2010), mental arithmetic (Amalric and Dehaene 2018), syllable temporal or-dering (Moser et al. 2009), and various kinds of visuomotor tasks, such as pointing or moving a cursor to a target (Culham and Kanwisher 2001). Thus taken together, the con-clusion to draw from these findings is not that musical transformations merely reflect visual processing, but rather that the dorsal stream in general, and the IPS in particular, carries out similar computations over diverse inputs.

All these visual and visuomotor tasks bear a family resemblance to our musical tasks in so far as they require some kind of mental transformation, reorganizing the elements of a pattern in some new reference frame—sometimes in a clear-cut coordinate system that corresponds directly to the real world (visual rotation, grasping) and other times on a more abstract level of cognitive representation (arithmetic, language). Similarly in the auditory domain, the relevant "space" on which the IPS operates may be based on temporal and spectral dimensions (the fundamental sensory dimensions in the auditory system, as discussed in Chapter 2). But sometimes the transformation may be applied within a much higher-order reference frame that may depend on learning, as shown by the sensitivity to key distance, rather than simple pitch height, as in the IPS response. It remains an open question whether neural populations that carry out these different operations are identical, partially overlapping, or merely located in close proximity to, or intermixed with, one another. The parietal cortex is very heterogenous and contains specialized cell populations for many visuospatial and motor subtasks (Gallivan and Culham 2015). As such, it will be interesting for future work to systematically explore this question in the auditory and music domains.

Working Memory Manipulation

We saw in Chapter 3 that the ventral stream is crucial for the maintenance of information within working memory, which is a critical function in audition. This maintenance mechanism allows us to concatenate sounds together, to determine their relationships to one another, and to access knowledge stored in long-term memory. But there is also a close link here to the transformational ability just discussed, which is the province of the dorsal stream. This is because we have a great deal of flexibility in manipulating these patterns and are not confined to only representing them statically. Thus in the context of working memory, the ventral and dorsal stream mechanisms work together—the first encoding and maintaining information, the second providing the means to manipulate the information.

In the studies reviewed in the previous section, the process of manipulation was described as pertaining to perception or imagery. But in fact, all these functions—perception, imagery, and working memory—appear to form a continuum and depend largely on similar distributed circuitry. For example, Phlippe Albouy carried out a magnetoencephalography (MEG) study in our lab (Albouy et al. 2017) that took our mental musical reversal task several steps further: listeners were presented with a simplified version of the task, using a three-tone pattern which they had to mentally reverse (last-to-first) during a two-second retention period, before comparing it to a second pattern. A nice feature of this study is that we analyzed the data from this silent time period between stimuli, when there is no stimulus-driven activity, thus allowing us to focus on purely mentally driven activity. As shown in Figure 4.3, MEG data collected within this timeframe showed a particular oscillatory signature: enhanced power in the theta (5 Hz) band, which was specifically associated with the listener reversing the tones in working memory. This theta frequency of oscillation in distributed frontoparietal regions has been previously implicated in working memory processing in various other contexts (Eriksson et al. 2015, Fell and Axmacher 2011). These findings also fit well with the role of the IPS, and related dorsal structures, in manipulating information in visual working memory (Champod and Petrides 2007, 2010). Therefore, although the task of reversing items in the mind's ear may be viewed as an imagery task, and

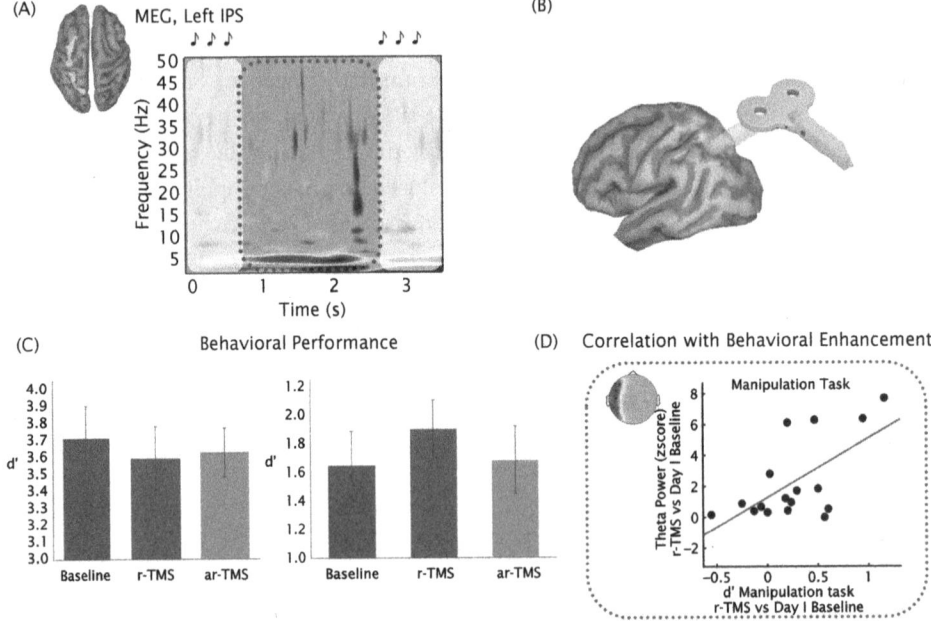

FIGURE 4.3 Causal evidence for involvement of parietal cortex in tonal transformation. A: Time-frequency plot of MEG data in tonal reversal task. Increase in power in the 5-Hz (Theta) frequency band is observed during the silent period (dotted red line) in between the presentation of the target and comparison tones (top), when the listener is performing the mental reversal. Data taken from the left frontoparietal region (inset). B: Target region for Transcranial Magnetic Stimulation (TMS) in left superior parietal cortex. C: effects of 5-Hz rhythmic stimulation (r-TMS) compared to arhythmic (ar-TMS) and baseline conditions. No effect is seen during simple comparison (left), but enhanced performance is seen specifically during rhythmic 5-Hz stimulation (green bar, right). D: Individual differences in degree of reversal task enhancement correlate with individual differences in degree of theta band power increase induced by the stimulation. Adapted from (Albouy et al. 2017) with permission from Elsevier.

the phenomenology associated with doing it is consistent with that description, the underlying circuitry is essentially identical to that involved in working memory manipulation.

The most remarkable finding of Philippe's study was that working memory manipulation ability could be selectively enhanced by targeting the IPS with rhythmic transcranial magnetic stimulation (TMS), specifically at the 5 Hz frequency but not in a control arrhythmic condition. The improvement was specific to the reversal task (not the simple retention task), and the degree to which each individual improved their performance was correlated with the degree to which they each showed enhanced theta power induced by the TMS (see Figure 4.3D). This set of findings, therefore, causally links the dorsal stream structures (IPS and dorsal premotor/prefrontal areas), and their particular oscillatory signature, to the ability to reorder information in auditory working memory, thereby providing direct evidence for the role of this circuitry in active manipulation of auditory information.

One might object that this kind of mental reversal is artificial and rarely used in real music; in fact, although some people with musical training can perceive such transformations (Krumhansl et al. 1987), most cannot. However, that was not really the point of this set of studies. Rather, the idea was to explore the mechanisms that allow us to reorganize perceptual elements into new patterns that we have never heard before. This general ability to change things around (whether reordering in time or transposing in pitch) can operate on sounds online as they are experienced, which is more in keeping with a working memory kind of description. But it could also operate on stored representations of sounds we may have previously experienced, which would be more in keeping with an imagery kind of description. Thus, whether one considers the function to be related to working memory or to imagery may depend more on the task one is performing than on the actual neural circuitry involved.

But just as manipulating symbols in mathematics or in language is critical to solving problems, one could also argue that, in the context of music, this ability can serve equivalent functions. For instance, a composer or arranger, who is trying to think of what the order of a chord sequence should be, might mentally manipulate the sounds involved to decide on the best outcome. Perhaps this ability could even represent one substrate of musical creativity, in so far as the ability to generate new sound patterns based on known ones seems like a prime example of what a composer or improvising performer would need to do. This conclusion is consistent with the finding that a network of several dorsal regions is engaged during improvisation (Berkowitz and Ansari 2008), and that its connectivity is modulated specifically by the performer's past experience with improvisation (Pinho et al. 2014).

Auditory-Motor Processing: Music Production

Given the prominent role of the dorsal stream in visuomotor processes, as outlined above, and given the fact that this dorsal route may be considered a multimodal convergence zone, receiving not only visual but also auditory (Molholm et al. 2006, Regenbogen et al. 2018) and sensory-motor inputs (Lewis and Van Essen 2000), it is not surprising that it should also prove to be critical for musical production. However, studying music production with the usual tools of cognitive neuroscience is not so simple, especially given the confines of an MRI machine. But scientists are nothing if not resourceful, and some have figured out ways to overcome the constraints in order to measure brain activity during music production. Several groups have adapted a keyboard so that it can be played within the scanner (Parsons et al. 2005, Bengtsson et al. 2007, Chen et al. 2012, Brown et al. 2013, Pfordresher et al. 2014). Others have used a variety of different interfaces, including trumpet (Gebel et al. 2013), cello (Segado et al. 2018, Wollman et al. 2018), or a guitar neck used for simulated playing (Higuchi et al. 2012). Of course, another way to look at music production without using a physical instrument, is to study singing (Kleber et al. 2007, Zarate and Zatorre 2008, Zarate et al. 2010, Belyk et al. 2015), about which there will be more discussion later.

Broadly speaking, these studies with different instruments converge in describing the basic auditory-motor network involved in musical production, which typically comprises not only the classic auditory and motor systems (the latter including basal ganglia and cerebellum) but also many of the dorsal stream structures, especially the dorsal premotor and the parietal cortex. Variations within this network do arise however, depending on the specifics of the instrument and also on the paradigm used (playing/singing simple tones/scales versus complex music, playing from memory versus reading a score versus improvising, learning from imitation, etc.). The challenge, as well as the interest, comes from the fact that music-making can take many forms and hence is likely to involve many different cognitive operations.

As a means of addressing this complexity, let us deconstruct a seemingly simple musical production process, such as playing a short sequence of tones on a piano, in terms of the components that may be involved. At a minimum, we can think of three motion-related aspects that must be regulated in order to play correctly: each finger must find its correct spatial target on the keyboard, the sequence of finger actions must occur in the correct order, and the movements of the fingers must be accurately timed (Zatorre et al. 2007). In addition, the sounds produced must be monitored in order to make corrections as necessary. What roles do the various components of the dorsal pathway play in these different sensory-motor processes?

Fractionating and Integrating Movement Parameters

The dorsal stream is known to be important for many aspects of movements to spatial targets in a correct sequence when cued by visual inputs as discussed in the first part of this chapter (Gallivan and Culham 2015). In the case of musical production, though, the spatial targets are unreliably related to visual cues because although they may sometimes be clearly visible (such as a piano keyboard or a drum kit), they are only partly visible in other cases (a saxophone or a flute, for instance) or even completely invisible (as in the case of singing). Besides, skilled musicians don't typically look much at their instruments, as they are busy looking at a score, or the audience, or each other; and sometimes, they even close their eyes when playing. Therefore, movement trajectories cannot be accounted for by the presence of visual cues in music playing.

But, even if the motor targets (keys, strings, etc.) are not always visible, movements organized in spatial patterns of the kind relevant for music do engage dorsal premotor cortex, parietal areas, and other components of the dorsal stream when cued only by the sound pattern to be executed (Brown et al. 2013) and also when performing from memory (Bengtsson et al. 2004). Sound production may be cued by visual input when people learn to play music by reading a score, which also engages the parietal cortex (Stewart et al. 2003); but in that case, the cue is not present in the physical position in space that the fingers must move to, but rather acts as a symbol to be interpreted. All these examples can be subsumed within the transformation operation described earlier, such that an action is generated based on an abstract transform from one coordinate system to another.

One factor that makes the study of movements challenging is that the different dimensions of movement are inherently linked to one another. It's difficult to dissociate the sequencing of movements and their trajectories from timing and spatial position, for instance, because any given sequence necessarily has to occur in space and extend over time. This issue is especially relevant for music performance in which the right actions have to be performed at exactly the right time (Palmer 1997). Such considerations and empirical data (Shin and Ivry 2002), have led some investigators in the motor domain to develop mathematical models in which movement trajectories and their timing are represented in an integrated manner in the brain (e.g., Laje and Buonomano 2013).

In music cognition research, a related question has been extensively debated because behavioral evidence points to a close perceptual interaction between pitch and rhythm patterns (Jones 1987), whereas other evidence, including from amusic persons (Peretz and Kolinsky 1993, Hyde and Peretz 2004), suggests that they are separable. An fMRI study specifically addressing this question (Brown et al. 2013) had pianists reproduce melodies using an fMRI piano and reported that variation in both pitch and rhythmic structure modulated most of the dorsal stream structures in a similar manner, implicating this network in both processes (see also de Manzano and Ullén 2012). However, the intraparietal sulcus region was more sensitive to pitch variation (compared to rhythm), which translates into a spatial code for playing on a keyboard, consistent with its role in transforming an auditory code into a motor/spatial code.

Although these studies of motor performance have consistently reported the involvement of the dorsal stream, it remained unclear, until recently, how the different parameters of movement were represented. New methods applied to neuroimaging data on the basic organization of the motor system have shed light on this question, showing that both segregated and integrated representations of movement parameters exist but within different neural substrates. In a series of studies, Jörn Diedrichsen and his team used multivariate pattern analysis of fMRI data as people executed a sequence of finger movements (without sound) and showed that well-known movement sequences were represented differently than novel ones in several dorsal-stream regions (Wiestler and Diedrichsen 2013). The most relevant finding, in terms of temporal versus spatial sequence representation, was the demonstration that these two features were jointly coded in primary motor cortex but could be separated within the dorsal premotor cortex (Kornysheva and Diedrichsen 2014), as shown in Figure 4.4. In other words, the premotor cortex can represent each dimension independently, thus allowing great flexibility in how they might be combined. In the case

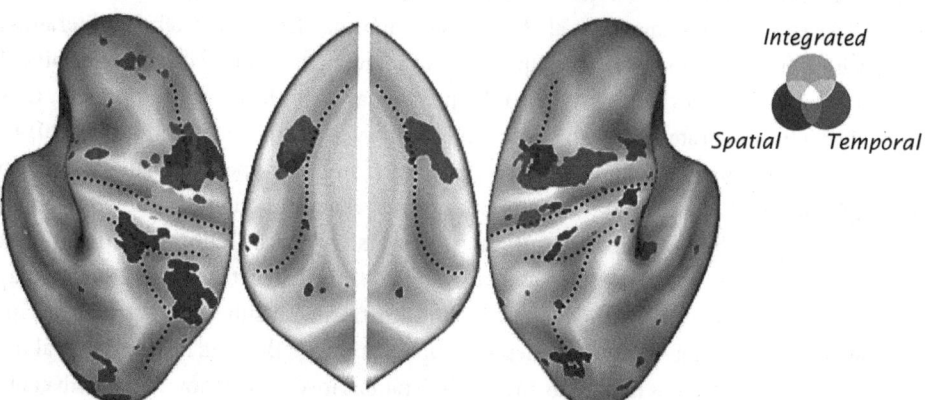

FIGURE 4.4 Separate and joint representation of spatial and temporal motor features in the dorsal stream. Functional MRI classifier accuracy is displayed on inflated cortical surface (lateral views on left and right, medial views in center). Distinct regions are associated with the spatial components of movements (blue), the temporal components (red), or the joint representation of both components (green). Separate spatial or temporal processing occurs in parietal and dorsal premotor regions; integrated processing occurs only in primary motor cortex. Reproduced under the terms of a CC BY 4.0 license from (Kornysheva and Diedrichsen 2014).

of music, this might be relevant to the commonplace situation where different rhythmic patterns are applied to the same melody (for instance, as in a simple jazz improvisation or in a theme and variation situation). Conversely, the primary motor area, that directly sends motor commands to the muscles, integrates these distinct features so that movements can be coordinated and executed on the desired instrument optimally, in both space and time.

Movement Timing and Metrical Structure

There is a large and diverse literature in relation to the timing of movements. But among the key structures that emerge are not only several dorsal-stream structures (especially dorsal premotor cortex and supplementary motor area) but also the cerebellum and basal ganglia. Cerebellar mechanisms are thought to be important for many distinct cognitive functions (Diedrichsen et al. 2019), some of which may contribute, directly or indirectly, to timing of movements (Ivry et al. 2002), including error correction, feedforward control, and optimization of movement trajectories. According to some prominent models in which actions are based on a forward model (with little or no feedback), cerebellar mechanisms are critical for the predictive aspects of movement (Bastian 2006), thus implicating the cerebellum and its connections as part of the broader predictive coding theme that is so relevant for music in general, and for music production in particular.

Patients with cerebellar damage show inaccurate neural encoding of rhythmic patterns, as shown by patterns of electroencephalography (EEG) beat-tracking (Nozaradan et al. 2017), which implicates cerebellar circuitry not only in the output aspect of timing but also in the representation of the temporal relationships of the pattern itself. Such findings agree with many neuroimaging studies that report cerebellar activity during passive auditory perception (Petacchi et al. 2005) and with the observation that individual differences in cerebellar anatomy correlate with performance on beat discrimination and timing tasks (Baer et al. 2015, Paquette et al. 2017). These findings from the cerebellar literature, along with others described in this chapter, show that it is often impossible to determine if a given structure is contributing more to the perceptual or to the action component, thus highlighting the integrated nature of motor and auditory networks (Patel and Iversen 2014).

Along with the cerebellum, the deep gray matter nuclei in the basal ganglia are also strongly implicated for various aspects of timing and related functions, as shown in an elegant set of studies by my colleague at Western University, Jessica Grahn (Grahn and Brett 2007, Grahn and Rowe 2009, 2012). But the dorsal cortical structures also seem to play an especially important role—most likely in coordination with subcortical structures, including the basal ganglia—when it comes to complex timing of the kind involved in making music. In a series of studies in our lab, my former student Joyce Chen, now at the University of Toronto, demonstrated this phenomenon quite clearly. First, she measured fMRI brain activity while people tapped to an isochronous sequence of tones, which just recruits auditory, primary motor areas, and cerebellum, as expected from prior work (Jäncke et al. 2000) but not any of the dorsal-stream regions. But when the same tones were accented by making them louder in such a way as to create a regular metrical structure (ONE-two-three, ONE-two-three), activity in the dorsal premotor cortex and its connectivity to the auditory cortex both increased as the stimulus became more clearly metrical (Chen et al. 2006). This effect did not occur if the accents were distributed in a random way.

A related finding was observed with more complex rhythms that varied in metrical complexity: as the sequence became more metrically complex, tapping behavior became less synchronized (Povel and Essens 1985), and the entire dorsal stream was engaged— including dorsal premotor cortex, SMA, and the intraparietal region, together with the lateral cerebellum (Chen et al. 2008b), forming a functionally connected network for the interface of auditory and motor processes, as shown in Figure 4.5. In this latter study, the

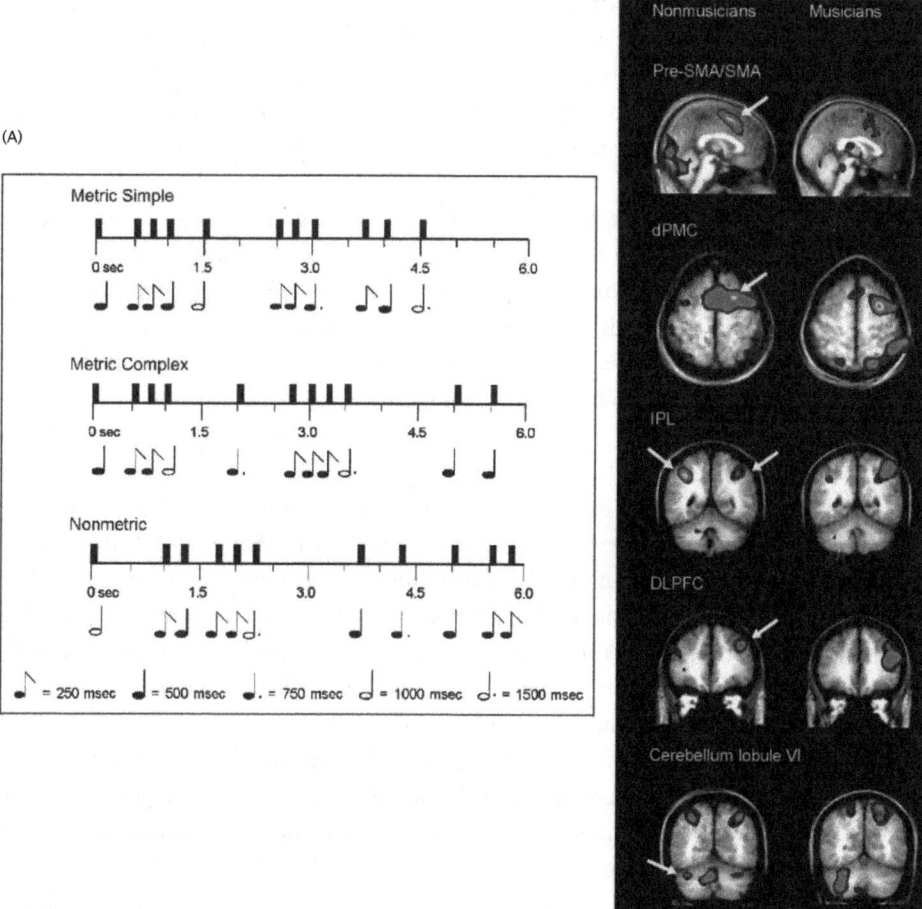

FIGURE 4.5 Dorsal stream involvement in metrical organization. A: Rhythmic stimuli consisting of an identical set of durations (shown both in terms of onsets in timeline and in music notation) are permuted across conditions, resulting in either a simple metrical organization (top), a complex metrical organization (middle), or a nonmetrical organization (bottom). B: Functional MRI data from nonmusicians (left column) and musicians (right column) showing modulation of activity in both groups as a function of metrical organization in dorsal stream areas and cerebellum. (Abbreviations: SMA: supplementary motor area; dPMC: dorsal premotor cortex; IPL: inferior parietal lobe; DLPFC: dorsolateral prefrontal cortex). Reproduced with permission from (Chen et al. 2008).

actual elements making up the various rhythmic sequences were the same; they were simply permuted to make them more or less metrically organized. This means that the dorsal regions are sensitive to an abstract feature of the stimulus—its metricality, or the relationships between elements, rather than the values (durations) of the elements themselves, mirroring what we saw earlier for transposition or mental reversal.

Do these findings indicate that the dorsal premotor cortex and other dorsal stream structures encode the metrical structure of a rhythm per se? This may not be the right question, because these areas work in tandem with auditory sensory regions. Perhaps a better interpretation of the role of the dorsal premotor cortex and its connected network, is that it helps to optimize the movements based on higher-order organizational features of the stimulus (in this case, its meter), as also shown by the increased connectivity to auditory areas for more complex meters. This interpretation is in line with the understanding drawn from neurophysiology (Hoshi and Tanji 2007) that the dorsal premotor cortex is implicated in indirect associations between a stimulus and an action. Thus, the durations of the stimulus sequence do not cue the motor response directly, but rather the motor action is cued by the abstract feature of meter.

This observation also links to the concept of prediction because metrical structure can be thought of as a providing a framework—a temporal grid—to enable predictions about when sounds are likely to occur (most likely at the strong beat, less likely on weak beats, and least likely at time intervals that don't correspond to integer subdivisions of the meter at all). When events fall on weak beats or when there is no event on a strong beat (a musical phenomenon called syncopation, as shown in Figure 4.6), it creates a mismatch because of the discrepancy between what is expected and what is heard. So without an expectation, or prediction provided by the meter, there could be no syncopation.

Tapping in synchrony to syncopated or nonmetrical rhythms is hard to do, although of course, musicians outperform nonmusicians on such tasks (Smith 1983). Therefore, it is of special interest to note that musicians also show greater responses in dorsolateral and inferior frontal cortex in tapping tasks (Chen et al. 2008b), which can be interpreted as representing prediction signals since musically trained individuals possess better internal models of when to expect events to occur. Here, again, we see how the brain's functional organization conforms to a concept derived from music theory (metricality), showing that such formal constructs not only capture aspects of the musical structure (which is what music theorists typically care about), but that they also have explanatory power in terms of musical behaviors, such as movements to music and the neural functions that support them. As we shall see in Chapter 8, this feature of temporal predictability and its violation plays an important role in the sensation of "groove" (Witek et al. 2014), which in turn, is linked to how pleasure can be driven by rhythmic structure.

FIGURE 4.6 Musical syncopation. The traditional song *Down by the Riverside* contains many elements falling off the beat (for example at the words "down" and "shield"), resulting in syncopation. Public Domain.

Interactions Between Auditory and Motor Systems

Role of Learning

Activation in various components of the motor system (premotor cortex, basal ganglia, cerebellum) is a common finding in neuroimaging studies in which people merely listen to musical sounds without any overt movement (Chen et al. 2008a, Bengtsson et al. 2009, Grahn and Rowe 2009, Gordon et al. 2018). This effect is not typically seen with other complex stimuli, for example, during observation of movies, except when certain specific stimuli are present, such as moving body parts (Hasson et al. 2004). Listening to a narrative story can also elicit responses within ventral premotor regions (S. M. Wilson et al. 2007), and hearing speech can also change the excitability of the motor system (Watkins et al. 2003). However, in the case of action observation or speech, the link to the motor action seems more direct—in that the visual action or the speech could be readily imitated by executing the correct motor commands; whereas in the case of music, the pattern of sounds that elicits the motor response may be quite abstract or may be beyond the capacity of the listener to imitate (I can listen to Glenn Gould all I want, but that doesn't mean I can play like him). This difference is most likely related to findings that suggest that, for visual actions or speech, ventral premotor regions are more consistently recruited; whereas for music, both ventral and dorsal networks are reported. The activity in the motor system that results from music listening can be interpreted as further evidence for the close coupling between music processing and the motor system. But is this coupling a consequence of explicit auditory-motor mapping that is established during training, or is it more general and thus intrinsic to music in some sense? We can derive a better understanding of this relationship by looking more closely at the circumstances in which these interactions occur, turning first to the role of learned actions.

Several studies of motor engagement during music listening reveal a specific relationship to learning to play an instrument (and hence probe a direct link between action and sound). Amir Lahav and colleagues, from Gottfried Schlaug's group (Lahav et al. 2007), investigated this idea by scanning volunteers who had no prior musical background before and after learning to play simple tone patterns on a keyboard (five tones, one assigned to each finger). The interesting aspect of the study is that, during scanning, people were merely listening (without movement) to the identical sounds, both before and after training; therefore, any observed differences must be related to the training itself, and not to the stimuli. The findings showed that just listening to the trained musical sequences was sufficient to engage both dorsal and ventral premotor cortex but only once people had learned to play those sequences (Figure 4.7A). Auditory responses were, of course, prominent but remained unchanged between the two conditions. An additional control condition is worth mentioning: Lahav and colleagues compared neural responses when listening to the trained pattern with responses when listening to an untrained pattern that consisted of the same tones but in a different order. They found that the ventral premotor region was responsive (albeit to a smaller extent) even to the untrained pattern, thus implicating this region in the basic mapping of each tone to an action, a point we shall return to further on.

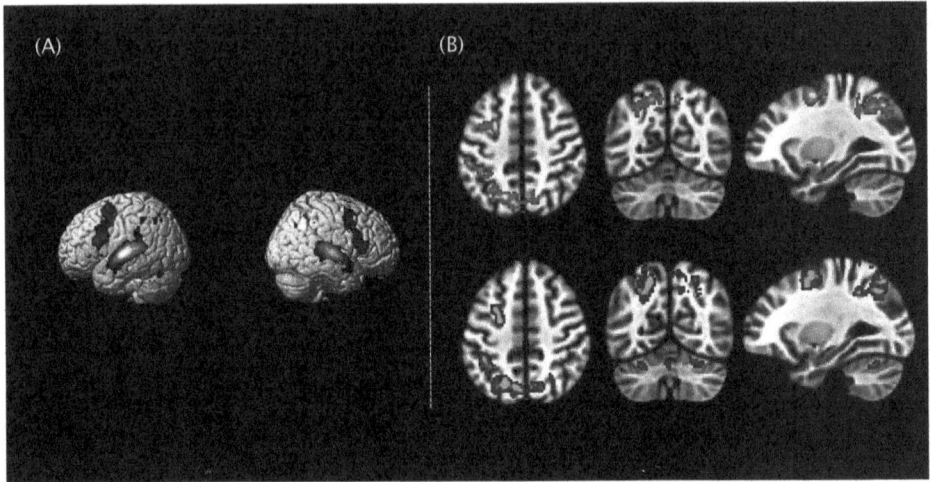

FIGURE 4.7 Recruitment of dorsal stream regions after learning to play a musical pattern.
A: Functional MRI results showing activity in dorsal and ventral motor regions when comparing listening to a pattern after versus before learning to play it. Used with permission of the Society for Neuroscience, from (Lahav et al. 2007); permission conveyed through Copyright Clearance Center, Inc.

B: Functional MRI results showing activity in dorsal premotor cortex and intraparietal area after versus before training when comparing listening to melodies that volunteers had learned to play versus equally familiar ones that had not been trained. Top: Orange areas indicate responses in a condition in which the melodies were played to the listener. Bottom: Blue areas indicate responses in a condition in which melodies were imagined by the listener. Similar dorsal-stream regions were active in both conditions. Reproduced with permission from (Herholz et al. 2016).

This finding of motor recruitment to sound patterns after learning is consistent with other related reports of increased motor cortex excitability when merely listening to music that musicians had learned to play (D'Ausilio et al. 2006). It also aligns with a cross-sectional study comparing pianists to nonmusicians as they listened to piano sequences, which showed greater fMRI responses in ventral and dorsal premotor areas in the musicians (Bangert et al. 2006). Similarly, there is a greater evoked EEG response from premotor areas of pianists to an altered tone in a melody that they had learned to play than to a melody they had memorized only by hearing (Mathias et al. 2014). Even macaque monkeys show activity in motor and premotor areas when they hear sounds they learned to produce (Archakov et al. 2020).

All these studies support the idea that the representation of actions and their associated sounds are integrated after learning. A very direct demonstration of the predictive nature of this linkage was provided by a study showing that the motor excitability of the particular muscle that would be used to play the next tone in a simple musical sequence was facilitated while hearing the previous tones in the pattern at least 50 milliseconds before the target tone actually sounded (Stephan et al. 2018). In other words, predictable auditory information can activate motor representations in an anticipatory manner, with high specificity to the finger to be used, even in a passive condition without movement. However, this effect was only observed for trained patterns, indicating once again that this type of coupling pertains to a learned association between sounds and actions.

An fMRI study with a six-week piano training protocol conducted by Sibylle Herholz and Emily Coffey in our lab, also supported the conclusion that dorsal premotor and parietal regions (as well as cerebellum) are recruited while merely listening to familiar melodies after (but not before) one has learned to play them (Herholz et al. 2016). But remarkably, as shown on Figure 4.7B, we also found that this entire dorsal system is equally active even when the melodies were merely imagined, not heard. Thus, this finding expands covert imaginal processes that we know involve the auditory cortex (Chapter 2) to incorporate the auditory-motor networks that are characteristic of overt musical processing.

The parallel activation patterns in dorsal stream regions for both perception and imagery are also consistent with their role in representing complex internal models, predictions, and transformations and argue in favor of an abstract representational code rather than a fixed sound-to-action association. The latter conclusion was specifically supported by a recent fMRI study by Orjän de Manzano working in Fredrik Ullén's lab in Stockholm, which used a similar approach of scanning nonmusicians after they had been trained to play piano melodies (de Manzano et al. 2020). Using a multivariate decoding approach, these authors were able to demonstrate that activity in auditory and premotor cortical areas could be used to classify melodies after training. But the novelty was that some degree of classification was also possible after training even for new, untrained melodies, indicating that the training may have generated more abstract sound-action correspondences, too.

An important detail that may appear contradictory is that, in our piano learning study, (Herholz et al. 2016) we saw dorsal but not ventral premotor responses when listening to melodies that had been trained; whereas these ventral regions were most prominent in the study by Lahav and colleagues (Lahav et al. 2007). In the latter study, however, the training was strictly limited such that each finger was associated with one and only one key, and the hand never moved from its fixed position on the keyboard. In contrast, in the study by Herholz et al, the melodies were bimanual, more complex, spanning a larger range, and hence required variable fingerings that were left up to each person to learn. Thus, for each sound, the task required selection from among several different possible movements, a situation known to involve the dorsal rather than the ventral pathway (Hoshi and Tanji 2006, Amiez et al. 2012), including during mental rehearsal of stimulus-action couplings (Cisek and Kalaska 2004). This dorsal-ventral distinction again emphasizes the more abstract nature of encoding in the dorsal pathway.

These findings of responses in motor regions to sound have often been interpreted in the context of the mirror-neuron idea (Rizzolatti and Craighero 2004), which explicitly links motor systems to the observation of actions and also to sounds produced by actions (Kohler et al. 2002). Experiments in which motor recruitment is seen only after training can fit this narrative (Lahav et al. 2007). However, this class of explanation does not readily account for the more abstract types of responses just described. In particular, it seems difficult to explain phenomena like neural responses in motor areas to the beat of a pattern which may occur at time-points that have no sound at all (notably, during syncopation), so there is no obvious imitation of a stimulus, since the stimulus is absent. Also, the more spontaneous recruitment of motor-related regions that are not linked to explicit learned actions, such as was reported by de Manzano et al. (2020), are harder to explain via a mirror-neuron account.

As mentioned above, recruitment of the motor system, and other distributed components of the dorsal stream, has been frequently reported in the absence of any training or explicit sound-movement association. Notably, both naturalistic listening to real music (Alluri et al. 2012, Abrams et al. 2013), as well as listening to artificial rhythmic patterns of varying complexity (Chen et al. 2008a, Grahn and Rowe 2009, Kung et al. 2013), elicit activity in premotor regions and the IPS. In those experiments, the activity occurs despite the lack of any specific real or imagined action, nor any association to action, nor any kind of training prior to the experiment—it's a spontaneous phenomenon. Therefore, there must be a different (or at least an additional) reason, other than the mirror-neuron idea, to explain this intrinsic sound-action coupling.

Temporal Predictions

A number of different concepts have been developed that can help to explain the apparently intrinsic link between rhythmically organized musical sound patterns and actions, in particular the recruitment of motor regions of the brain during exposure to complex rhythms and music (for a recent review, see Damm et al. 2020). One class of models uses the idea of endogenous neural oscillations to explain motor entrainment to periodic sounds, including music (Large and Palmer 2002), via nonlinear interactions between putative auditory and motor oscillating networks (Large and Snyder 2009). A role for cerebellar mechanisms has also been proposed in a predictive timing context (Schwartze and Kotz 2013). More recent theoretical work has emphasized how oscillatory mechanisms originating in motor cortices interact with ongoing processing in auditory cortices (Morillon et al. 2019). The idea has also been extended beyond rhythmically organized patterns to encompass the processing of aperiodic events (or quasi-periodic ones, like speech) via a phase-reset mechanism originating in dorsal structures that can change ongoing oscillations in auditory areas (Rimmele et al. 2018). However, we should keep in mind that spontaneous synchronization to sounds is almost entirely confined to music, due to its temporal regularity compared to speech. This phenomenon was shown empirically by my friend and BRAMS co-director Simone Dalla Bella and colleagues using tapping tasks; they concluded that "Music's peculiar and regular temporal structure is likely to be the main factor fostering tight coupling between sound and movement" (Dalla Bella et al. 2013).

A complementary theoretical approach to the link between auditory and motor systems, focused specifically on music processing, was developed by Ani Patel and John Iversen. In their "Action Simulation for Auditory Prediction," or ASAP hypothesis, Patel and Iversen (2014) propose that

> "the motor planning system uses a simulation of body movement . . . to entrain its neural activity patterns to the beat period, and that these patterns are communicated from motor planning regions to auditory regions where they serve as a predictive signal." (p. 4)

This model may be thought of as a specific version of the more general concept that action simulation underlies many cognitive processes (Jeannerod 2001). It is also directly relevant to a principal idea of this chapter: that auditory cortex–dorsal pathway interactions can be understood in the context of prediction and thereby contribute to both music production and perception (and eventually, to pleasure).

In addition to addressing the spontaneous recruitment of motor-related regions to temporally structured rhythms, ASAP was explicitly designed to account for several other interesting behavioral phenomena that are worth discussing in further detail. One such feature that was described long ago (Fraisse 1948), is the greater capacity of auditory signals than visual ones to induce rhythmic movement and the greater accuracy of such movements to auditory than to visual stimuli (Repp and Penel 2002). Although moving visual stimuli (like a ball bouncing against a wall) can sometimes entrain movements as accurately as auditory inputs (Gan et al. 2015), there is no evidence that this effect extends beyond simple repetitive rhythms to encompass complex, hierarchical beat-based movements of the kind most relevant for music (Ammirante et al. 2016). Besides, there is little spontaneous urge to move in such circumstances: nobody dances to a tennis match, no matter how regular the serves and strokes are. This discrepancy, between the influence of auditory versus visual inputs on the motor system has been linked to the basal ganglia (Hove et al. 2013), in agreement with prior studies on beat perception (Grahn and Rowe 2012); but the modality difference would seem to imply some broader underlying special connection between auditory and motor networks, as we shall see below.

Another important behavioral phenomenon is that beat processing is predictive; that is, motor actions typically occur either precisely on the beat or very slightly before it (Repp and Su 2013). Therefore, the motor action must be planned in advance of the beat point rather than being a reaction to a stimulus; otherwise, it would arrive too late. Most people, even those without musical training, are able to produce these predictive movements quite accurately (Repp 2010); although, not surprisingly, trained percussionists are the most accurate (Krause et al. 2010). This common facility to tap to a beat seems to be part of the general predisposition to move to temporally organized sounds, which doesn't require much more than exposure, since it manifests itself spontaneously early in life (Zentner and Eerola 2010). In fact, in infants, the link between sound and motion is a two-way street, as moving to a particular beat also influences subsequent beat perception (Phillips-Silver and Trainor 2005).

This mostly experience-independent ability of moving predictively to the beat also implies a fairly fundamental link of some kind between auditory and motor systems. The ASAP idea, that motor structures influence the auditory system, would be a good candidate mechanism for this feature as well. Evidence in favor of this idea comes from one of Jessica Grahn's fMRI studies that showed engagement of putamen and supplementary motor and premotor cortices during perception of rhythms containing beat structure. They also showed a greater contribution from the putamen in a condition of beat continuation (tapping after a sound sequence ends), which involves a greater predictive component than beat finding (Grahn and Rowe 2012). We will revisit these ideas in the context of pleasure and groove (Matthews et al. 2020) in Chapter 8.

Tapping to a beat, however, is not merely a matter of finding an underlying pulse and moving at that rate. In real music, meter is often organized hierarchically (Large and Palmer 2002), such that several different levels may be differentiated according to subdivisions that are most often related by integer ratios (for review of these phenomena in a musical context see [London 2012]). Listeners can then choose to focus on lower or higher levels of the hierarchy (Toiviainen and Snyder 2003), which in turn may be related to training or experience with the relevant style of music. They may also perceive beats at different phases of the meter, for example on the stronger or the weaker beats (i.e., downbeats or upbeats) and will sometimes shift from one to the other, as shown in Figure 4.8, indicating flexibility in the sound to action relation (Patel and Iversen 2014).

In some musical styles, multiple meters may even coexist simultaneously (Vuust et al. 2014). There are of course also some musical systems that largely eschew a regular meter (such as Gregorian chant, or much contemporary Western art music) and others in which metrical structure waxes and wanes throughout the piece. However, integer ratios seem to characterize temporal structure of music across many different societies (Jacoby and Mcdcrmott 2017). The key point here is that even for metrically well-structured music, which is also common across cultures (Savage et al. 2015), there is not necessarily a one-to-one relationship between the timing of sounds and the actions elicited by them, which is why a straightforward mapping of one to the other is not a sufficient explanation.

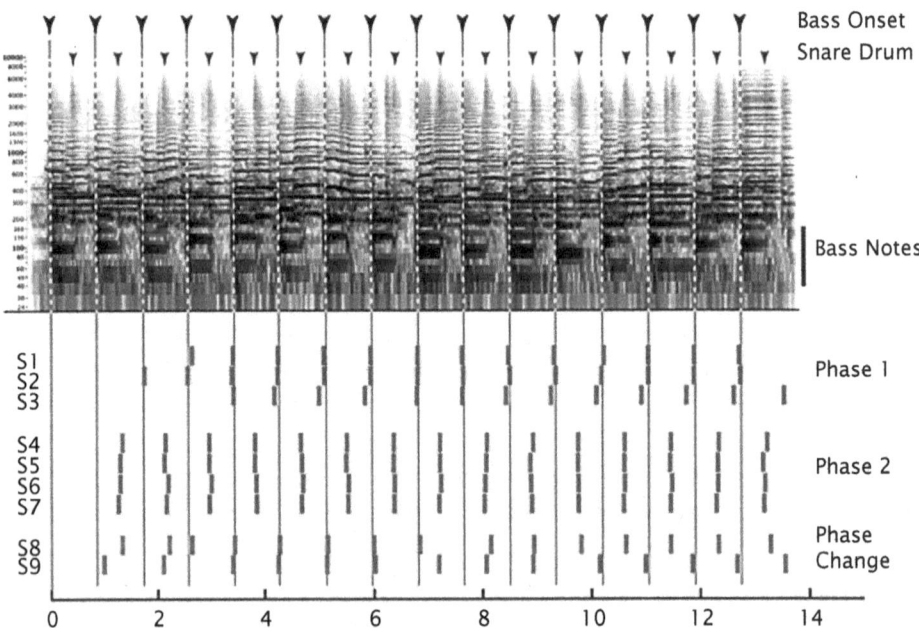

FIGURE 4.8 Individual differences in beat perception. Top: spectrogram (frequency as a function of time) of an excerpt of jazz music (*Stompin' at the Savoy*, by Benny Goodman). The inverted arrows at the top show the onsets of the double bass and the snare drum, which always occur in alternation. Bottom: Red lines indicate position of taps made by listeners when asked to tap to the beat of this music. Subjects 1–3 tap primarily to the beat indicated by the bass, subjects 4–7 tap to the beat indicated by the snare drum (phase shifted with respect to the bass), and subjects 8 and 9 switch from one percept to another over time. Reproduced under the terms of a CC BY 3.0 license from (Patel and Iversen 2014).

Moreover, movement to real music that contains multiple levels of metrical organization can involve engagement of different body parts for each of the different levels (e.g., hips for the slower levels, head or arms for intermediate, and fingers for the faster levels), as shown by a study from Petri Toiviainen's lab in Finland that measured motion patterns as people danced to Motown hits (Burger et al. 2018). This finding makes sense in terms of the biomechanics of the different body parts (it's hard to wiggle your hips as fast as you can wiggle your fingers!). But, more importantly, it suggests that the nervous system can track multiple metrical levels simultaneously. These phenomena further underline that perception of beat structure is not merely passive but can change, depending on the listener's conscious influence.

Another example of this top-down control is when a beat can be imposed on a stimulus that contains no such higher-order structures (which is why a ticking clock, for instance, can be made to sound, at will, like it's playing in duple or triple time—especially in the middle of the night). Naturally, musicians and composers exploit this phenomenon to create metrical ambiguity that leads to the music being heard in different ways, depending on various musical structural cues and the listener's knowledge and experience (Hannon et al. 2004). This common experience of imposing a beat on a pattern (Repp et al. 2008) strongly implies that there is some kind of top-down modulation of the incoming auditory information, as opposed to the pattern simply generating entrainment in the perceptual/motor system in a passive way. This kind of "constructive" view of metrical structure emphasizes its abstract nature: it is a mental representation of a stimulus pattern, but not the stimulus pattern itself.

An important neural correlate of this top-down metrical construction phenomenon has been provided by a few investigators who measured MEG responses to physically identical rhythmic patterns, as listeners were instructed to mentally impose a beat on one or another element of the rhythm (Iversen et al. 2009, Fujioka et al. 2015). The results showed a distinct modulation of induced oscillatory activity to each sound in the beta band (~20 Hz) such that the amplitude was greater for the item that was mentally "accented," compared to those that were not. Even more interesting, John Iversen and colleagues were able to show that with syncopated rhythms there is an induced beta-band response, even in the absence of any actual sound at that moment in time, as long as the listener perceives a beat to be present. This means that these sorts of responses are driven more by the internal representation of the beat pattern than by the physical sound itself, in keeping with the argument that beat patterns are abstract entities.

In related MEG work, Takako Fujioka and her colleagues showed that even under passive conditions, while listeners were distracted by watching a silent movie, there is still a modulation of beta-band activity just before each tone in a sequence is sounded—but only if it is predictable and not random. This effect shows that the prediction mechanism can operate in a kind of automatic mode, even if it is also clearly subject to volitional control (Fujioka et al. 2012). But where does the modulation come from? According to ASAP, and other arguments presented in this chapter, predictive signals should originate in dorsal-stream structures. Some support for this idea comes from experiments showing that inhibition of posterior parietal regions, via TMS, caused behavioral disruption of beat perception but not of simple interval timing (Ross et al. 2018). Similarly, temporally correlated beta-band modulations can be localized to auditory areas and motor regions (Fujioka et al. 2012). So, there are clearly interactions between these systems. But how exactly might they work together?

How Auditory and Motor Systems Influence One Another

As we have seen, temporal acuity—essential for both music production and perception—critically depends on some kind of predictive mechanism. Behavioral studies show that tapping, or moving in time to rhythmic sequences, improves the accuracy of their perception (Su and Pöppel 2012, Manning and Schutz 2013). There are now good empirical results that point to descending influences of the motor system on the auditory system that directly support this perceptual enhancement. Luc Arnal and colleagues used MEG to show that the accuracy of deciding whether a target tone was delayed or on time—in an otherwise temporally regular sequence—was related to the amplitude of oscillations in the beta band in coupled auditory-motor networks, which could be measured a few hundred milliseconds before the occurrence of the target (Arnal et al. 2014). This clearly predictive oscillatory signature in an active task is compatible with the passive beta-band activity to repeated tones just discussed (Fujioka et al. 2012). It is also reminiscent of the predictive modulation of specific muscles prior to playing a tone from a learned sequence (Stephan et al. 2018), indicating that these motor-to-auditory influences manifest themselves in many different contexts, including both spontaneously and in circumstances where there is a learned mapping.

Further specific evidence for the influence of motor activity on auditory processing comes from experiments by Benjamin Morillon, who demonstrated behaviorally that perception of a temporally regular tone sequence embedded in a distractor sequence could be enhanced by silently tapping along to the rhythm of the target sequence (Morillon et al. 2014). Subsequently, this paradigm was applied in the context of MEG (Morillon and Baillet 2017). In this experiment, shown in Figure 4.9, the target and distractor sequences were presented quasi-periodically and in an interleaved fashion such that every other tone corresponded to the target sequence. Thus, although there was a sound every 333 milliseconds (3 Hz), the target pattern was presented at the slower rate of 667 milliseconds (1.5 Hz). Analysis of the MEG responses, from both active tapping and listening conditions, showed a response in the auditory cortex at 3 Hz, corresponding to the rate of incoming sound presentation (a mixture of targets and distractors), and a response in the dorsal sensory-motor cortex at 1.5 Hz, corresponding to the rate of the target melody. This latter motor response also corresponded to bursts of beta-band oscillations, supporting the conclusion from prior studies that beta activity in motor areas modulates perceptual analysis occurring in auditory regions. That interpretation is also supported by the analysis of directed connectivity between these two regions: information corresponding to the rate of the total sound input (3 Hz) flowed from auditory areas to dorsal regions; whereas information corresponding to the attended targets (1.5 Hz) flowed in the opposite direction from the dorsal regions to the auditory cortex (Morillon and Baillet 2017).

Taken together, the strong bidirectionally functional interactions between motor and auditory systems are relevant to several interesting aspects of music processing. At a general

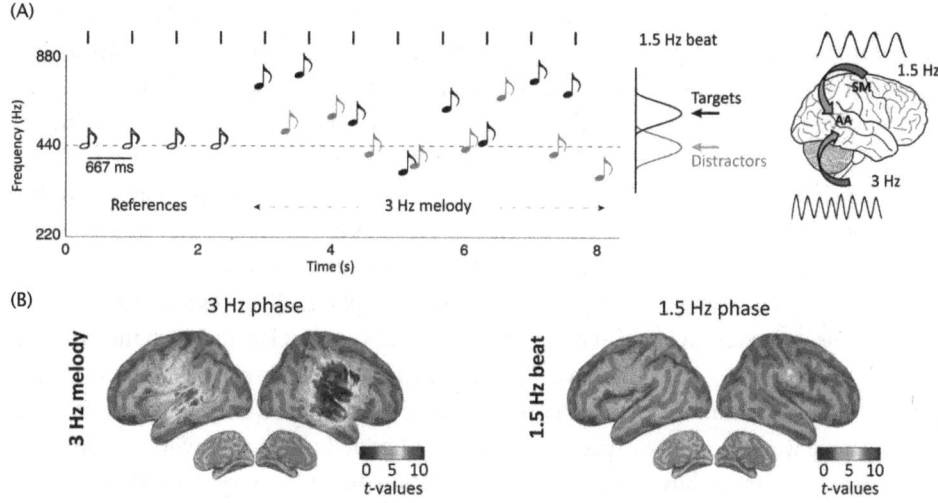

FIGURE 4.9 Role of dorsal regions in temporal predictions. A (left): Stimulus sequence used, starting with four pacing tones (open symbols) presented at a rate of 1.5 Hz (667 ms), following which target tones (black symbols) were presented at this same rate but interleaved with distractor tones (gray symbols). The combination of all tones resulted in a 3 Hz rate of tone presentation, but attention had to be directed only to the target tones, whose pitch differed on average from the distractors. A (right): Depiction of the hypothesis that temporal predictions for the target tones would occur at 1.5 Hz and originate in motor regions, while stimulus-driven activity would occur in auditory areas and occur at 3 Hz. B (left): MEG data showing that phase locking with the 3 Hz stimulus rate originates from auditory and adjacent regions. B (right): MEG data showing that phase locking with the 1.5 Hz target stimulus rate originates in dorsal and sensory-motor regions. Adapted from (Morillon and Baillet 2017).

level, they help to explain the utility of spontaneous movements to music since they are likely to shape the listening experience and help to organize and disambiguate sound patterns. A direct demonstration of this phenomenon extended into the imagery domain was provided in Mor Regev's experiment, described in Chapter 2 (Regev et al. 2021): intersubject correlation of fMRI activity for imagined melodies was higher in the auditory cortex when people were asked to tap to the beat than when they performed the task without movement. Therefore, even though the tapping did not add any information about the sound (indeed, there was no sound—it was all imaginary!), it generated a more accurate and/or highly synchronized internal representation, such that the dynamic changes in brain activity in the auditory cortex became more tightly aligned across listeners.

At a more specific level and as every musician knows, movements such as foot tapping are useful to the performance of music for the same reason, especially during learning: they help to impose a metrical structure which improves the representation of the musical pattern. Importantly, motor influences can be covert since motor-to-sensory interactions appear (as we have seen) even in passive listening, imagined listening, or in active listening without overt movement. Similarly, my McGill colleagues Rachel Brown and Caroline Palmer showed that individual differences in motor imagery skill correlate with the ability to learn to play musical sequences, demonstrating that the ability to imagine movement is helpful to musical learning (Brown and Palmer 2013).

In an important theoretical paper, Giacomo Novembre and Peter Keller argued that, in addition to these aspects just mentioned, common coding of perception and action can support the integration of actions in group music-making. Predictive mechanisms can be used to interpret the actions of others as well as oneself. They proposed that "coupling of perception and action might scaffold the human ability to represent complex (structured) actions and to entrain multiple agents—via reciprocal prediction and adaptation—in the pursuit of shared goals" (Novembre and Keller 2014), thus broadening the concepts beyond mere perception and motor control to encompass social aspects as well.

A nice demonstration of how this kind of group coupling influences real music performance comes from an experiment conducted in Laurel Trainor's lab, in which they studied the body movements of musicians in a piano and string trio as they played together (Chang et al. 2019). They found that the better the coupling of the movements of the members of the ensemble were, the more expressive the music sounded to listeners who merely heard the performance (and thus had no visual information about the movements). Therefore, synchronization seems to be related to how well musicians communicate with one another, leading to higher-quality musical performance.

Connectivity Between Auditory and Motor Systems

All these phenomena linking sounds to action require anatomical connections between auditory and motor systems as a substrate for the reciprocal flow of information that is assumed to underlie such interactions. At the start of the chapter, we reviewed some of these anatomical pathways, so now it is relevant to consider evidence that their structural and functional organization is indeed linked to auditory-motor integration in musical contexts. If the reciprocal influence of motor and auditory systems is mediated via these pathways, then we would expect that their organization ought to be related to musical functions.

One source of information on this question comes from cross-sectional experiments in which measures of anatomical connectivity in white-matter tracts were found to differ as a function of musical training. For example, various studies have used diffusion MRI to suggest that the anatomical organization of fiber tracts that interconnect auditory and motor areas—targeting both dorsal and ventral motor structures—is enhanced in those with musical training, compared to those without (Halwani et al. 2011). A similar finding has also been reported in children, as a function of the amount of training received (Loui et al. 2019). These findings from analysis of white matter complement other work that shows enhancements in gray matter structure in auditory and dorsal regions of musicians (Schneider et al. 2002), including cerebellum (Baer et al. 2015), and that correlate with music task performance (Foster and Zatorre 2010a). Of particular relevance is the finding that the intercorrelation of cortical thickness between auditory and ventral premotor regions is greater in musicians compared to controls (Bermudez et al. 2009), implying that changes in gray-matter structural features between these structures are linked by their frequent coactivation.

These group comparisons suggest that musical training changes the structural connections between auditory and motor systems. However, more direct evidence for a causal influence comes from longitudinal studies. For example, Figure 4.10 shows a study by

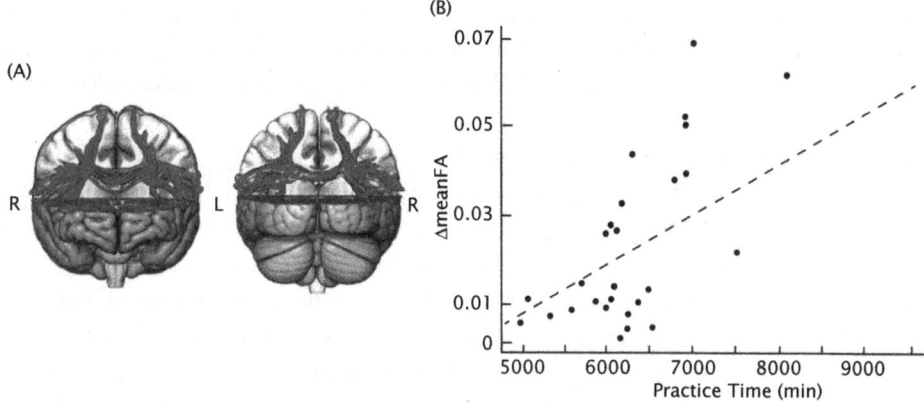

FIGURE 4.10 Effect of music training on white-matter organization. A: White-matter tracts interconnecting auditory with motor regions measured with diffusion MRI. B: Change in fractional anisotropy of white matter in the tracts interconnecting auditory and motor areas as a function of total practice time for each individual. Reproduced with permission from (Li et al. 2018).

Qiongling Li and colleagues from Beijing University, who analyzed white-matter structure of auditory-motor connections in groups of people before and after receiving musical training for 24 weeks. They reported an increase in connectivity associated with training (and amount of time spent practicing) but not in a control group. The effect largely disappeared after 12 weeks without further practice, directly showing that it was experience-dependent (Li et al. 2018). Similarly, in one of the first longitudinal studies of music training in children, Krista Hyde and colleagues reported anatomical changes in both auditory and motor areas after training (Hyde et al. 2009).

Complementary findings to these structural connectivity studies come from analyses of functional connectivity: for example, network analyses of functional imaging data during naturalistic music listening show that musicians automatically engage neural networks that are more action-based, comprising sensory-motor regions, compared to nonmusicians (Alluri et al. 2017). Enhanced interactions between sensory and motor regions in musicians have also been shown in the absence of any stimulus during resting-state scanning. Increased fMRI-based connectivity between auditory cortex and ventral premotor cortex (the same region involved in direct auditory-motor mapping) is greater in musicians (Palomar-Garcia et al. 2017) as is the case between the insular cortex and auditory/motor areas (Zamorano et al. 2017). Using EEG, enhanced connectivity in musicians has also been observed between auditory cortex and sensory-motor/frontal regions, without any task being performed (Klein et al. 2016). These resting-state results suggest that auditory-motor interactions associated with training manifest themselves even in the absence of any musical task and, as such, imply a more permanent rebalancing of how these systems interact, due to the coupling of motor actions with sounds that a musician has repeatedly experienced.

All these findings, that musical training is associated with enhancements in the connectivity between auditory and motor systems, constitute a necessary condition for the claim that these systems are important for music production. But as we have seen, the spontaneous engagement of auditory and motor systems need not be related to musical training at all. So how does the connectivity of these systems relate, more generally, to the claim that there is some kind of special link between sound and action, independent of the consequences of training?

There are several ways to address this question. First, a number of studies have indicated that there are individual differences in the degree to which auditory and motor systems are linked, even in the absence of any musical training, and that this variability is predictive of the speed or ease with which people can learn musical tasks. Specifically, there are significant correlations between measures of the structural organization of auditory-motor fiber tracts obtained prior to learning and the outcome of musical artificial grammar learning (Loui et al. 2011). Even more directly relevant are two studies that have shown that the speed of learning to reproduce musical patterns can be predicted by diffusion MRI measures of tracts that interconnect auditory with dorsal regions (Engel et al. 2014, Vaquero et al. 2018). As shown in Figure 4.11, the better the connectivity, the more predisposed people will be to learn auditory-motor mappings quickly and efficiently.

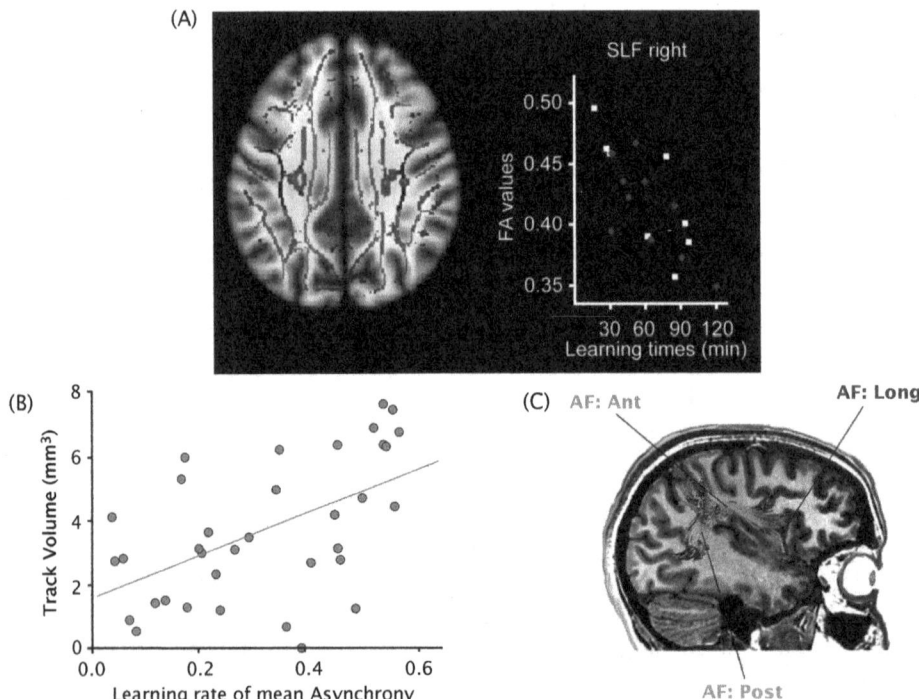

FIGURE 4.11 White-matter organization predicts speed of learning. A: Brain image showing white-matter skeleton (green) of the superior longitudinal fasciculus measured with diffusion MRI and the regions within this tract whose functional anisotropy correlated with learning speed (yellow-orange). Scatterplot shows fractional anisotropy as a function of speed of learning for each individual: faster performance is associated with higher values of white-matter organization. Reproduced with permission from Engel et al. 2014.

B: Scatterplot showing volume of the anterior segment of the arcuate fasciculus as a function of learning rate in a rhythm task: better learning is associated with higher volume. C: Anatomical depiction of the arcuate fasciculus with the anterior segment shown in green. Adapted from (Vaquero et al. 2018) with permission from Elsevier.

The microstructure of the white-matter tracts connecting auditory and ventral frontal regions also predicts individual variation in the degree of EEG mismatch response to rhythmic deviants (Vaquero et al. 2021), consistent with the role of frontotemporal interactions in generating these responses as discussed in Chapter 3. These findings are also in line with other studies showing that activity in auditory areas prior to learning predicts how well people will learn to discriminate pitch patterns (Zatorre et al. 2012) and learn to play piano melodies (Herholz et al. 2016). Evoked MEG responses in auditory cortex of nonmusicians predict melody discrimination as well (Schneider et al. 2002). However, it is not only the quality of the auditory encoding that is predictive but also how those representations interact with dorsal-stream structures that is important in many aspects of musical perception and learning.

Also very relevant is a recent study in which our former postdoc Indiana Wollman, now doing research at the Philharmonie de Paris, trained people to play our MRI-compatible cello (to be further discussed later in this chapter) and measured their brain activity before and after the training (Wollman et al. 2018). Cello performance scores obtained after training were correlated with functional connectivity between auditory cortex and the pre-SMA, which is part of the dorsal pathway, measured prior to training. In other words, those who, for whatever reason, have higher levels of functional interaction between their auditory and dorsal motor systems learned to play more efficiently over the four-week period. Related findings come from experiments carried out in the lab of Caroline Palmer and her associates, comparing memory for melodies that were produced on a keyboard versus those that were only passively perceived. Not only were the former better remembered, but cortical motor regions showed greater EEG responses to altered tones in the previously produced melodies (Mathias et al. 2014, 2016). All these examples point in the same direction: interactions between auditory and motor systems influence learning of auditory-motor tasks, and individual differences in connectivity between these systems are predictive of the efficacy of such learning.

Another important point about auditory-motor connectivity is that, independently of training or individual differences, intrinsic interactions between these two systems seem to be very prominent. We have already reviewed the top-down oscillatory modulation from dorsal motor structures to auditory systems, which are thought to optimize the parsing of auditory information or impose temporal regularities on incoming events (Morillon et al. 2019). But there is additional intriguing information that suggests a specific coupling between the two. For example, in a combined brain stimulation and fMRI study, my former postdoc Jamila Andoh found that transcranial magnetic stimulation applied to the right auditory cortex influenced two resting-state functional networks: the auditory network, as expected, but also the somato-motor network, including much of the dorsal auditory-motor pathway. Therefore, modulating the auditory system has an influence on the motor system.

This effect may be related to the more extensive, network-level anatomical connections from right auditory cortex to dorsal areas (Misic et al. 2018) that we will consider in greater detail in Chapter 5. Another piece of relevant evidence is that a region of the dorsal premotor cortex close to the laryngeal motor representation can be characterized in terms of auditory spectrotemporal modulations, just like the auditory cortex itself, and is functionally connected to the auditory cortex (Venezia et al. 2021). These results taken together, hint at an intrinsic, possibly privileged, auditory-motor coupling, most likely particular to the temporal domain, which is at the heart of many musically relevant behaviors. Furthermore, this coupling likely only requires exposure, rather than specific training, in order to develop (although it likely develops even further with specific training).

Singing

The ability to sing may be thought of as a particularly interesting and relevant musical function to examine in the context of auditory-motor interactions. Singing appears in all human cultures and has near-universal acoustical and behavioral features (Mehr et al. 2019). It also emerges early in development in a spontaneous fashion, even if it takes a long time to mature fully (Tsang et al. 2011). Furthermore, contrary to what is often assumed, the ability to sing simple songs reasonably in tune is widespread in the population. This ability was shown by Simone Dalla Bella and his collaborators, who asked random people in a park to sing a well-known tune, and found generally good performance, despite the lack of any specific musical training (Dalla Bella et al. 2007).

All these observations support the idea that the human nervous system possesses the natural capacity to encode musically relevant information from the environment and engage the vocal motor system to reproduce it. Like for all music (and speech) production skills that humans possess, it's important to realize the computational difficulty of the task, which requires a complete transformation: starting from sound waves hitting the eardrum, all the way to the exact combination of air pressure and muscle tension being exerted on the vocal cords, in coordination with breathing, to reproduce those original sound waves. Or rather, not simply to reproduce the original waves, but to reinterpret them at a more abstract level, because singing preserves the global features of the song—the relationships between elements—but not necessarily the exact pitches, durations, or other low-level features. A great deal of research has emerged in recent years on many aspects of singing (Welch et al. 2019). But here I shall limit the discussion to the role of the dorsal-stream structures in the auditory-motor integration processes needed to sing.

Many of the principles of sensory-motor control that are relevant for singing are, of course, also relevant for speech. Extensive neurophysiological evidence has identified the brain network involved in cognitive control of vocalizations—including ventral premotor areas, the cingulate cortex, and the SMA—which interact with primary motor and brainstem nuclei to coordinate breathing, activation of the larynx, and the articulators (Loh et al. 2017). Sophisticated computational models also exist of how the brain's vocal control system integrates auditory and somatosensory feedback to produce spoken words (Houde and Chang 2015, Kearney and Guenther 2019). It is not surprising that much of the neural hardware for speech and song overlap, given that both require coordinating the same muscle groups and integrating auditory and sensory feedback. But the demands of singing differ from those for producing speech, particularly in terms of the increased requirement for accuracy of pitch control (Zatorre and Baum 2012) as well as other differences, such as breath control. Consistent with this observation, several studies have indicated that singing involves additional structures beyond those implicated in speaking, especially in the right hemisphere (Özdemir et al. 2006, Kleber et al. 2007).

The so-called singing network, shown in Figure 4.12, encompasses neural structures within the dorsal stream, together with auditory and motor cortical regions. In particular, it includes the cortical representation of the larynx, an area which, in humans, has direct projections to the brainstem nuclei that control the vocal musculature and allows volitional control over our vocalizations (Simonyan and Horwitz 2011). A very interesting detail about the laryngeal motor representation is that, in humans, the precentral sulcus contains not one but two regions that control the larynx, one more ventral and one more dorsal (Eichert et al. 2020). This organization contrasts with that of other primates, whose larynx is primarily controlled via a single ventral cortical region. The more dorsal larynx region would seem to be a human specialization that allows conscious, fine motor control over laryngeal muscles and may explain why humans learn to speak and sing effortlessly, whereas apes don't do either (Belyk and Brown 2017). In addition to these structures, the singing network includes basal ganglia, cerebellum, and certain specific brainstem nuclei. However, the insula seems to play an especially important role, because of how it integrates feedback signals coming from both the ear and the body (Kleber et al. 2013).

While the idea of predictive coding is highly relevant here, it plays out somewhat differently than in other domains, because of how the sensory-motor system is modified via training. Specifically, although auditory feedback is normally needed to sing in tune (because it provides an error signal), trained singers are less dependent on it. This effect was shown in an experiment run by my former student Jean Zarate, in which she asked non-singers and trained singers simply to sing back a target tone, which everyone could do adequately. But on certain trials, the pitch of the tone was unexpectedly shifted up or down, and people were asked to ignore such perturbations and continue singing the original tone. Non-singers found it difficult to do so and often partly compensated for the shift, whereas trained singers were better able to ignore what they heard and continue singing the intended pitch (Zarate and Zatorre 2008). Similarly, trained singers were not affected at all when auditory feedback was entirely blocked by noise, whereas the accuracy of non-singers

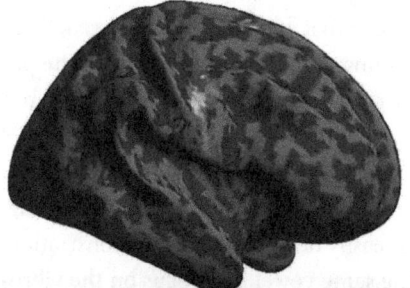
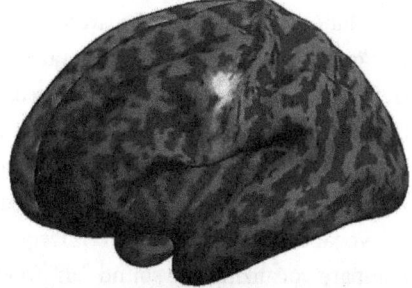

FIGURE 4.12 Singing network. Inflated brain representation of functional MRI responses associated with singing. Prominent activity is seen in the dorsal laryngeal motor region, together with more ventral motor areas. Other important nodes in this network include auditory cortex, insula, and dorsal parietal and premotor regions. Used with permission of the Society for Neuroscience, from (Kleber et al. 2013); permission conveyed through Copyright Clearance Center, Inc.

was much worse under that condition (Kleber et al. 2017). Conversely, in other conditions of these experiments, when singers or non-singers were asked to correct for the shift in pitch that was artificially introduced, they were both able to do so reasonably well by shifting their voice pitch in the opposite direction of the perturbation. This compensation had already been shown in related speech production experiments (Larson et al. 1999).

These behavioral findings are consistent with reports from singers (as well as from some instrumental musicians) that they have a fair amount of flexibility in deciding when, and how, to incorporate input from the environment: if the auditory input is reliable or relevant (for instance, matching the pitch to be sung to that produced by another singer or instrument), then it is integrated and adjustments are made; whereas if it is interfering (e.g., if the person next to you in the choir is off-key), then the input can be ignored or suppressed. In other words, musical training enhances the predictive model, which means that singers can better control their vocal system when needed, even without feedback via top-down control; whereas non-singers are more dependent on feedback (input signals), and hence are more affected by error signals, when the sound they hear is absent or incorrect. This conclusion is consistent with the fact that when we speak we are only partly dependent on feedback, adjusting if necessary (e.g., speaking more loudly in a noisy room) but otherwise being able to speak fluently, even without any sound input at all (Lane and Tranel 1971). That's because we are all expert speakers of our language, even if we are not all expert singers or instrumentalists.

The neural substrates of these abilities to integrate feedback as needed are in line with what we would expect from the roles of the dorsal-stream structures and their interactions with auditory regions. When sound information is either to be ignored or masked, auditory cortex activity is either upregulated or downregulated, respectively (Zarate and Zatorre 2008, Kleber et al. 2017), especially in trained singers. At the same time, there is recruitment of dorsal-stream structures that are usually considered part of the mechanism that supports implementation of motor actions independently of feedback (i.e., the top-down prediction signal). In the condition where listeners were asked to compensate for the pitch perturbation, there was an even greater reliance on dorsal-stream structures, especially the intraparietal area in singers. This result is most likely related to its role in processing pitch shifts and transpositions, as discussed earlier in this chapter. A similar effect was reported in a piano study by Peter Pfordresher and colleagues such that when pitch was altered, there was greater brain activity in several structures, including dorsal stream regions in the parietal lobe, supplementary motor area, and anterior cingulate; so the phenomenon is not unique to singing (Pfordresher et al. 2014).

However, there is another important factor to consider in explaining how singers can maintain accurate singing without feedback: somatosensory information. When we sing, the vocal cords vibrate at a given frequency, which is easily detectable (as a demonstration, compare vocalizing the sound "ah" to whispering the same vowel, and focus on the vibro-tactile sensation coming from the throat). This somatosensory information, together with other proprioceptive cues, such as muscle tension, can be used by singers when more direct feedback (like the actual sound produced) is not available. This kind of sensory information may provide important feedback for other instruments too, including string instruments, where the vibration sensed by the fingers plays a prominent role (Wollman et al. 2014), and also kinesthetic (mechanical touch) feedback on the fingers for piano playing (Goebl and Palmer 2008).

To investigate the neural basis of vibrotactile feedback in singing, Boris Kleber, now a research scientist at the University of Aarhus in Denmark, carried out a somewhat heroic study when he was a postdoc in our lab, in which he applied topical anesthesia to the vocal cords of trained singers while they sang in the fMRI scanner (Kleber et al. 2013). The anesthesia numbs the vocal cords, thus temporarily suppressing vibrotactile information, without interfering with the ability to vocalize, which is controlled by muscles that are unaffected by the anesthesia. Singers (but not non-singers) were thus able to maintain accurate pitch under anesthesia (this also explains why the famed Italian tenor Enrico Caruso was known to spray ether in his throat, to numb any pain, and continue singing [Sataloff et al. 1993]).

In the trained singers that Boris tested, the neural effect of the anesthesia was to downregulate the activity in the laryngeal sensory-motor cortex, along with the insula, reflecting a decreased reliance on sensory and interoceptive information in favor of a feed-forward, or predictive, model. In the non-singers, there was instead an increase in insula activity, likely reflecting increased reliance on the missing or degraded input. The increased importance of cortical sensory-motor regions representing the larynx in singers is further supported by the findings that these regions are active in trained opera singers, when they sing an aria (Kleber et al. 2009), and that there is an enhanced anatomical structure in these same regions among singers (Kleber et al. 2016).

It's worth mentioning that imagined singing, like other aspects of imagery we covered earlier, also engages most of the same singing network, including both auditory and dorsal-stream areas as during overt singing (Kleber et al. 2007). Interestingly, self-reported auditory imagery ability is also related to accuracy of overt singing (Pfordresher and Halpern 2013). These types of findings extend the link between imagery and motor systems and have led to theoretical proposals that imagery entails access to an integrated sensory-motor representation (Keller 2012, Pfordresher et al. 2015). This concept, in turn, helps to explain how imagery is useful both as mental practice and as a means of generating novel musical ideas.

Some of the foregoing findings seem to paint a picture whereby singing has some special status. Yet, we have also already seen evidence that controlling different kinds of instruments may engage some similar mechanisms. To what extent is singing unique in terms of its neural substrates? This has been a hard question to address in any detail because existing studies have focused either on singing or instrumental playing, but not both, and not with comparable paradigms. Furthermore, most neuroscience research on playing instruments has been carried out with keyboards, which differ from singing in many ways. Not only are there the obvious differences in terms of controlling hands/fingers versus larynx/mouth, but the keyboard also incorporates a discrete mapping between action and sound, as each key corresponds to one and only one pitch—there is no way to play the pitches "in the cracks" of the keyboard. This feature also means that pitch correction in real time is impossible: if you hit a wrong note on a keyboard, you cannot correct its pitch by wiggling your finger a bit. But with singing, as with fretless string instruments, which have continuous mapping between muscle action and pitch, feedback-based, on-line adjustment is possible.

To address these questions directly, we devised the world's first (and, so far, only) cello that can be played inside an MRI scanner. With help from our colleagues Avrum Hollinger and Marcelo Wanderley, from the music tech team at McGill's Faculty of Music, we were able to build an electro-optical cello with the right dimensions on the fingerboard and strings, so as to allow the fingers to play as usual, and with a tiny bow that can fit inside the bore of the magnet (Hollinger et al. 2007). Furthermore, it has gut strings, which are not only compatible with the high magnetic field of the scanner but also, we like to joke, lend period-instrument historical accuracy. The advantage of using this device is that both singing and cello playing have continuous mapping of action to pitch (as opposed to discrete mapping), allowing us to test the same individuals with the same tasks in both modalities and look for similarities or differences.

This is precisely what Melanie Segado, working with Virginia Penhune and me, did for her PhD thesis—compare brain activity and connectivity patterns, as cellists reproduced individual tones on the cello or sang them out loud. The result was that the previously identified singing network emerged during singing and largely overlapped with the pattern that resulted from cello playing (Segado et al. 2018). As shown in Figure 4.13, the principal difference was that the cortical motor area for controlling the arm and fingers was active only during cello playing; whereas singing recruited the dorsal larynx representation in the motor cortex, as expected.

But we wanted to go further than merely documenting overlap to see if the same kind of auditory-motor integration, as discussed above with respect to auditory feedback in singing, also operates in cello playing. To do so Melanie implemented a pitch-shift paradigm with cello playing and singing, identical to that previously applied only with singing in Jean Zarate's studies: as the volunteers reproduce a target tone, the pitch is shifted unexpectedly, and they are asked to either correct for it or ignore it (Segado et al. 2021). The findings nicely replicated prior data, with dorsal-stream regions—such as the IPS and premotor areas—being most active for the compensation task, and auditory areas being most active during the ignore task, regardless of instrument (voice or cello). Moreover, connectivity analyses showed that, for both compensate and ignore tasks, there was increased coupling between auditory cortex and SMA and between auditory cortex and primary motor regions for both instruments (see Figure 4.13).

These findings would suggest that the so-called singing network is, in fact, a more general system capable of controlling other sound-producing devices besides the voice. It's only at the final point, where the direct motor control is enabled, that the networks diverge for singing versus other instruments in order to control the relevant muscles. This set of results makes sense because the voice and each instrument require not only different groups of muscles but also very different types of actions for the coordination of limbs and/or breathing. But, if this is so, it still leaves the question open—how did such fine motor control for producing sound arise? It is fairly obvious that the vocal motor control system is phylogenetically ancient, presumably being present in humans since we were human and certainly so in comparison to the cello motor control system (which can't date back earlier

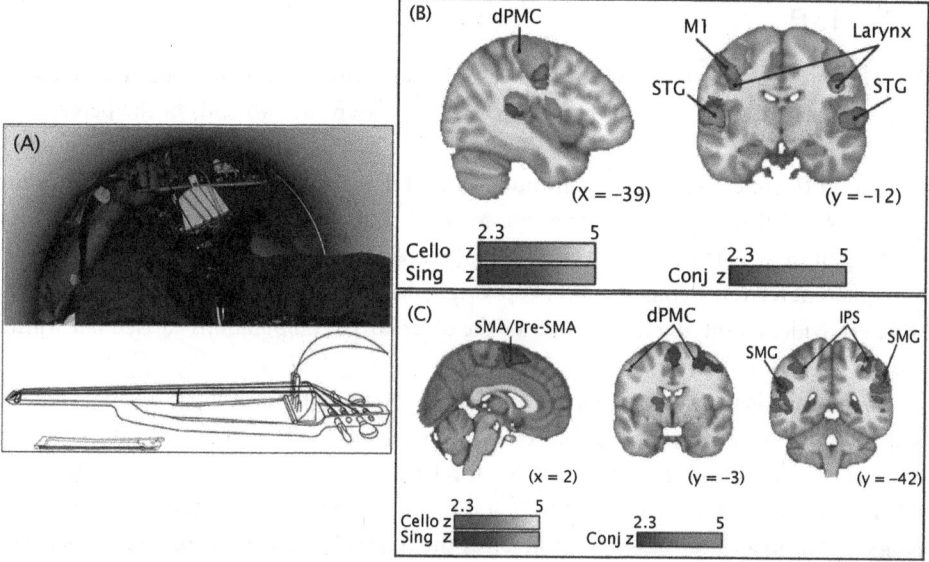

FIGURE 4.13 Functional MRI-compatible cello. A (top) Photograph of cello being played in the scanner; (bottom) line drawing of MRI-compatible cello and short bow. B: Brain activity during simple tone reproduction with cello (orange), singing (blue), and their conjunction (green). Auditory and motor areas (primary motor, M1; superior temporal gyrus, STG) overlap except for the larynx region which is active only for singing. C: Brain activity during compensatory pitch-shift task for cello (orange), singing (blue), and their conjunction (green). Areas of overlap are seen in the entire dorsal pathway. Abbreviations: supplementary motor area (SMA), dorsal premotor cortex (dPMC), primary motor cortex (M1), intraparietal sulcus (IPS), and supramarginal gyrus (SMG). Adapted from (Segado et al. 2021) with permission from Elsevier.

than around the 16th century, when cellos were first manufactured). And nobody—not even Yo-Yo Ma—is born already knowing how to play the cello, whereas everyone is born knowing how to vocalize (indeed, it's usually the first thing we do upon emerging from the womb!).

Therefore, an attractive hypothesis for our ability to harness our motor system in the service of musical instruments is that it represents an instance of neural "recycling" (Dehaene and Cohen 2007). According to this line of thinking—which is related to Stephen Jay Gould's notion of exaptation (Gould and Vrba 1982)—if a neural network already exists to carry out the relevant computations for vocalization (including the feedforward and feedback aspects that depend heavily on the dorsal stream, as we have seen in this chapter), then this same network can be repurposed to act upon different inputs and outputs. Thus, the fact that playing a musical instrument is possible at all may be due to having much of the neural hardware necessary for the right cognitive operations already in place. This circuitry can then be used to control different effectors than the ones used to vocalize and to produce many more different sounds than the ones that can be produced by a human vocal tract. This view that instrument playing piggybacks onto mechanisms for song control also implies that song may be a fundamental, primordial form of musical communication, a conclusion that is in accord with findings that songs across many diverse societies share common features (Mehr and Krasnow 2017).

Reprise

The multiple functional roles of the dorsal stream are critical for many cognitive functions that are essential for music. In this chapter, we have seen the two ends of the perceptual-motor continuum that are especially linked to these brain circuits: manipulation of information on the one hand and organization of motor sequences in time and space on the other. Both of these operations, working memory and auditory-motor interactions, may be thought of as part of a broader concept, involving the transformation of information into different reference frames—a concept originally developed in the visuomotor domain, which provides a unifying framework for the diversity of computations carried out within the dorsal stream.

I suggest that similar principles apply in the domain of musical processing. One example unique to music that depends on dorsal structures is transposition, which may be thought of as conserving a pattern of tonal relationships, but shifted in pitch space to a different reference frame. Importantly, in this case, the reference frame is not a simple mapping of pitch in a linear fashion but involves the more elaborate construct of musical key space, at least in Western tonal music. Temporal manipulation, such as reversing a tone sequence, also involves some of the same dorsal structures and may be thought of as analogous to pitch-based transformations, except that they occur in a temporal reference frame. Although reversal is a somewhat artificial operation, the dorsal pathway is also important for other, more musically commonplace temporal transformations, given the role of premotor circuitry in the representation of metrical structure. Examples of temporal transformations might include tempo changes, systematic modifications to a rhythm, such as emphasizing "swing" or syncopation, or switching a musical theme from duple to triple meter, as is often done in many styles of music. The abstract nature of the metrical representation in the dorsal stream would enable the perceptual stability needed for a listener to understand and follow shifting musical patterns that generate diversity and maintain interest.

Those examples are on the input (perception) side. On the output (production) side, acting upon an instrument or the vocal musculature to produce sound may also be considered in terms of a transformation: in this case, a sound pattern is transformed into a set of motor actions to reproduce the sound. This operation is analogous to the visuomotor transform that enables a visual spatial position to guide an action to that coordinate in space. The dorsal stream provides the sensory-motor interface that allows precise actions on the environment and can do so in both visual and auditory domains. Similar concepts have been developed in the speech domain as well. The same dorsal-stream system is critical for creating internal models of what the sound patterns should be, given a certain set of actions to be executed; or conversely, what the actions must have been that generated a given sound pattern. Indeed, these models may be thought of as the internal representations of the sensory-motor transforms that allow for the two-way input/output information flow. Such models can also be constructed without any action being executed, and without any sound being emitted, which relates to the auditory and motor components of musical imagery. This feature of covert sensory-motor processing allows enormous flexibility in our ability to think about sound: it frees us from the constraints of dealing only with what we have previously experienced or can physically produce, enabling us to create new sound patterns never heard before, which is the essence of musical creativity.

This concept of internal sensory-motor models is, of course, central to the capacity to make predictions and evaluate the outcome of those predictions. Without a roadmap, it would not be possible to know how what is happening now is related to what came before or to what should come next. In the context of the perceptual and sensory-motor operations subserved by the dorsal stream, the predictions largely relate to enhancing the ability to follow the organization of a musical pattern (for instance, by using metrical structure to enable predictions of relevant events and thereby process them more efficiently). Furthermore, they also serve to structure movement patterns to allow accurate musical production (for instance, to prepare muscle actions in anticipation of upcoming events while playing a sequence). Importantly, predictive mechanisms in the dorsal stream tend to operate on the temporal level, emphasizing *when* events are expected in contrast to the ventral stream which is more concerned with predictions about *what* events to expect. Predictions need not be confined only to concrete physical dimensions; rather the dorsal stream is able to represent abstract features and indirect mappings, meaning that predictions can operate at higher-order levels both of perceptual and motoric organization. Given the complex nested hierarchies that are commonplace in music, such a system, with its abstract predictive models, is well-suited to represent these cognitive relationships.

5

Hemispheric Specialization

Two Brains Are Better Than One

I remember very well the day in 1981, when Roger Sperry won the Nobel prize "for his discoveries concerning the functional specialization of the cerebral hemispheres," as articulated by the Nobel committee. I had just started my postdoctoral fellowship in Montreal a month or so earlier, when Brenda Milner burst into our office with great excitement to announce the news. It was especially nice for her, as she had worked with Sperry at one point (see Milner et al. 1968). It was also one of the few times a psychologist had won the Nobel. But beyond that, it served as a recognition of the reality of how the two halves of our brains differ. It also highlighted the importance of research on the topic, one which has fascinated me for a long time in the context of music processing and is the subject of this chapter.

The idea that the two cerebral hemispheres carry out different functions antedates Sperry's work, of course, by more than a century. The French neurologist Paul Broca is well-known to have identified, in the 1860s, that left hemisphere damage results in language disorders. But the same phenomenon was noted as early as 1836 by Marc Dax, a physician who had treated soldiers injured in the Napoleonic wars (Manning and Thomas-Antérion 2011). Since language was the first function to be described in terms of asymmetries—undoubtedly because of the salient symptoms of aphasic patients—much of the subsequent research on other cognitive functions, including music, was linked to language. The concept of a "dominant" hemisphere, which was a common term for a number of years, implied that cognition mostly depended on language processes, which occurred in the left hemisphere.

Some of the most interesting historical observations about music and its lateralization come from reports of dissociations between musical functions and language. A remarkable example of sparing of musical function, described by Alexander Luria, is that of the Soviet composer Vissarion Shebalin (Luria et al. 1965), who, despite profound aphasia caused by a massive left-hemisphere stroke late in his life, was nonetheless able to compose many new works, including sonatas, quartets, and symphonies, which were praised by the likes of Shostakovich. This case study is remarkable as it suggests that advanced musical function

From Perception to Pleasure. Robert Zatorre, Oxford University Press. © Oxford University Press 2024.
DOI: 10.1093/oso/9780197558287.003.0005

can exist independently of language. Other similar dissociations have also been described, such as occurred with the famous French organist Jean Langlais, who continued to play and compose even after extensive left-hemisphere damage (Signoret et al. 1987). But ultimately, such idiosyncratic anecdotal reports are of limited value, because of many complex and un-controlled individual factors that may influence the outcomes. Taken as a whole, this case-study literature does not lead to much of a theoretical framework with predictive validity, since many different combinations of aphasia can occur, with or without musical difficulties (Stewart et al. 2006). As such, it's hard to know where to go next, experimentally, based on such one-off observations.

This situation led to a bit of a dead end in the many years that followed of individual patient-based studies on music and lateralization. These studies either concluded that lan-guage and music were lateralized to the left and right hemispheres, respectively, or that both functions were intermixed (Critchley and Henson 1977). However, neither of these conclusions is satisfactory, because they do not probe into the subtle and complex ways in which the various cognitive functions involved in each domain may coexist or differ within various processing pathways, as was emphasized in an important theoretical paper two dec-ades ago (Peretz and Coltheart 2003). A notable exception to the failure to propose mech-anisms in this literature comes from a review paper by Dennis Phillips and Mary Farmer, who were among the first to propose that speech perception problems following brain damage were specifically linked to a deficit in the processing of rapidly changing acoustical inputs, rather than to any domain-related factor (Phillips and Farmer 1990). Related ideas were also developed by Paula Tallal from the developmental perspective (Tallal et al. 1993). Although these ideas were important, they only dealt with the speech side of the equation and did not address music.

There have also been various attempts to categorize the differences in hemispheric functions more generally, in terms of functional capacities or even cognitive styles. For example, the concept that the left hemisphere was "analytical," while the right engaged in "holistic" processing, was popular for a while (Bradshaw and Nettleton 1981). Another viewpoint, particularly based on lesion data, held that the left hemisphere was organized in a more focal manner compared to the right, whose functional organization was more diffuse (Semmes 1968). These ideas were formulated in such a way that it became difficult to generate distinct, falsifiable hypotheses; but, at least, they did refocus the question to-ward the idea that each hemisphere makes unique contributions. This reconceptualization of the problem was also part of the legacy of Sperry and his followers—to emphasize that the two hemispheres had separate, and likely complementary, capacities that needed to be analyzed carefully, and that the old notion of a dominant hemisphere that not only did all the talking but basically did all the thinking too had to be abandoned (Sperry et al. 1969).

Coming back to comparisons between the processing of music versus language, they are very instructive in many ways, as discussed in detail in Ani Patel's excellent book on the topic (Patel 2010). In what follows for the rest of this chapter, rather than dealing with music and language as unitary, encapsulated domains, I will focus more on the neural substrates of specific features that are important for musical processing, and contrast them with those most relevant for speech, to see how each hemisphere may be specialized for particular aspects of processing.

Experimental Lesion Studies of Perception of Tonal Sound Patterns

A breakthrough for our understanding of hemispheric effects in music-related perceptual functions came in 1962 with Brenda Milner's experimental study of auditory abilities, measured before and after temporal-lobe excision in epilepsy patients (Milner 1962). This paper avoided the ad-hoc, descriptive, single-case study approach that had dominated the literature for a long time. It also promoted the idea that one could fractionate different aspects of musically relevant function by testing for abilities such as pitch, timbre, and melody perception separately, based on one of the few tests of musical skill available then, the Seashore test of musical talent (Saetveit et al. 1940). Although limited in many ways, this test did yield clear results that showed that resection within the right, but not the left, temporal lobe disturbed the ability to distinguish melodies and timbres, as shown in Figure 5.1. As such, after a lot of conflicting and heterogenous accounts, this study set the stage for trying to understand why the right temporal cortex would be more important than the left in these specific tasks that involved analysis of tonal patterns.

In my lab's early days, we continued in Brenda's footsteps, working with similar patients, and consistently observed that right temporal excision disrupted many different musical tasks, especially those requiring analysis of spectral patterns. This result was observed in melody discrimination tasks (Zatorre 1985, Samson and Zatorre 1988), with judgments of pitch direction (Johnsrude et al. 2000), with missing fundamental stimuli (Zatorre 1988),

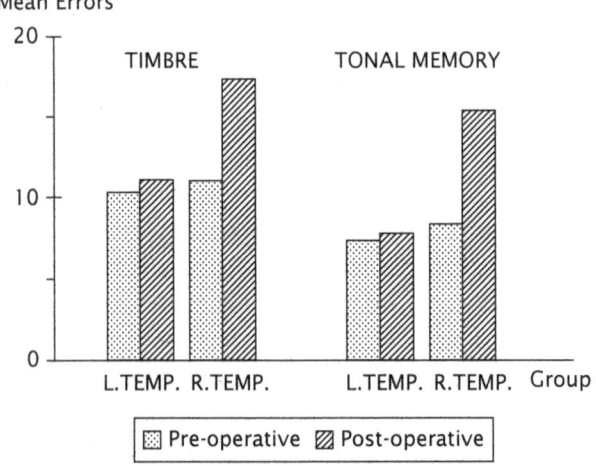

FIGURE 5.1 **Temporal-lobe excisions and tonal functions.** Error rates on timbre discrimination and tonal memory tasks before and after temporal-lobe excisions in epilepsy patients. Significant elevation in errors was observed after right temporal-lobe excision, but there was no effect of similar excision on the left side. Reproduced with permission from (Milner 1962).

in pitch judgments of tones differing in timbre (Warrier and Zatorre 2004), and in melodic imagery (Zatorre and Halpern 1993), as well as short-term and long-term pitch memory tasks (Zatorre and Samson 1991, Samson and Zatorre 1992). Additionally, we also observed deficits in the reproduction of rhythmic patterns in this same population (Penhune et al. 1999). Similar right-side lesion effects on melodic processing were obtained by other groups as well (Liégeois-Chauvel et al. 1998).

It should be noted that, as a general rule, the effects in many of these deficits were relative rather than absolute, meaning that left-sided lesions did sometimes produce impairments but usually not as severe as those on the right. The surgical excisions in this epilepsy population were circumscribed, unlike vascular lesions which tend to be diffuse. On the one hand, this feature was especially important in the days prior to anatomical imaging because it meant that we could be fairly certain about the location of the lesion. But, on the other hand, it was more difficult to be sure that the epileptic discharges, which were common in that population, had not affected both hemispheres to some extent. Furthermore, because the lesions were confined to the temporal cortex, it allowed us to study the role of the auditory areas within the temporal lobe. However, studying the role of ventral and dorsal streams, outside the temporal cortex, was more difficult in this group.

More recent studies that used different types of patients with more widespread vascular lesions, and also took advantage of newer brain imaging methods to map the location and extent of lesion (Stewart et al. 2006, Sihvonen et al. 2019), provide valuable new information and largely uphold the asymmetries seen with the surgical excision population. The recent meta-analysis by Sihvonen et al., shown in Figure 5.2, is particularly relevant in this regard: when looking at the lesion sites that resulted in the greatest deficits on music-related tasks, they found that the damage was concentrated in right-hemisphere auditory dorsal and ventral pathways; but they observed essentially no overlap with aphasia-producing lesions, which, instead, consistently encompassed the left hemisphere. Analysis of the white-matter fiber tracts affected by the lesions also pointed to the importance of right-hemisphere connections between auditory regions and other structures, especially in the ventral stream. This conclusion is highly consistent with the various mechanisms that support musical functions, discussed in previous chapters. These authors were also able to document the pattern associated with spontaneous recovery from music deficits. They found that such recovery was more common in those with less damage to right-hemisphere anatomical connections and was also associated with more bilateral recruitment of undamaged areas in the left hemisphere, suggesting some possibility of compensation from the other side.

In addition to this information from acquired lesions, as reviewed in these papers, there is also the large body of evidence in relation to congenital amusia that was already discussed in detail in Chapter 3. The vast majority of studies on congenital amusia demonstrate a disruption in the organization of the auditory cortex and inferior frontal cortex and/ or in their connections within the right hemisphere rather than the left. This asymmetry has been observed in studies of functional activation (Albouy et al. 2013), functional connectivity (Hyde et al. 2011), directed functional connectivity (Albouy et al. 2015), analysis of gray-matter anomalies (Hyde et al. 2007), analysis of white-matter fiber connections (Loui

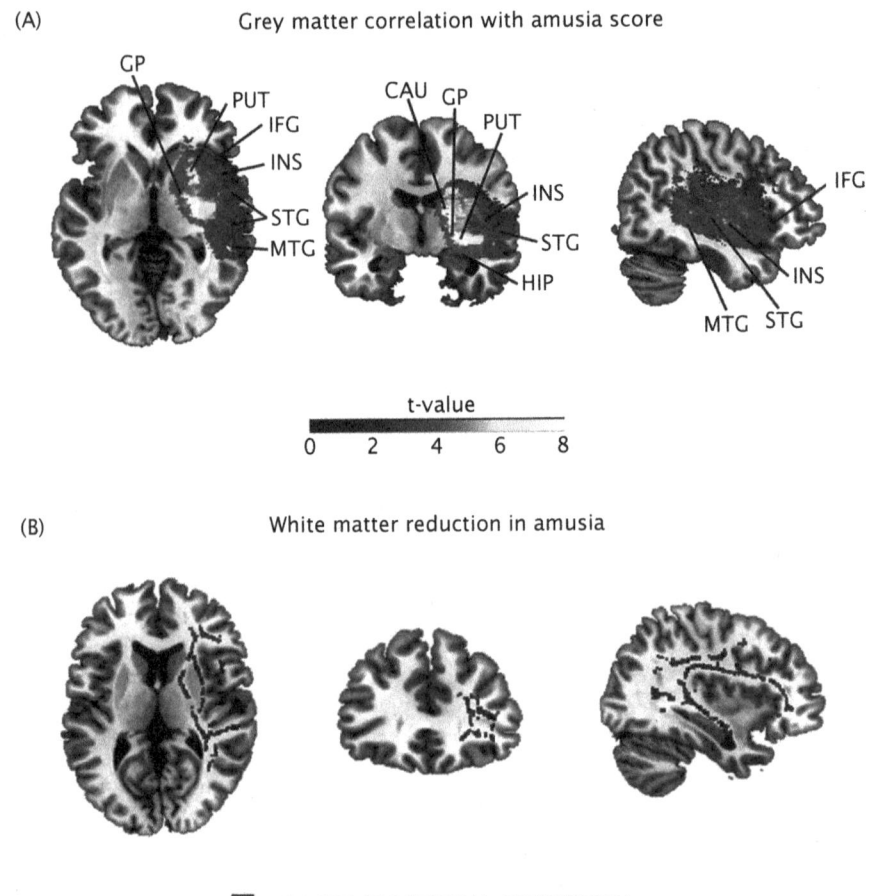

FIGURE 5.2 Vascular lesions and musical functions. A: Map of lesion locations associated with deficits in musical tasks in a population of stroke patients in the acute stage. Maximum overlap is centered within the right superior temporal cortex and adjacent deep structures. B: Comparison of white-matter structure in stroke patients who did not recover musical functions to those who did. Reductions in fractional anisotropy are observed throughout the right-hemisphere fiber tracts interconnecting auditory areas with dorsal and ventral regions. Adapted from (Sihvonen et al. 2019) with permission from Elsevier.

et al. 2009), and machine-learning-based classification analysis of functional and structural data (Albouy et al. 2019). The fact that this very selective deficit for musical processing—with its particular behavioral manifestation in terms of fine-grained pitch processing (Hyde and Peretz 2004, Tillmann et al. 2016)—shows such consistent right-hemisphere lateralization again points to the critical role of right lateralized structures in important aspects of music processing.

All of these lesion studies conclusively demonstrate the presence of hemispheric asymmetries and prove a direct causal link between the functioning of structures within the right hemisphere and certain musically relevant tasks. In fact, they constitute a kind of a ground truth in the sense that any model or theory, ultimately, has to be able to explain why lesions have asymmetric effects. However, they still leave open the mechanistic reason for the observed asymmetries. Why is there this highly consistent result in study after study? What is it about the right hemisphere—or at least the right auditory cortex—that makes it critical for at least some aspects of music processing? Or conversely, what is it about music that makes it dependent on processes carried out, in large part, by systems within the right hemisphere? And exactly which bits of the right hemisphere are relevant? Can we address these questions by contrasting the demands of music with those of speech?

Music Versus Speech

Similarities and Differences Between Music and Speech

Comparing music and speech yields differences or similarities, depending on whether you prefer half-full or half-empty glasses of water. There are many ways in which the cognitive demands of music and speech are similar (Patel 2010). Perhaps the most fundamental similarity is that both speech and music essentially depend upon modulating sounds in various ways to convey information between individuals or groups of people. As such, as I suggested in Chapter 1, both may be considered as species-specific communication signals. These similarities may, therefore, result in some sharing of cognitive and/or neural resources (see [Peretz et al. 2015] for review of neural overlap between the two domains).

Another important similarity is that speech depends on fine motor control of the vocal musculature, and music makes use of the voice too. One component of vocal motor control used in both domains concerns the ability to consciously modulate the tension on the vocal cords, resulting in changes to vocal fundamental frequency and perceived changes of pitch. There may even be a link between the acoustics of vocalized speech sounds and the origins of musical scales. Recent studies analyzed the amplitudes of harmonics present in human speech (from both English and Mandarin) and found that the probability distribution of the harmonics that are amplified while speaking various vowels can be used to predict the structure of musical scales. The musical intervals that are formed from these harmonics preferentially fall into ratios that occur in a chromatic scale, which is commonly used in several (though not all) cultures (Ross et al. 2007). Therefore, it is possible that the tonal structure of musical scales may derive, in part, from the acoustics of vocal speech sounds, providing a fundamental link between vocal production and musical structure. This commonality, however, is more about the physics of sound and perhaps reflects the evolutionary adaptation of the auditory nervous system to exposure to these frequency relations. As such, it does not directly imply anything about the neural substrates of music versus speech, nor does it inform us about hemispheric specialization.

An argument against the exclusive importance of the voice to music is that people everywhere can and do make music by manipulating basically anything that can be made to produce sound, including parts of their bodies (clapping hands, stomping feet) and objects readily found in the natural environment (sticks, logs, rocks). As noted in Chapter 1, the existence of very ancient musical instruments suggests that making music with objects is most likely a fundamental aspect of music in our species. Neurally speaking, this means that there is not as tight a mapping between motor control and musical sounds, as there is for speech. The greater degrees of freedom in the ability to use different effectors to produce

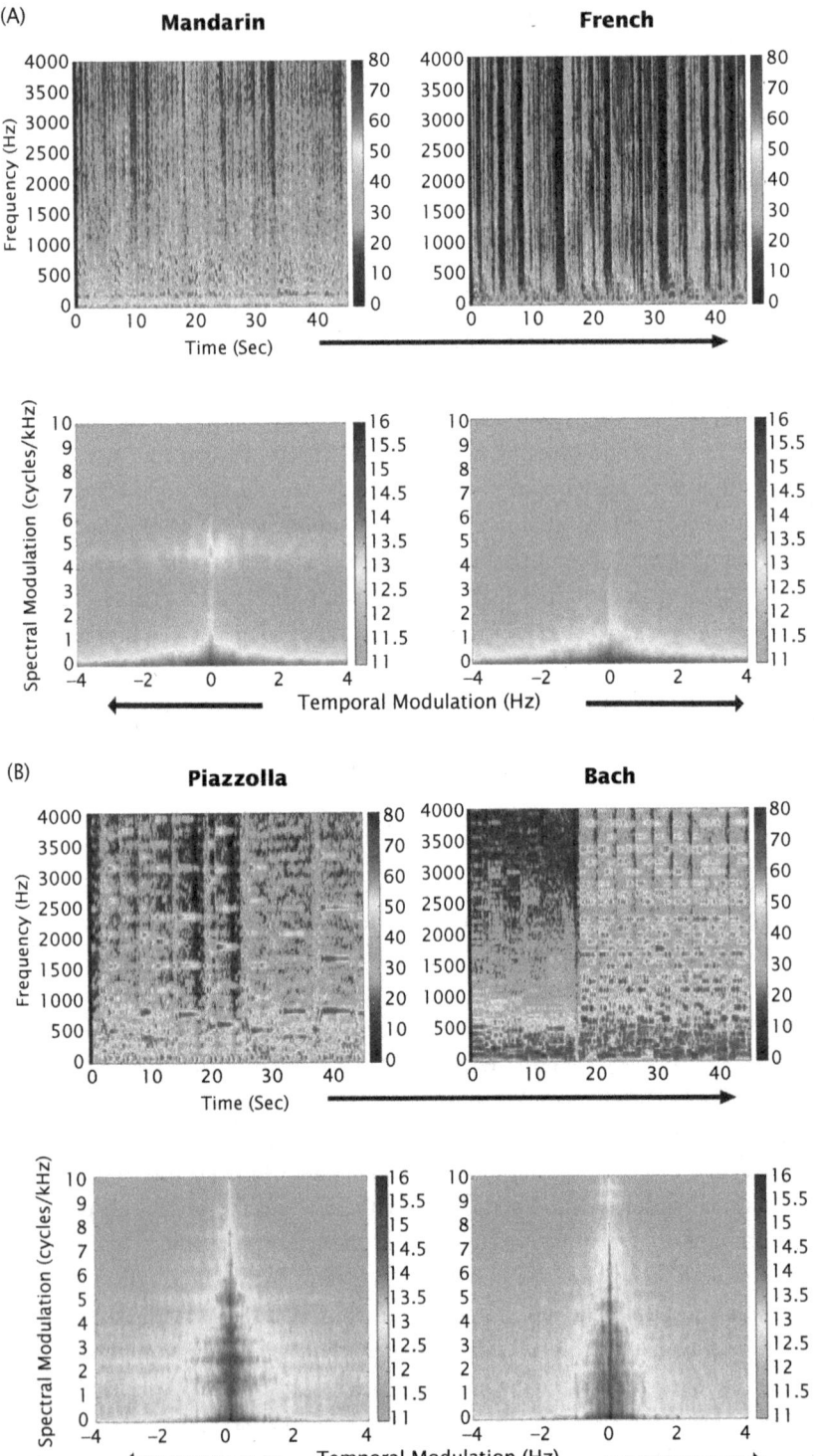

FIGURE 5.3 Spectral and temporal cues in music and speech. A: Top: Spectrograms (frequency as a function of time; amplitude shown in color scale) for a 50 s sample of speech in Mandarin and in French. Note the many onsets (vertical striations) for both languages, corresponding roughly to the syllable rate;

music is consistent with the many different types of flexible mappings between auditory and motor systems that exist in the dorsal stream, as described in Chapter 4. This is true even if playing an instrument may utilize some of the same neural substrates as singing (Segado et al. 2021).

On a more cognitive level, it is relevant to consider that both systems can be described as generative, in the sense that complex structures are built up from simpler elements, at various hierarchical timescales: syllables are combined to create words, which are combined to create sentences, speeches, or stories; tones are combined to create melodies, which are combined to create songs, sambas, or symphonies. These complex patterns are generated by the application of recursive syntactic rules that describe the permutations that occur in both domains (Patel 2010). These, and the other parallels across domains, suggest that some of the same neural mechanisms may be at play. However, to some extent, they might also arise from parallel circuitry within each hemisphere.

These interesting similarities notwithstanding, in this chapter we are most interested in understanding how and why there seem to be hemispheric differences in the way that speech-relevant or music-relevant sounds are processed. Hence, we will focus more on the way that the processing of speech and musical sounds may differ (Zatorre and Baum 2012). Our group's theoretical position is that, whereas music particularly (but not exclusively) exploits spectral elements of sound, speech (to a greater degree) depends on temporal elements.

Based on material presented in Chapter 2, we know that neurons throughout the auditory nervous system are sensitive to spectral and temporal modulations of sound and that they explicitly encode these features. To get an intuitive sense of how speech and music exploit these parameters differentially, we may simply observe how temporal and spectral sound energy are distributed in representative music versus speech samples. Plotting the distribution of spectral versus temporal modulation (see Figure 5.3) in two music samples (in this case, an extract from Astor Piazzolla's *Oblivion* scored for strings and bandoneón and the opening bars of JS Bach's *Passcaglia and Fugue in C Minor* for organ) shows that they both have a wide range of spectral modulation but a relatively narrow range of temporal modulation clustered toward the slow end of the scale (vertical areas on the display). In contrast, speech samples (here, spoken French and Mandarin) both show the opposite pattern, with a higher rates of temporal modulation (horizontal areas on the display) but

Bottom: Spectrotemporal modulation plots (Spectral modulation on the y-axis, Temporal modulation on the x-axis; power shown in color scale) of the same sound clips shown in A. Note the generally horizontal distribution of energy along the bottom of the plot, indicating that the sounds contain mostly temporal modulations, without much spectral modulation (Mandarin contains some additional spectral modulations compared to French, corresponding to F0 changes in this tonal language). B: Similar layout as in A for two 50 s music clips, an excerpt from Astor Piazzola's *Oblivion* scored for bandoneón and strings, and the start of JS Bach's *Passacaglia in C Minor* for organ. Note that the spectrograms (top) contain more horizontal striations, indicative of rich harmonic content, as compared to the speech. The spectro-temporal plots (bottom) show a much more vertical distribution of energy than the speech, indicating that spectral modulation is very prominent and spans a wide range in these music samples, but that the temporal modulation is more limited (0–2 Hz). Images courtesy of P. Albouy; used with permission.

lower rates of spectral modulation compared to music (note the small additional spectral modulation band for Mandarin, a tone language).

The model my colleagues and I have been developing integrates this knowledge in the context of hemispheric specialization. We propose that the processing asymmetries arise from differential resolution in the spectral versus temporal domains, within right and left auditory areas, respectively (Zatorre et al. 2002a, Zatorre and Gandour 2008). We termed the idea the "tradeoff hypothesis" because specialization of one dimension would trade off with that of the other. A complementary hypothesis was proposed by David Poeppel and his team from New York University, based on the idea of different time integration windows in each hemisphere (Poeppel 2003). According to this view, the right and left auditory cortices integrate auditory information over longer (~150–300 ms) versus shorter (~20–80 ms) timescales, respectively. This idea is more anchored to specialization for speech in the left hemisphere, without a strong prediction about music, but it is entirely compatible with our own model. Both of these models are related to earlier ideas also developed from lesion studies (Phillips and Farmer 1990). However, before dealing with neural evidence of hemispheric specialization, I will first discuss the ways in which the use of spectral information differs for music versus speech, followed by a discussion of the same issue for temporal features, in order to better understand the nature of the processing in each domain for which the brain must optimize.

Differences in the Use of Spectral Information

One of the most relevant differences between speech and music, for our present purposes, is related to how spectral information, and in particular pitch, is utilized in each domain. Pitch variation in speech is part of prosody, which also involves modulation of other parameters, such as intensity and duration. Prosodic pitch has important linguistic functions, both at the word level, to distinguish one word from another in tonal languages such as Mandarin or Thai, and also as part of sentence-level intonation, which is found in most languages and typically signals grammatical differences, such as question versus statement (Botinis et al. 2001). It also communicates affect. There is a rich literature showing that prosody seems to depend more on dorsal and ventral pathways in the right rather than the left hemisphere (Witteman et al. 2011, Sammler et al. 2015), providing further arguments for lateralization being based on acoustical cues rather than on linguistic status of the stimulus. But even though pitch is most definitely used in speech communication, how it is used seems to fundamentally differ from its use in music.

One critical difference is that pitch variation in speech (at both syllable and sentence levels) usually follows continuous contours, whereas in music, it more commonly follows discrete scale steps (Sato et al. 2020). There exist some tonal languages that use differences in the pitch of level tones (as opposed to contours) linguistically, although these tend to be limited to a small number (most frequently two or three tones, unlike music where the entire spectrum is divided into many intervals). Furthermore, these tones do not really correspond to fixed frequency ratios, as in most music, since they vary in pitch height and interval size across different speakers, depending on their vocal pitch range (for discussion see Patel (2010)). There is no particular limitation in vocal production of pitches along a continuum—anyone can produce a glide with their voice from a low to a high pitch. Yet, a majority of musical cultures make use of scales, where the octave is subdivided into several (most often 5, 7, or 12) discrete pitch values (Savage et al. 2015).

These scale tones are related by the ratios of the fundamental frequency between them rather than by absolute pitch values. The specific pitch ratios that correspond to these scale tones are accurately represented in memory, as shown years ago in pitch-matching experiments (Attneave and Olson 1971). Small deviations from the correct values in melodies are readily detected by listeners, even when the melodies are transposed to different keys across trials (Trainor 1996) or when spectral timbre differs across tones (Warrier and Zatorre 2002). Anyone who has played an instrument that requires control over intonation (fretless string instruments like the violin and woodwinds, where pitch depends both on air pressure and fingering, like the clarinet and, of course, the voice) will be acutely aware of how slight pitch errors can be noticed by the audience (otherwise, the "auto-tune" tool would not be so popular!). Systematic deviations from nominal values certainly do occur (Rakowski 1990), and they are often used for expressive purposes, which, in fact, is quite common in music performance (Gabrielsson 1999). But this only works because there is an implicit understanding of what the scale pitch values ought to be, allowing the musician to deviate from it for a desired effect. Hence, it just as much depends on the precision of the underlying representation of scales in the memory of the musician.

There is no real equivalent to precise intonation in speech: it would be odd to say that a news reporter's speech was out of tune, for example, in the way that playing music can sound out of tune. Although prosody contributes to speech comprehension in various important ways, prosodic modulation of pitch does not appear to depend on specific pitch targets, but rather on directional contour. To illustrate this difference, in a 2012 paper we prepared a demonstration of the different ways in which precise pitch values affect speech versus music, by altering sung and spoken speech in exactly the same way (Zatorre and Baum 2012). When the pitch values of the voice fundamental frequency are stretched or compressed by 50%, the change introduced in the speech is quite subtle and is not rendered unnatural. In contrast, the identical manipulation applied to the song makes it sound so badly out of tune that it's comical. Even removing all pitch information in speech, by flattening the fundamental frequency of sentences, has rather limited effects on comprehension of meaning, unless the listening conditions are degraded, for example, by noisy backgrounds (Miller et al. 2010). This is even true for tonal languages, such as Mandarin (Xu et al. 2013).

Not only are musical pitch values discrete and organized according to scales, but in Western music (and other musical systems as well), there is also a hierarchical structure within scales, such that some tones are considered more stable than others. This hierarchical organization leads to many interesting perceptual phenomena, including, for example, voice-leading and key relationships, which may be subsumed under the term tonality (Krumhansl 1990). Tonality is a prominent feature of much Western music. But it's especially relevant for the central narrative of this book because it allows for predictive process to occur based on tonal structures (Huron 2006). As a very simple example of this hierarchical prediction, try playing a simple C major scale but stopping on the 7th scale degree, B natural, without playing the final C—it sounds incomplete, and you expect it to continue to the final tone. Such organization could not exist if particular pitch relationships were not accurately specified within musical scales that, in turn, determine melodic structure. The existence of scales is also critical to harmony, which results from the simultaneous sounding of multiple tones drawn from those scales, in particular relationships to one another. Harmony provides the framework for harmonic movement, key modulations, tension, and resolution in tonal music. Thus, discrete and accurate pitch representations are essential to the kinds of predictive processes that I am emphasizing throughout this book as critical for the generation of emotion and pleasure in music.

There is no obvious hierarchical organization of pitch variation in speech intonation contours that is comparable to tonality in music, and there is certainly no counterpart to harmony in speech. In music, it is common for different instruments or voices to play different but coordinated lines of music to generate both harmony and polyphony, but this does not happen in speech. Indeed, in the absence of turn-taking, when multiple people speak simultaneously, it results in the cocktail-party phenomenon discussed in Chapter 2, usually requiring selection of one speaker and suppression of the others. Or, even worse, it results in confusion and cacophony, such as happens when people speak over each other during an argument (just watch any heated political debate for a demonstration of this effect). One exception to this rule might be in quasi-musical speaking situations, where there is conscious alignment, both in time and often in pitch, across multiple speakers This situation is somewhat intermediate between normal speaking and actual singing and can be observed in some group activities, like collective prayer or patriotic pledges. While one may speculate about the relation between speech and music under these particular circumstances, and also in other intermediate speech/music art forms like rap or *sprechstimme*, these situations are, nonetheless, clearly different from everyday speaking.

These distinctions in the use of pitch, across speech and musical domains are important for purposes of understanding brain function because they provide possible clues as to why there is a hemispheric difference for processing tonal musical patterns. According to our view, the division of hemispheric labor rests, in part, on specialized right-hemisphere circuitry that enables a fine-grained representation of spectral information in general, and accurate pitch information in particular, both of which are essential for tonal music. Hence, this idea would explain the previously reviewed evidence for the importance of the right hemisphere in music. In contrast, left-hemisphere circuitry contains a coarser representation of pitch information, which is normally sufficient for prosodic speech processing, because prosodic contours do not usually depend on fine or exact frequency differences. Having discussed spectral cues, we now turn to how temporal cues are differentially used in music and in speech.

Differences in the Use of Temporal Information

The differences between music and speech, in terms of the use of temporal information, are present at various scales. Temporal cues for music are obviously essential, at a global level, for the organization of rhythm. As with tonality, there is also a hierarchical structure of rhythm and beat in many styles of music for which no obvious counterpart exists in speech. Speech is characterized by quasi-periodic temporal structure, and this organization exists at different scales (phonemes, syllables, phrases). It exhibits its own type of hierarchical structure (Patel 2010, Kello et al. 2017), but, for the most part, it does not contain the type of precise regular spacing of events that is found in many styles of music with high metrical structure and a clear beat, that often induce movement, as we saw in Chapter 4. Temporal cues at a local level (within individual tones) are also important to distinguish different instrumental timbres (McAdams et al. 1995), but similar cues are also relevant to distinguish individual sounds of speech from one another.

It is tricky to claim that any given class of temporal cues is critical for speech because, if nothing else, the human vocal tract can produce many different kinds of sounds of varying temporal grain and of different amplitude and frequency modulation rates. These include sounds that can last either a very brief moment or an indefinite amount of time (or until you run out of breath, anyway) and sounds that lack any periodicity (say, "shhh") or have very clear periodicity (say, "aaaah"). But there is a class of sounds—the stop consonants—which are apparently found in every language and are dependent on relatively rapid changes in acoustical events. The differences between a voiced and unvoiced consonant (e.g., bad versus pad) or between consonants pronounced with closure of the lips or with the tongue (e.g., bad versus dad) depend a great deal, although not exclusively, on how the acoustics of these sounds evolve over short time periods. This idea led some investigators to propose that speech is critically dependent on a mechanism that can resolve these rapidly changing events (e.g., Tallal et al. 1993). However, this view has been challenged on the grounds that speech depends on many cues over different timescales and that stop consonants play no special role (McGettigan and Scott 2012). One way around this issue is to avoid a focus on any one element of speech but, instead, to systematically manipulate the temporal cues that are globally available in natural speech and determine to what extent they are important in comprehension.

This is precisely the approach taken by Bob Shannon and collaborators in Los Angeles, whose seminal study, more than a quarter century ago, provided an important piece of empirical evidence supporting the primacy of temporal cues for speech perception (Shannon et al. 1995). They were interested in the degree to which speech could be understood despite degradation in the spectral domain. This question is very relevant in the context of developing cochlear implants, because they only transmit information into the cochlea via a limited number of electrodes, each one of which is placed in a different frequency region of the basilar membrane. This arrangement means that, although the spectral resolution is quite poor, temporal information is relayed relatively accurately. To test how this might affect perception of speech, Shannon distorted normal speech by removing all the normal spectral information and replacing it with amplitude-modulated noise passed through 1–4 filter banks centered at different frequencies, thus mimicking how information might be transmitted via a small number of channels in an implant. The results showed that sentence perception was essentially perfect with four spectral channels and was 90% correct with only three. The conclusion, as reflected in the title of the study ("Speech recognition with primarily temporal cues"), focused on how preservation of the timing was sufficient to understand speech, a finding that later contributed to the widespread use of cochlear implants for speech perception. But it would be equally correct to conclude from this study that spectral cues are not so critical for speech perception. A later study, using the modulation transfer framework that was introduced in Chapter 2, found compatible results in terms of dependence on temporal modulations for speech comprehension (Elliott and Theunissen 2009). Music was not tested in these studies, although the latter one pointed out that distinguishing the voices of males and females required more spectral modulation information. We shall return to this point below, to show that melody and harmony require higher spectral resolution than is needed for speech.

There is another related way to look at how speech and music differ objectively in terms of their temporal structure, which involves extracting the temporal envelope information from the acoustical stream of sound. Any sound that is not perfectly steady will contain fluctuations in amplitude over time. These fluctuations can be characterized by computing the temporal modulation spectrum (Singh and Theunissen 2003). This technique represents a way of deriving an unbiased, descriptive measure of how quickly sound events are changing, without making any particular theoretical assumptions about which cues are important. For speech, this metric is largely related to the speed with which the mouth opens and closes when speaking. For music, on the other hand, it mostly corresponds to the onsets of different sounds (tones, chords, percussive sounds, etc.) and is hence closely linked to rhythm and beat.

Nai Ding and colleagues applied this method to systematically compare music and speech (Ding et al. 2017). As shown in Figure 5.4, they observed an interesting difference: speech samples, from nine different languages (see also Poeppel and Assaneo 2020 for further evidence and discussion), all showed a consistent peak in the temporal modulation distribution at about 4–6 Hz, reflecting a typical syllable rate (it feels natural to say "blah blah blah blah blah" in one second). The equivalent temporal modulation rate for a variety of Western musical genres (classical, rock, jazz) was generally around 2 Hz, less than half the speed of speech. There is, of course, a great deal of variability within each domain on this metric: some people speak more quickly than others, some music has very fast passages or quite slow sections. Despite this variability, when looking at the broader trend, the difference is striking. This study only examined Western musical genres; but in a very recent study (Albouy et al. 2023), our group was able to show that the distinction between speech and music, based solely on temporal versus spectral modulations, holds across a wide range of speech and song samples from 21 different societies across the planet, such that a machine-learning classifier can distinguish the two based solely on these modulation rates. The reason that these findings are especially important in our context is that they point to another clue about how the processing of the two domains may differ in the brain: since speech tends to be faster than most music—that is, it has faster temporal modulations—it requires a processor that is well-tuned to its speed, while music requires higher spectral resolution. As we shall see below, auditory cortical functional organization across the two hemispheres provides just this mechanism.

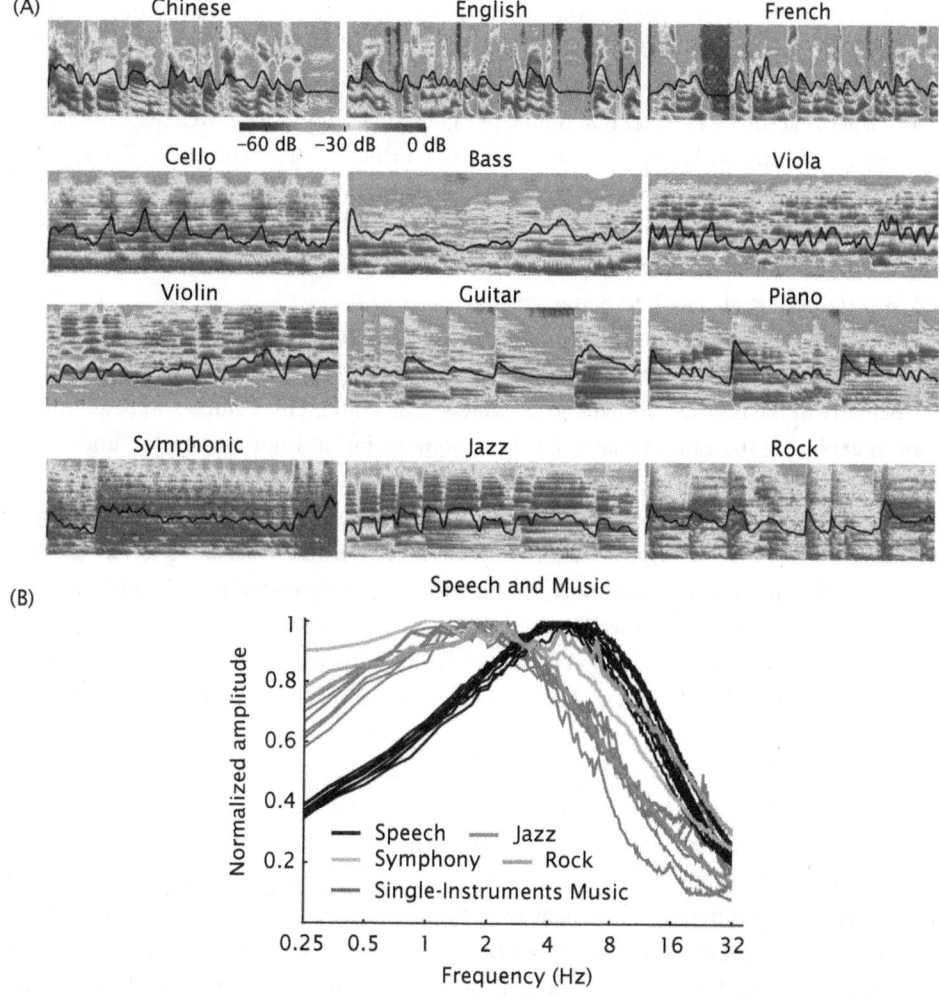

FIGURE 5.4 Temporal modulations in music and speech. A: Spectrograms of various examples of speech and music. Dark line corresponds to overall amplitude envelope. B: Temporal modulation spectra of different samples of Western music (colors) and speech (black). Average modulation rate for music is found around 2 Hz while for speech the average is around 4–6 Hz. Adapted from (Ding et al. 2017) with permission from Elsevier.

Neuroimaging Evidence for a Hemispheric Difference in Pitch Processing

In parallel with the many lesion studies reviewed above that showed greater disturbance from right-sided lesions for many musical tasks (particularly those involving analysis of pitch or harmonic patterns), many functional neuroimaging studies have been carried out that show asymmetries in the same direction. In earlier years, the research was more a matter of documenting their existence, and the circumstances under which they were observed, rather than testing specific hypotheses about the source of the asymmetry. Many of the relevant studies have been mentioned in previous chapters, in the context of specialization of auditory cortex (Chapter 2) or the auditory ventral stream (Chapter 3) for the processing of tonal components of music. These include studies using straightforward contrast approaches that showed greater magnitude of right than left auditory cortex responses when comparing listening to simple tone patterns versus various control conditions (Zatorre et al. 1994, Griffiths et al. 1999, Patterson et al. 2002), as well as a variety of more active tasks—for example, requiring deviance detection of tones or chords (Tervaniemi et al. 2000), processing of dichotic instrument timbres (Hugdahl et al. 1999), or imagery for instrumental timbre (Halpern et al. 2004), to name only a few.

Higher-order aspects of tonal processing, such as abstract deviance detection of the sort discussed in Chapter 3 have also been shown to yield right-hemisphere asymmetries, particularly in the inferior frontal cortex (Maess et al. 2001). These studies were all important in exploring the range of lateralized phenomena but did not explain the underlying reason for the asymmetric responses. To get at this question, subsequent studies manipulated acoustical features more systematically, in order to test the hypothesis emerging from the discussion above: that right auditory cortical mechanisms are sensitive to fine-grained spectral differences, which are most relevant for tonal processing and, in turn, are the reason why right-sided asymmetries often emerge with tonal tasks.

To obtain evidence for this idea, Pascal Belin and I designed a functional imaging experiment (Zatorre and Belin 2001) in which we manipulated the size of pitch intervals in a sequence of tones in a parametric manner, from a coarser to a finer-grained pattern, as shown in Figure 5.5 (top). At one end of the continuum, we presented tones that were separated by a large, constant pitch distance of one octave. Then, keeping pitch range and number of tones constant, we progressively subdivided the octave into finer and finer intervals, so that, at the other end of the continuum, successive tones were separated by much smaller frequency differences (as small as 37.5 cents, or roughly a third of a semitone). We reasoned that if right auditory cortex neurons are sensitive to fine spectral differences, then they should respond more as the number of resolvable tones increases; whereas the left auditory cortex, being sensitive only to coarser frequency differences, would not increase its activity as much, since it would tend to treat such tones as equivalent. This pattern is precisely what we observed: the slope of the activity increased more in two regions of the right auditory cortex, one close to the pitch-sensitive region (see Chapter 2) and another in a more ventral temporal cortical region; whereas, in the left hemisphere, there was also a response to pitch variation in a homologous area, but it was less pronounced. The results of this study were replicated, in all essential respects, by Kate Watkins' team from Oxford University, who also showed that the results were consistent at an individual level (Jamison et al. 2006). Both these studies also showed complementary responses on the left side for temporal modulation, as we will discuss in the next section.

The prediction that the right auditory cortex is more sensitive to fine-grained frequency differences than the left auditory cortex continued to be supported by a number of additional studies. For example, Krista Hyde, Isabelle Peretz and I (Hyde et al. 2008) carried out a functional imaging study in which we presented patterns of ascending and descending scales, varying in the size of successive pitch intervals from very small (6 cents) to very large

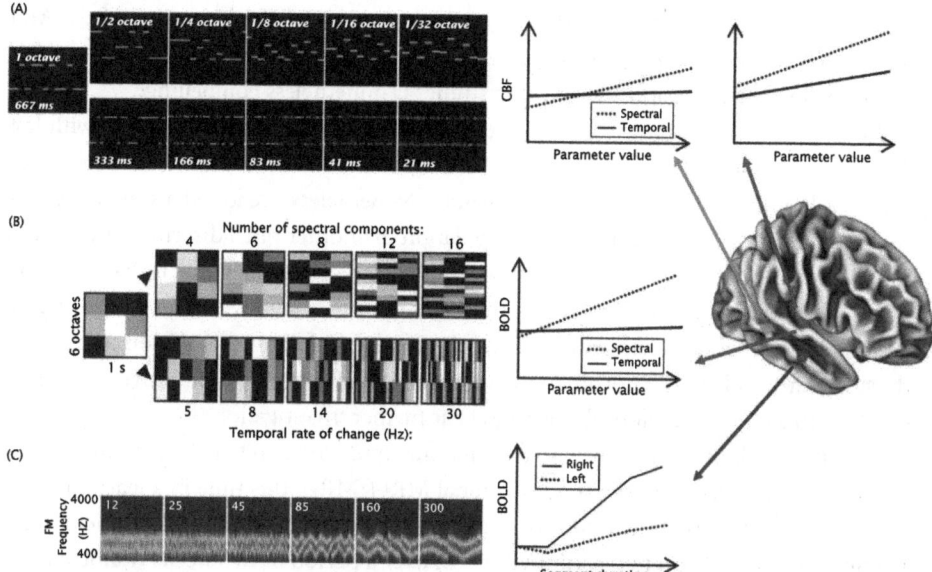

FIGURE 5.5 Right auditory cortex response to acoustical modulation. A (left): Parametric variation of tone sequence resulting in progressively finer spectral content (top) or progressively faster temporal rate (bottom); (right): Schematic depiction of functional MRI responses from right superior temporal gyrus (red) and sulcus (green), showing progressively greater response to spectral than to temporal changes. Adapted with permission from (Zatorre and Belin 2001).

B (left): Parametric variation of noise sequences resulting in progressively finer spectral content (top) or progressively faster rate (bottom); (right): Schematic depiction of functional MRI responses from right superior temporal gyrus (red), showing progressively greater response to spectral than to temporal changes. Adapted with permission from (Schönwiesner et al. 2006).

C (left): Parametric variation in concatenated narrow-band noise stimuli resulting in progressively longer segment durations (indicated on each panel, in ms); (right): Schematic depiction of functional MRI responses from right superior temporal sulcus (blue), showing progressively greater responses to longer segments. Adapted with permission from (Boemio et al. 2005).

(200 cents or, in musical terminology, one whole tone). We found that, while the activity in a region in right posterior auditory cortex scaled in a continuous manner with the entire range of variation, the equivalent region on the left did not respond much until the largest frequency difference was reached. In other words, it treated the smaller pitch differences as if they were not present, which is exactly what one would expect if the left auditory cortex lacks resolution in the spectral domain.

This result was particularly well-aligned with the result of one of our earlier lesion studies, in which we reported that pitch discrimination thresholds were larger in patients with excisions encroaching onto right Heschl's gyrus than in those with similar damage on the left side (Johnsrude et al. 2000). Critically, the right auditory cortex-lesioned patients, who scored poorly in that study, were still mostly able to perform the task—they just needed a much larger difference in frequency between the tones to be able to do so, which makes sense if they were having to rely on their left auditory cortex with its coarser frequency representation. The threshold that these patients with right-sided excision needed to perform the task was on the order of 200 cents, which is exactly the point at which the brain activity started to increase in the left auditory cortex in Krista's imaging study. And while the precise pitch value corresponding to the threshold may depend on several variables, the pattern across lesion and imaging studies is nonetheless compelling.

In passing, it is also interesting to note that in the Johnsrude study, patients with left auditory cortex lesions were marginally better than the controls. It's rather unusual to see that a brain lesion leads to *enhanced* performance. Nonetheless, we found a similar significant enhancement with the same patient population in another pitch discrimination study as well (Warrier and Zatorre 2004). Our admittedly speculative explanation is that disabling the cortical system with the coarser spectral resolution may remove some interfering or inaccurate functional response, while leaving the system with better resolution to respond to the task. And while we do not advocate undergoing brain surgery to improve one's pitch perception, this kind of result probably does bear further investigation.

We undertook another study with a similar manipulation of pitch distance, in order to look at hemispheric differences using functional MRI (fMRI), this time in a learning context. We created a series of melodies consisting of microtonal intervals ("micromelodies") and trained people to learn to discriminate them over a period of two weeks (Zatorre et al. 2012). We were inspired by a very old study that suggested that people could learn to discriminate very small pitch intervals (Werner 1940) such that, after learning, they report a kind of perceptual expansion, allowing them to perceive melodic patterns that they could not hear before. We found a similar behavioral enhancement in our experiment: after two weeks of training, most listeners were better able to perceive the micromelodies even with very small interval sizes (on the order of 10 cents). The fMRI data prior to training replicated the effect observed by Hyde et al. (2008): the activity pattern in the right auditory cortex increased more with increasing pitch interval size, compared to the left side. Following training, the maximum modulation in brain activity was also located in the right auditory cortex, showing that the improved pitch perception ability was mediated by changes in this region.

To obtain causal evidence for the importance of right auditory cortex to fine-grained pitch learning, my former PhD student Reiko Matsushita performed a brain stimulation study in which she showed that perturbation of the right auditory cortex blocked pitch learning of micromelodies (Matsushita et al. 2015). In a second study, she found the same behavioral effect after right-sided stimulation (but not left) and also observed that the degree to which the learning was affected in each individual was correlated with reduction of the amplitude of evoked MEG responses from the right auditory cortex after stimulation (Matsushita et al. 2021). These findings indicate that the right auditory cortex is critical for learning-based enhancement of micromelody discrimination and that the effect is linked to a physiological index of right auditory cortical response.

Additional evidence in favor of the conclusion that the right auditory cortex encodes frequency information more accurately—leading, in turn, to a finer-grained representation of pitch—comes from Emily Coffey's study of the cortical frequency-following response already mentioned in Chapter 2 (Coffey et al. 2016, Coffey et al. 2017a). The amplitude of both the neuromagnetic as well as the hemodynamic signal, corresponding to the frequency-following response, was significantly larger in the right than the left hemisphere, although it was present bilaterally (see Figure 2.12). This right > left asymmetry was also subsequently confirmed in other laboratories (Hartmann and Weisz 2019, Gorina-Careta et al. 2021). The reason this asymmetric measure is especially relevant is because the frequency-following response specifically represents the periodicity in the stimulus, unlike, for example, the fMRI response, which only reflects the overall magnitude of blood oxygenation.

Perhaps most importantly, in the MEG study of 2016, Emily showed that the amplitude of the brain signal coming from the right (but not the left) auditory cortex was correlated with each individual's ability to discriminate pitch. Thus, those individuals whose right auditory cortex represented the stimulus frequency more strongly were also better at distinguishing small pitch differences, thereby providing a critical brain–behavior link. A similar link between behavioral pitch ability and neural responses was also uncovered with fMRI by Federica Bianchi and colleagues in Denmark. They showed that for musically trained individuals, the activation level in the right (but not the left) auditory cortex was correlated with their pitch discrimination thresholds, once more supporting the importance of right auditory cortex function in relation to fine-grained pitch processing (Bianchi et al. 2017).

A possible neurophysiological mechanism for this enhanced frequency resolution in the right auditory cortex was suggested by Kuwook Cha, working with Marc Schönwiesner and me (Cha et al. 2016), who examined local functional connectivity patterns in fMRI data collected during passive listening to tones of various frequencies. Recall from Chapter 2 that core areas of auditory cortex receive sharply tuned lemniscal inputs and have the highest frequency selectivity, with more distal regions showing more integration across frequencies (Bitterman et al. 2008, Schönwiesner et al. 2014). We also know from neurophysiological recordings that sharp tuning of neurons to frequency is dependent on excitatory intracortical inputs more than on thalamic inputs (Liu et al. 2007) and that this happens more in earlier than later cortical regions.

In Kuwook's experiment, he identified the preferred frequency of each voxel and then computed the functional connectivity between pairs of voxels as a function of their difference in peak frequency tuning, as shown in Figure 5.6. The result showed that, overall, voxels with similar tuning (closer in frequency preference) had higher connectivity than those with more distant frequency tuning preferences. In other words, voxels that are more similar to one another are more interconnected. But the most relevant finding was that, in the right core region, connectivity was very strongly influenced by the frequency difference between each pair of voxels—voxels that preferred the same frequency were strongly interacting, but voxels with distant frequency preference were weakly interacting. On the left side, this trend was still present but much less prominently. One could say that the left core auditory cortex doesn't care as much about frequency selectivity, in terms of its functional connectivity pattern, whereas the right core area does.

Put another way, the left auditory cortex integrates more across frequencies, while the right tends to segregate frequency information. These observations provide a plausible mechanism for hemispheric differences in frequency resolution at early stages of cortical processing: the enhanced functional connectivity could represent the intracortical inputs from neural populations with similar tuning, which amplify and sharpen frequency

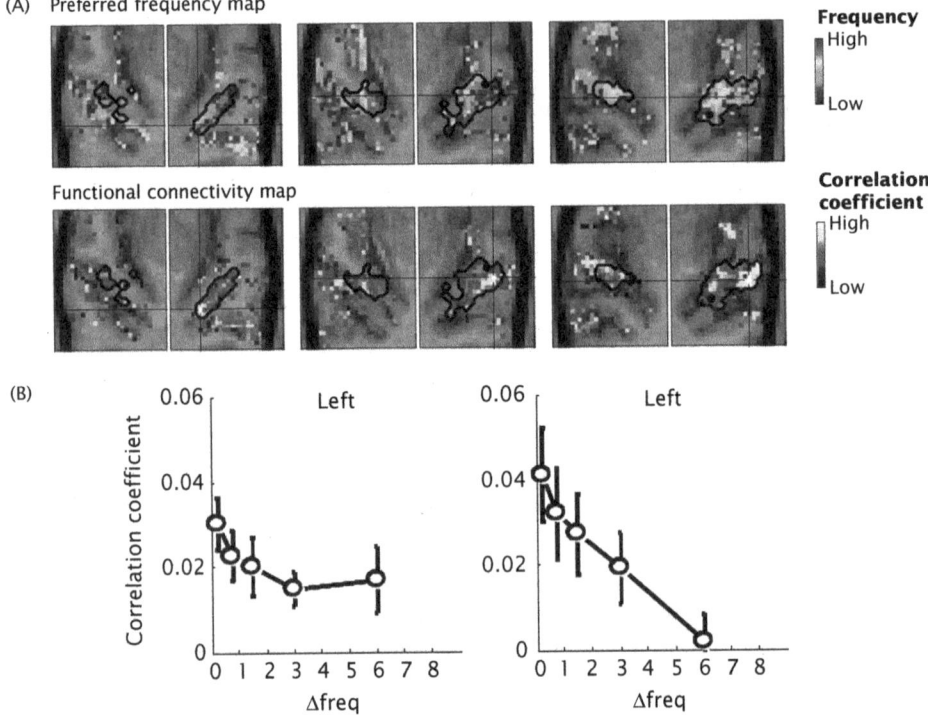

FIGURE 5.6 Voxe-by-voxel functional connectivity in auditory cortex. A (top row): Preferred frequency maps based on functional MRI selectivity for each voxel in three horizontal slices from one representative individual. Black outlines indicate core areas. (bottom row): functional connectivity (correlation value) maps of three individual target voxels (marked with crosshairs) with all other voxels. B: Combining preferred frequency and functional connectivity maps yields plots of connectivity for each voxel as a function of frequency difference between target and all other voxels. Mean functional connectivity across all listeners for voxels in left core area shows only a modest dependence on frequency, while functional connectivity for voxels in the right core area shows a steep drop-off in connectivity as the frequency difference between target and other voxels becomes larger. Adapted with permission from (Cha et al. 2016).

resolution in the right core area, leading to the asymmetries observed in imaging and lesion experiments presented above.

Neuroimaging Evidence for a Hemispheric Tradeoff in Temporal Versus Spectral Processing

All the studies that were just reviewed in the prior section used tones or sounds with distinct pitch since they were mostly focused on tonal or melody perception. But the model we espouse proposes two broader ideas: first, that the right auditory cortex has a better spectral resolution in general, that is, not only for periodic sounds with distinct pitch; and second and more importantly, the model proposes that the complementary nature of the hemispheric specialization would be reflected in enhanced temporal resolution within left auditory cortical areas. We turn next to that issue.

An fMRI study by Marc Schönwiesner and his colleagues in Leipzig showed evidence in favor of both of these propositions (Schönwiesner et al. 2005). It followed a similar logic to that of Zatorre and Belin (2001) in that parametric variation of a stimulus set was used to elicit activity in brain regions related to the variable in question. But instead of using tones, Marc manipulated a series of noise bands, as shown in Figure 5.5 (middle), by parametrically varying the number of independently modulated frequency bands (spectral complexity) and the rate of intensity change within each band (temporal complexity). Although there were bilateral responses to both parameters, part of the right superior temporal gyrus anterior and lateral to Heschl's gyrus, covaried more with the spectral parameter; whereas activity in the equivalent region on the left superior temporal gyrus covaried more with the temporal parameter. This finding supports the idea of a complementary hemispheric specialization for spectral and temporal processing and shows that this effect does not depend on the presence of periodicity in the stimuli (since noise bands were used), nor does it depend on the stimuli being semantically meaningful or familiar in any way (since they just sound like strange noises).

The finding that more rapid changes over time recruited left auditory cortical areas was also in line with the studies already mentioned using tonal patterns (Zatorre and Belin 2001, Jamison et al. 2006). In those experiments, in addition to the manipulation of pitch intervals, the speed at which such changes occurred was also varied. Both studies reported that left auditory cortex responses were more sensitive to the temporal rate of stimulation, contrasting with greater sensitivity to frequency variation on the right. A similar conclusion was reached in an MEG study carried out in Japan, which took advantage of the mismatch negativity phenomenon (discussed in Chapter 3): when a deviant stimulus involved a modification of a temporal feature, the mismatch response was greater on the left; but when the deviant involved a spectral feature, the response was greater on the right (Okamoto and Kakigi 2015).

All these studies using nonspeech stimuli support the concept of complementary hemispheric specialization driven by acoustical cues. An additional study also supported this concept but used real speech stimuli, which were manipulated by altering their temporal or spectral structure (Obleser et al. 2008). This study reported that increasing temporal or spectral detail in speech elicited responses from the left and right anterolateral auditory cortex, respectively. A more recent study has extended the importance of hemispheric differences for speech sounds by showing that spectral versus temporal feedback cues are processed differentially within right and left auditory cortex during speech production (Floegel et al. 2020). These outcomes show that even for speech sounds, the hemispheric difference at the level of auditory cortex is related to specialization in processing of acoustical cues rather than sensitivity to higher-order lexical or semantic features, which, instead, tend to emerge at hierarchically later levels (Davis and Johnsrude 2003).

Several other studies have specifically explored in more detail the temporal resolution of the left auditory cortex. A sophisticated approach to this problem comes from an fMRI study, which used a set of concatenated segments of noise stimuli, varying the durations of each segment across a range, from rapidly changing (12 ms) to more slowly changing (300 ms) (Boemio et al. 2005). As shown in Figure 5.5 (bottom), sensitivity to this parameter was bilateral and symmetrical in primary auditory cortices. However, the more slowly modulated signals preferentially drove activity in more antero-ventral portions of the right, but not the left, auditory cortex. A similar result from yet another study emerged by varying the correlation between successive time frames of a complex stimulus, resulting in greater right than left response in ventral auditory cortical areas as the stimuli were more smeared in the time domain (Overath et al. 2008).

Causal evidence for the role of left auditory cortex in temporal resolution, comes from brain stimulation data showing increased thresholds for gap detection after left, but not right, disruption (Heimrath et al. 2014). Finally, an older study with electrical recordings from Heschl's gyrus in epilepsy patients showed that short temporal events, occurring both in speech and nonspeech sounds, are better resolved in the left auditory cortex compared to the right (Liégeois-Chauvel et al. 1999). The results of all these studies favor the interpretation that right and left auditory cortical areas process information on longer and shorter timescales, respectively, and demonstrate that these effects are not dependent on the stimuli being linguistic in any way. Hence, these outcomes support the proposal that left auditory cortex is relatively more specialized for high temporal resolution (shorter segments, more rapidly changing events), whereas right auditory cortex operates at a lower temporal resolution (longer segments, slower events).

A complementary way of looking at these asymmetries in the time domain is by considering oscillatory properties of the brain. If it is true that auditory processes in each hemisphere have different timescales, then one might expect that feature to be reflected in intrinsic patterns of oscillations. This idea was explored by Anne-Lise Giraud and colleagues, who measured oscillations with EEG and hemodynamic activity with fMRI, under rest conditions (Giraud et al. 2007). They found that spontaneous EEG power variations correlated best with left auditory cortex fMRI activity, within a faster frequency range (28–40 Hz, corresponding to events occurring on a timescale of 25–35 ms), whereas these fluctuations correlated best with right auditory cortex in a slower frequency range (3–6 Hz, corresponding to events on a timescale of 150–300 ms). Since there was no stimulus or task, these results reveal intrinsic coupling between particular ranges of spontaneous oscillation frequencies and local brain activity levels, within auditory areas of each hemisphere. Yet another converging approach to the asymmetry question is to "ping" the brain and see how it resonates—that is, probe the response of left and right auditory cortical regions to a very brief tone and measure their response profiles in terms of spectral content. Measures taken from implanted electrodes in epilepsy patients after such stimulation show that right auditory cortex responds most strongly in the slower 3–5 Hz range, compared to the left, whose response is more in the 25–50 Hz range (Giroud et al. 2020).

All these findings clearly align with the hypothesis that hemispheric asymmetries reflect different timescales, as proposed by the asymmetric sampling model (Poeppel 2003). But they also make sense in terms of the spectral/temporal hemispheric specialization model and, more specifically with the concept we proposed, that these two dimensions are differentially represented within each hemisphere because enhancement of one dimension comes at the expense of another—hence the term tradeoff (Zatorre et al. 2002a). This idea of a reciprocal relationship between time and frequency, sometimes referred to as the Heisenberg-Gabor uncertainty principle, was codified mathematically in the mid-20th century (Gabor 1946), and sets out fundamental limits of information representation, which must be respected by any biological system. According to this principle, both spectral and temporal information cannot be known with precision simultaneously; specifying one entails a loss of detail in the other. In nontechnical terms, the idea, as applied to our model, is that the left auditory cortex processes events on a more rapid timescale, which consequently entails less accurate encoding of spectral information because speed means fewer samples of the changing information are obtained over time. Conversely, better spectral resolution in right auditory cortex requires integration of information over a longer time, to get an accurate representation of the frequency content of a stimulus. The asymmetric sampling idea focuses only on the temporal specialization because it is intended to explain the neural basis for speech. But if there is complementary specialization for temporal and spectral information processing, then this would present a more complete picture, not only relevant for speech but also for music processing.

A Few Discrepancies and a Search for Mechanisms

Up to this point, I have focused on the large body of converging evidence from many different research approaches that seems to paint a very clear picture of what underlies hemispheric specialization for auditory signals. However, I have glossed over a few points that are less clear. One issue is that, even as the left–right asymmetry is very consistent in the above literature, the precise location of the auditory regions that show asymmetry is not as clear. In particular, although some studies show clear-cut asymmetries in core or closely surrounding belt areas (Liégeois-Chauvel et al. 1999, Zatorre and Belin 2001, Giraud et al. 2007, Cha et al. 2016), others show such effects in more downstream regions only, in portions of the superior temporal gyrus, anterior (Boemio et al. 2005, Giroud et al. 2020) or posterior (Hyde et al. 2008, Zatorre et al. 2012, Coffey et al. 2017c) to core regions. Indeed, in some studies, we can see asymmetries in both core/belt and more distal regions simultaneously (Zatorre and Belin 2001), suggesting that more than one aspect of the processing has an asymmetric component.

Perhaps these findings are best understood not as discrepancies but rather as reflecting the many different types of stimuli and measurement approaches that have been used in these studies, which, in turn, engage somewhat different asymmetric processing pathways or mechanisms. We know from the work reviewed in Chapters 2 and 3, that there is a hierarchical organization of auditory regions. Therefore, more elementary aspects of spectral or temporal processing may take place in earlier portions of the cortical processing stream, while more complex or abstract aspects might engage asymmetric mechanisms in later processing stages. However, from the available evidence, it does not seem likely that asymmetries are absent in core areas and only emerge in later processing stages, as has been proposed (Hickok and Poeppel 2007). Rather, it seems more likely that different aspects of asymmetric processing manifest themselves at earlier or later levels of the hierarchy. What exactly those aspects may be remains to be better understood.

The other potential discrepancy is that a difference in sensitivity for processing spectral versus temporal features has frequently been reported in terms of anterior versus posterior auditory cortical regions within one hemisphere. In Chapter 2, several studies were mentioned that reported a gradient along the superior temporal cortex such that posterior regions showed greater response to higher temporal modulation, while anterior regions showed greater response to spectral modulations, in both functional imaging and intracranial recording data (Santoro et al. 2014, Hullett et al. 2016); see also (Herdener et al. 2013). This anterior–posterior distribution appears, at first glance, to contradict the left–right distribution of sensitivity. But the two organizational principles likely coexist in the brain. Indeed, in our first paper examining spectral versus temporal sensitivity, we noted both a hemispheric difference, as described above, as well as an anterior–posterior difference, exactly consistent with that reported by later studies (Zatorre and Belin 2001).

The anterior–posterior distinction can be best understood in terms of the organization of the ventral and dorsal pathways and their relative specialization for time-invariant versus time-sensitive processes, respectively. But how this organization interacts with hemispheric differences remains poorly understood and seems like a major area for progress in the coming years. In the case of music, we may speculate that for the aspects of music processing that require perceptual analysis and/or production of pitch patterns, the more important pathways might lie within right-hemisphere ventral and dorsal streams, respectively. Whereas for the aspects of music that require high temporal precision, in either perception or action, the ventral and dorsal systems in the left side might contribute more. Among the many open questions, is exactly how the complementary specialization plays out in terms of interhemispheric communication and cooperation, which would clearly be required in many real-life conditions.

It should also be noted that not every relevant study has uncovered asymmetries of the sort reviewed above (although none that I know of have reported the opposite effect). For example, the voxel decomposition method described in Chapter 3 (Figure 3.7), which nicely segregates speech-sensitive and music-sensitive cortical regions within the temporal lobe, shows no evidence for a left–right difference (Norman-Haignere et al. 2015, Norman-Haignere et al. 2019). It's not entirely clear why this otherwise powerful method does not appear sensitive to asymmetries that we know to exist, not even for speech, where it would be even more expected than for music. It could, however, be related to the fact that the naturalistic stimuli used contain modulations in both spectral and temporal parameters, thus not allowing the differences elicited by those low-level cues to emerge. Indeed, one might argue that the lack of asymmetry with this approach shows that hemispheric differences are not driven by domain-specific factors (i.e., the categories of speech and music themselves), else the domain-specific regions that were identified should have been lateralized.

The next step concerns understanding the neural mechanisms that might actually underlie the hemispheric differences that have been detailed in this chapter. The proposals we have been entertaining, including spectral versus temporal tradeoff, differential time windows, or oscillations, are all valid at some level of description but are mostly underspecified at the level of the neural computations that would implement them. We turn next therefore to recent studies that provide some answers to this challenge.

Spectrotemporal Modulations and Hemispheric Specialization

We saw in Chapter 2 how the responses of auditory neurons, in humans and other species, can be described in terms of joint sensitivity to temporal and spectral modulations in the stimulus (Figure 2.14). Such a mechanism provides an efficient way of encoding properties of complex, real-world sounds (Singh and Theunissen 2003) and seems to be the way that evolution has opted to go, at least in many species that have been studied. We also saw earlier in the present chapter how musical and speech sounds differ in their temporal and spectral properties and that these acoustical cues elicit different hemispheric responses. Therefore, it is natural to put all these ideas together, to see if the spectrotemporal modulation framework can help to explain hemispheric specialization. Two recent studies, one by Adeen Flinker and colleagues in David Poeppel's lab and the other by Philippe Albouy in my own group, have approached the issue using the spectrotemporal framework. Their findings not only seem to converge with one another, but also help to integrate a lot of the data from many other approaches.

In our study (Albouy et al. 2020), we wanted to examine the role of spectral and temporal modulations in speech and music, both behaviorally and in terms of brain responses. To get the right sound set, we first went to our friends in the Faculty of Music at McGill and found a composer who was willing to write some odd music for us: we had generated 10 short sentences that had an identical number of syllables, and we asked him to write 10 distinct melodies that all had the identical rhythm, and that would fit these sentences. We therefore ended up with 100 different short "songs," consisting of the 10 sentences crossed with the 10 melodies; this meant that understanding the words wouldn't tell you which melody it was sung to, or vice-versa. Next, we found a bilingual soprano willing to sing our silly songs (ultimately, there were actually 200 of them because, in true Montreal fashion, we created both French and English versions). We then proceeded to record them with the assistance of my McGill colleague Martha DeFrancisco, a professional recording engineer. Armed with this set of sung sentences, we then used an algorithm (Elliott and Theunissen 2009) to selectively alter them, so as to progressively remove either the temporal or spectral modulations that are naturally present in the stimulus. What's novel here is that because we used sung speech, both the verbal and melodic components are merged into a single acoustical sound wave. Therefore, when we manipulate the cues, we do so on the same physical stimulus and not on two separate stimuli, as in many previous studies, thus providing a more direct test of the hypothesis without any bias introduced by the stimuli.

Based on much of the work described above, we predicted that temporal cues would be more important for comprehension of the verbal content, whereas spectral cues would be more important for perception of the melodic content. This is precisely what we observed (for speakers of both languages), as shown in Figure 5.7: as the temporal cues became more

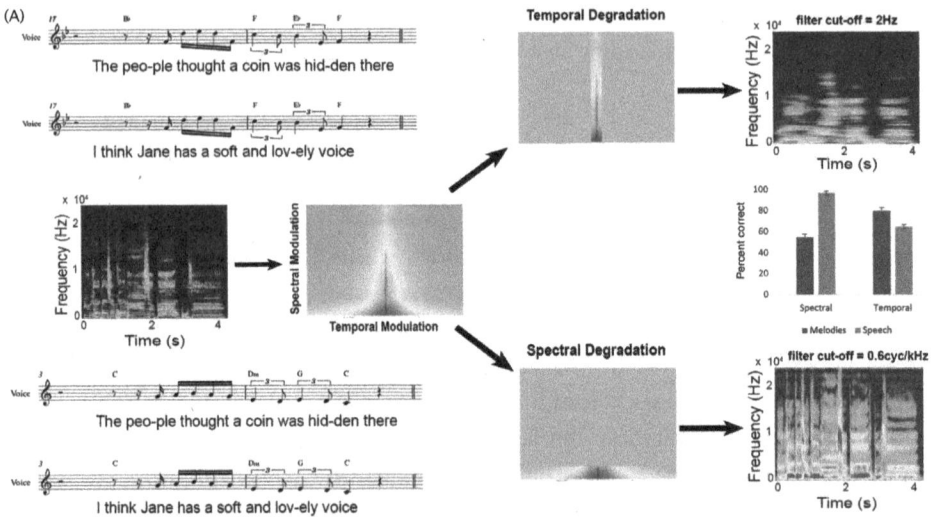

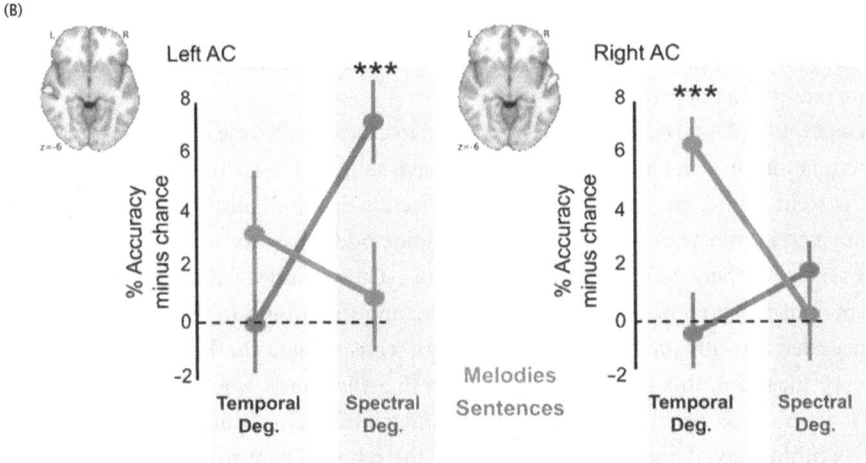

FIGURE 5.7 Behavioral and neural effects of spectrotemporal degradation in music and speech.
A: Sung stimuli (music notation) consisted of the same tunes sung to different phrases or vice-versa, yielding an orthogonal set of songs with matched melodic and speech content. These songs (spectrogram and spectrotemporal plots in middle panel) were then degraded either in the temporal domain, leaving spectral modulation intact (top) or vice-versa (bottom). The effect of this manipulation can be seen in the resulting spectrograms (right side) where the temporal degradation smears the temporal information but leaves spectral information unchanged, while spectral degradation smears the spectral information but leaves temporal information unchanged. The behavioral result (right panel bar graph) shows that behavioral performance for melodic content is severely reduced after spectral compared to temporal degradation (blue bars), while performance for speech is reduced after temporal compared to spectral degradation (orange bars). B: In the left auditory cortex, functional MRI classification performance for decoding speech content is reduced to chance only after temporal degradation; while in the right auditory cortex, functional MRI classification performance for decoding melodic content is reduced to chance only after spectral degradation, paralleling the behavioral effects shown in A. Adapted with permission from (Albouy et al. 2020).

degraded, the speech became incomprehensible, but the melody could still be discerned; whereas when the spectral cues were degraded, the melody was quickly lost, but the speech remained understandable, as was expected from Shannon's work (Shannon et al. 1995).

This clear double dissociation was also reflected in the brain data, which showed that the melodic information could be decoded from right, but not left, anterior auditory cortex; whereas the verbal content could be decoded from left, but not right, regions. To test the correspondence between the behavioral results and the brain data, Philippe computed a confusion matrix (showing which melodies/sentences were confused with each other) and found that the decoding data from the brain (right for melodies, left for sentences) matched that of the listeners. In other words, the brain activity made the same mistakes as the behavioral data, which reassures us that the way people heard these sounds was indeed based on what we were measuring in their brains. Finally, as seen in Figure 5.7, an analysis of how the degradations affected the brain responses also showed a parallel to behavior so that degrading in the spectral domain led to lower classification accuracy for melodies in the right auditory region, while degrading temporally led to lower accuracy for sentences on the left side.

The second study to explore this issue used a similar approach of filtering spectral versus temporal modulations (Flinker et al. 2019). The study is very complementary to our fMRI study, as they used MEG and intracranial recordings, rather than fMRI, to examine brain responses to either verbal content of sentences or to the timbre of the voice (male versus female, which is probably largely based on pitch and spectral cues) rather than to melodies, as we did. They observed that when temporal information is removed from speech, its comprehension was strongly altered, yet identification of voice timbre was unimpaired. Conversely, degradation of spectral information did not greatly affect verbal intelligibility, but it did make voice identity very difficult to perceive. So, the double dissociation is once again shown for temporal versus spectral information processing. In the MEG data, a clear left auditory cortex advantage emerged for the processing of the temporal cues in speech, with a weaker laterality effect for the spectral cues in the right auditory cortex. Intracranial data, which do not depend on modeling of sources, since the activity is recorded directly from the patient's brain, paralleled the MEG results.

These converging results across experiments with similar goals but different methods make us quite confident that hemispheric asymmetries are strongly dependent on the kind of temporal versus spectral processing advantages we had proposed earlier. The advantage of explaining asymmetries based on the principle of modulations is that it provides a neurally plausible mechanism, which was absent from prior accounts. It is plausible because of all the neurophysiological evidence from animal studies and imaging evidence from human studies, reviewed in Chapter 2, that indicate that neural tuning can be well described in terms of spectrotemporal receptive fields. We would additionally posit that there is segregation of tuning, not only in the anterior–posterior direction but also across hemispheres, with higher spectral tuning on the right and higher temporal tuning on the left. Therefore, this account of hemispheric specialization helps to explain why speech and music depend, at least to a significant degree, on different hemispheres: it's not because "speech is on the left, and music is on the right," as is sometimes simplistically concluded; but rather, it's because each of these two human communication systems tends to exploit opposite ends of the spectrotemporal continuum, and it is the specialization for spectral versus temporal processing that is lateralized.

These ideas also dovetail nicely with the broader concept that the brain is well-adapted to its environment and optimizes its representation thereof in the most efficient manner (Gervain and Geffen 2019). Just as neural responses in the visual system are well-matched to the statistical properties of visual scenes (Simoncelli and Olshausen 2001), so too are auditory responses similarly optimized to characteristics of sounds from the natural world (Smith and Lewicki 2006). Most importantly for our story, neurons of different species show particular enhancement to relevant communication signals. Thus, the brains of birds (Woolley et al. 2005), cats (Gehr et al. 2000), and monkeys (Wang et al. 1995) are tuned to the spectral and temporal features of bird songs, cat meows, and monkey calls, respectively, because these are the sounds that are highly relevant to the animal for its survival. It's not surprising, therefore, that a similar mechanism applies for speech (Gervain and Geffen 2019) and, I would argue, also for music.

Indeed, although the literature tends to display a hegemony in favor of speech (often arguing, at least implicitly, that it's the only human communication system that matters), the results discussed in this chapter suggest that humans have evolved not one, but two parallel communication systems, with complementary specialization in each hemisphere for the basic acoustical elements that are most relevant for each. Hemispheric optimization of spectral and temporal tuning not only provides an efficient match to the relevant features of each type of signal but also nicely solves the acoustical uncertainty problem: since higher resolution in one domain comes at the expense of the other, having two systems allows both to proceed in parallel.

Is Hemispheric Specialization Driven Exclusively by Sensitivity to Acoustical Cues?

I don't think so. The model of hemispheric specialization that I've championed emphasizes how lateralization emerges from bottom-up mechanisms, tied to particular acoustical cues. The evidence for lateralization with stimuli that are neither speech nor music nor carry any particular meaning is overwhelming, thereby invalidating the hypothesis that language-specific mechanisms are at the origin of lateralization. But this should not be taken to mean that such bottom-up features explain everything, as there is certainly a role for feedback systems in understanding how the two hemispheres work. Chapters 3 and 4 emphasized the hierarchical organization of auditory ventral and dorsal streams, with their feedforward and feedback loops enabling not only processing of incoming stimuli but also the modulation of the incoming information by prediction or other control signals that originate in higher-order structures in the pathway. This concept, no doubt, also applies to hemispherically distinct mechanisms.

Although much remains to be understood about the interaction between these top-down influences and specialized processes within the left and right auditory cortices, it seems plausible that higher-order neural systems that are related to more cognitive, motor, or even affective processing of the stimulus could amplify, modify, or perhaps even reverse specialization that already exists at earlier levels. For example, a musical passage containing prominent harmonic progressions might initially depend on specialized processing of the spectral information within right auditory cortex; but as the music evolves over time, top-down signals coming from inferior frontal cortex might refine the spectral representations in right auditory areas (as documented in Chapter 3 in the context of working memory), leading to greater asymmetries than if the stimulus were only a brief chord without any contextual embedding.

There is, at present, scant direct evidence for this idea. But it is notable that some studies using short meaningless stimuli, as opposed to long realistic stimuli, seem to find more evidence for anterior–posterior organization of spectral versus temporal cues, with a weaker right–left organization (Schonwiesner and Zatorre 2009, Santoro et al. 2014). This purely acoustically driven bottom-up process may be significantly influenced by iterative analysis of temporally structured stimuli, in regions more distant from early cortex along the ventral stream (Leaver and Rauschecker 2010), which we know integrates information over progressively longer timescales (Farbood et al. 2015, Overath et al. 2015). This interaction may lead to the more prominent hemispheric differences observed in the recent studies (Flinker et al. 2019, Albouy et al. 2020), which used relatively long, coherent, and well-structured musical or speech stimuli that would engage both auditory working memory and predictive mechanisms, especially because of the active tasks that were employed.

This idea that descending influences from hierarchically higher components of the pathway may influence hemispheric effects also helps to explain a number of interesting phenomena where different patterns of hemispheric involvement are seen under conditions where the stimulus remains identical, but some task or cognitive factor changes. For example, it was shown many years ago that an analogue of speech containing only sine waves is heard as nonsense by most people at first; but after some training, the semantic content of a sentence can be understood (Remez et al. 1981). Several functional imaging studies compared brain activity with such stimuli, and found that left auditory cortical regions were more strongly recruited after training, when the stimuli were perceived as speech, than prior to training (Liebenthal et al. 2003, Dehaene-Lambertz et al. 2005). This effect was present only in those individuals who learned to perceive the content (Möttönen et al. 2006). A related phenomenon can be seen with learning of new sounds of speech, as shown in an fMRI study by Narly Golestani: in this experiment, part of the left posterior auditory cortex responded more after listeners were trained to distinguish a speech contrast that is not present in their native language, compared to before training (Golestani and Zatorre 2004).

In all these cases, since the stimulus was unchanged, only exposure or training (but not acoustical cues) could have caused the change in lateralized activation. Selective attention to different stimulus features can also shift asymmetries, in a way that is consistent with our model: right auditory cortex is more active when listeners are asked to report the direction of pitch changes in frequency-modulated tones; but left auditory cortex is more active when listeners are asked to judge the duration of the identical stimuli (Brechmann and Scheich 2004). This effect indicates that top-down signals can influence earlier processing in an asymmetric manner.

For a musical example of this kind of phenomenon, it is interesting to look at the "speech-to-song" illusion, discovered by Diana Deutsch and her group in San Diego (Deutsch et al. 2011), whereby a brief segment of normal speech is looped and repeated many times, and, under some circumstances, people perceive it as singing (Falk et al. 2014). When this shift from perception as speech to perception as song was studied with functional imaging, it was found that the song percept was associated with greater activity in pitch-sensitive regions bilaterally and in right superior temporal and right premotor areas (Tierney et al. 2013). Since the acoustical cues had not changed across trials, these shifts in activity patterns cannot be solely explained by bottom-up influences. But the data do support the idea that as the illusion becomes stronger, perhaps music-related structures in more distant parts of the pathway (including motor-related areas) come into play and modulate the earlier responses.

These sorts of findings have sometimes been taken as evidence that brain activity in general, and hemispheric differences in particular, are driven only by higher-order factors (including, for instance, semantic processes) and that they cannot be explained by differences in early stages of processing. However, since the evidence for spectrotemporal tuning lateralization is so strong, a better explanation is that there are complex interactions between earlier and later stages and that feedback processes most likely explain the differential patterns that are observed when stimuli are identical, but knowledge or experience changes their processing. In those cases, predictive-type descending neural signals very likely influence earlier processes via the mechanisms described in more detail in Chapters 3 and 4, leading to an apparent domain specificity.

Anatomical Considerations

Local Features

In this chapter, I have focused on functional asymmetries, as they are the most relevant for our understanding of tonal, and hence musical, processes. But since function generally depends on structure in biology, it is worthwhile to examine some anatomical features that may underlie the functional differences. There is a long history describing differences in the gross anatomy of the two hemispheres, going back to early 20th century anatomists. Konstantinos von Economo (who, among other things, was a WWI flying ace) was one of the first to notice that some auditory cortical structures seemed to be larger on the left side than on the right (von Economo and Horn 1930). The majority of anatomical studies have focused on regions posterior to Heschl's gyrus, particularly the planum temporale, which was also reported to be larger on the left side in a famous anatomical study of gross morphology (Geschwind and Levitsky 1968). The definition of this structure has proven problematic, however, leading to different results depending on its boundaries, which are not always clearly identifiable in a consistent manner (Westbury et al. 1999, Zetzsche et al. 2001).

Not surprisingly, the vast majority of work examining the relationship of these asymmetries to function has focused on language, with relatively inconclusive results (Jäncke and Steinmetz 1993, Dorsaint-Pierre et al. 2006, Tzourio-Mazoyer et al. 2018). Moreover, many studies focused on the idea that only language functions are lateralized and, therefore, made the assumption that any anatomical asymmetries must be related to language, without considering that music (or other) functions might also be asymmetric. Worse yet, there is an implicit, and simplistic, bias that "bigger is better," and that since language is more important than anything else, the asymmetries in size must also be related to language. In any case, as I argued earlier, treating complex functions, such as language or music, as unitary wholes is not always very useful. So, I will instead concentrate more on anatomical relationships to asymmetries in spectral and temporal processing more relevant to our hypothesis (for reviews, see Shapleske et al. 1999, Chance 2014).

As we saw in Chapter 2, Heschl's gyrus, which usually contains most of the primary auditory cortex, is asymmetric in size. This observation has been made repeatedly, not only with the postmortem dissections of the last century, but also in large samples using modern neuroimaging techniques (Kong et al. 2018, Eckert and Vaden 2019). In addition, several studies have documented that not only volume but also cortical surface area is left asymmetric in auditory regions (Meyer et al. 2013, Dalboni da Rocha et al. 2020).

A closer look at the anatomy reveals that the leftward asymmetry of Heschl's gyrus may be as related to the white matter as to the gray matter. We observed this phenomenon in the first in-vivo anatomical imaging study of Heshchl's gyrus, carried out by Virginia Penhune (Penhune et al. 1996), who measured the volume of gray and white matter in this structure by (painstaking!) manual delineation of MRI scans. Results showed that the asymmetry was largely due to a greater volume of white matter underlying the gyrus on the left. This leftward white-matter effect was replicated using the same method in several independent samples (Dorsaint-Pierre et al. 2006, Warrier et al. 2009), including even in congenitally deaf individuals (Penhune et al. 2003). In-vivo imaging of myelin with MRI also suggests leftward asymmetries in several auditory temporal-lobe regions, including Heschl's gyrus (Sigalovsky et al. 2006). The same thing was observed in postmortem analysis of white-matter volume and also of the thickness of the myelin sheath surrounding axons in human posterior auditory cortex samples (Anderson et al. 1999). The reason that white matter, and especially myelin, is relevant to our understanding of auditory cortex function is that myelin acts as an electrical insulator for axonal communication and is, hence, critical for accurate temporal transmission of information (Fields 2008). Therefore, this finding could be consistent with the proposition of enhanced temporal resolution within left auditory cortex, although obviously much more needs to be done to solidify these ideas.

More detailed analysis of the cellular structure and connectivity of auditory areas in human postmortem tissue has revealed numerous asymmetries, especially in the organization of cortical columns (for review, see Hutsler and Galuske 2003). These columns are wider in diameter and spaced further apart in the left hemisphere (Seldon 1981a, 1981b) compared to the right, where they are closer together and more interconnected. In posterior auditory areas, there is a wider spacing of patches of intrinsic connections on the left than the right side (Galuske et al. 2000). Pyramidal cells within these columns are larger in size on the left as well (Hutsler 2003) and have longer axons (Hutsler and Gazzaniga 1996). However, the way in which these differences may relate to acoustical processing was not established by these studies.

To establish a link between anatomy and function requires that both be measured and that they be related to one another. For obvious reasons, this is hard to do with postmortem tissue analysis, so we must rely on less detailed in-vivo imaging. At the level of Heschl's gyrus, Narly Golestani showed that greater white and gray matter volume on the left, but not the right, was associated with learning to distinguish new speech sounds (Golestani et al. 2006). Nick Foster reported a converse finding for melody discrimination tasks, where performance was correlated with the volume and thickness of gray matter in lateral portions of the right, but not the left, Heschl's gyrus (Foster and Zatorre 2010a). Peter Schneider and colleagues reported that musical task performance can be related to anatomical features in both left and right auditory regions (Schneider et al. 2002); but in another study, this group also found that anatomical asymmetries favoring the right Heschl's gyrus were associated with greater reliance on spectral cues to pitch of complex tones (Schneider et al. 2005). Taken together, these studies are partly consistent with the idea that structural asymmetries map onto speech or music-relevant tasks but don't directly address the acoustical cues that may be behind the correlations.

In order to look more specifically at the relation between anatomical asymmetry and spectrotemporal processing, Catherine Warrier, working in Nina Kraus' lab, measured both Heschl's gyrus volume and functional activity to spectral and temporal stimulus variation, using Marc Schönwiesner's method (Schönwiesner et al. 2005). The most interesting finding was a double dissociation in the function–structure relation such that greater volume of Heschl's gyrus on the left led to greater functional response to temporal, but not spectral, stimuli and vice-versa on the right side (Warrier et al. 2009). This finding directly supports the contention that asymmetries in anatomical structure are related to the acoustical features for which each auditory cortex is specialized. But measures of tissue volume still leave open the question of how exactly anatomical features relate to functional responses.

To address this question, a more fine-grained understanding of neural microstructure is required. Recent developments in MRI that allow measurement of orientation of axons and dendrites in the cortex have revealed numerous asymmetries, particularly in auditory areas (Schmitz et al. 2019). Although a precise understanding of how these patterns map onto cellular and columnar organization remains to be worked out, a promising finding is that the density of axons and dendrites in left, but not right, posterior auditory areas correlates with latency measures of auditory responses to speech sounds from EEG (Ocklenburg et al. 2018). This finding is also consistent with a correlation between onset latency and white-matter diffusivity in left auditory cortex that our team observed (Coffey et al. 2017c). These observations support the idea that faster processing is associated with anatomical features in left auditory cortex, as shown in Figure 5.8.

In fact, we can take this concept further to explain not only temporal but also spectral processing. If the distribution of cortical columns is related to tonotopy, then the closer spacing and interconnectivity between them that is observed on the right is compatible with our finding that functional connectivity in right auditory areas is higher for more similar frequency tuning than it is on the left (Cha et al. 2016), as shown in Figure 5.6, because this sharpening of tuning arises from intracortical inputs (Liu et al. 2007). Consequently, such an arrangement would result in higher spectral resolution on the right. Conversely, the

Left hemispheric PT   Right hemispheric PT

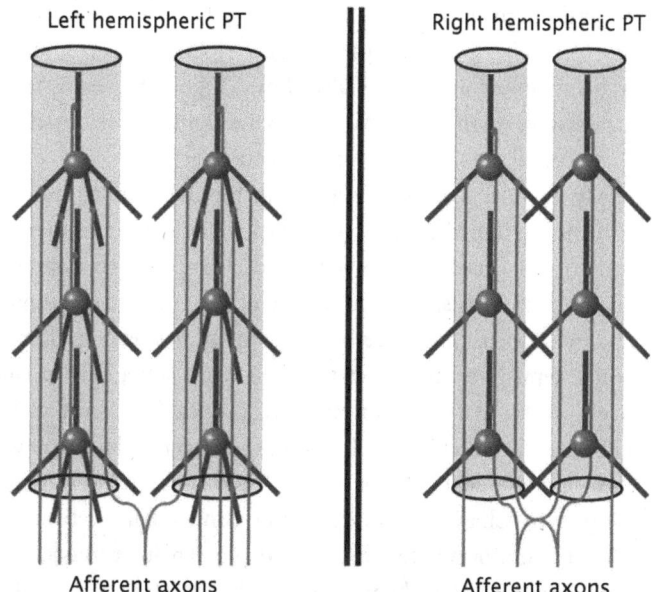

Afferent axons   Afferent axons

FIGURE 5.8 Diagrammatic depiction of microstructural asymmetries of the planum temporale.
Each panel shows two microcolumns in left and right hemispheres. The columns in the left hemisphere
display higher density of dendrites and axons, and neurons on the left show a higher degree of arboriza-
tion. Microcolumns are also wider and further apart from one another in the left hemisphere, while they are
spaced closer together on the right. Higher density of dendrites and afferents on the left side could enable
more synchronous firing, resulting in greater temporal resolution. Reprinted with permission of AAAS from
(Ocklenburg et al. 2016). © The Authors, some rights reserved; exclusive licensee AAAS. Distributed under
a CC BY-NC 4.0 License (http://creativecommons.org/licenses/by-nc/4.0/).

wider spacing and longer-range connections on the left auditory areas, together with larger
cell sizes and denser cellular organization, could be associated with more integration across
frequencies (lower spectral resolution) and faster responses (higher temporal resolution).
These ideas remain speculative at the moment but provide some useful testable hypotheses.

Long-Range Features

Differences between the hemispheres extend beyond the organization of particular auditory cortical regions, to also encompass network-level interactions. These are essential to
support all forms of higher-order cognitive processing, as we have repeatedly seen in earlier
chapters, so understanding some of the long-range connectivity asymmetries is highly relevant. We will consider both intra-hemispheric and inter-hemispheric connections, starting
with the former.

One of the most relevant intra-hemispheric fiber tracts that exhibits structural asymmetry is the arcuate fasciculus. Together with the superior longitudinal fasciculus, it interconnects auditory areas in the temporal lobe with motor and other regions in the frontal
lobe, as part of the dorsal auditory pathway (Petrides and Pandya 2009) (see Figure 4.1).
Numerous studies have repeatedly documented asymmetries in its structure, although
the details differ. Macrostructural measures derived from MRI scans of the white matter

show that the left-hemisphere arcuate fasciculus has a greater fiber density (Nucifora et al. 2005) and volume (Ocklenburg et al. 2014) than the right. Microstructural measures of these connections, using diffusion imaging, also show a higher degree of structural organization of the fiber pathways on the left compared to the right, whose organization tends to be more diffuse (Catani et al. 2007, Lebel and Beaulieu 2009).

Not surprisingly, there is a strong tendency in the literature to interpret these differences in terms of the need for better organization in the left hemisphere to handle the demands of speech processes, harking back to the old "dominant hemisphere" concept. But if we think of it in terms of hemispheric complementarity, then the findings make better sense. A greater, or more focused, degree of connectivity between auditory and anterior regions within the left hemisphere is indeed associated with better performance on certain language tasks, including, for instance, word learning (López-Barroso et al. 2013) and phonological tasks (Lebel and Beaulieu 2009). Looking exclusively at language will necessarily lead only to conclusions about language, of course. But the brain does other things too.

As touched upon in Chapter 4, white-matter connectivity between auditory and motor structures is also important for learning to play an instrument, or to sing, and these findings often implicate right-hemisphere pathways (Halwani et al. 2011). Lucía Vaquero, working with Virginia Penhune and Antoni Rodríguez-Fornells, demonstrated that learning of musical tasks is associated with both macrostructural and microstructural properties of the arcuate fasciculus in the right hemisphere only (Vaquero et al. 2018) (for more on this point, see also Engel et al. 2014). Similarly, Pysche Loui and her colleagues in Boston showed that anatomical measures of auditory-motor white-matter fibers in the right hemisphere were related to successful learning of a pitch-based artificial grammar task (Loui et al. 2011). These findings do not necessarily indicate that left-hemisphere white-matter pathways play no role whatsoever in music learning; they likely do. But what they do demonstrate is that it's not simply that the left-hemisphere connections are somehow more prominent or better but rather that networks within each hemisphere are specialized for solving certain tasks; and if the task involves spectral sound features, it is likely to engage right-hemisphere networks. We can get further clues to this differential organization by looking at interhemispheric structural network organization.

White-matter pathways interconnect structures across the two hemispheres, as well as within them. These interhemispheric connections primarily course through the corpus callosum and anterior commissure and are critical for many functions, as Sperry showed, including of course music-related ones. At a minimum, since many musical instruments require coordination of the two hands, which is mediated via the corpus callosum (Johansen-Berg et al. 2007), structural modifications with musical training are likely to occur in this structure (Schlaug et al. 1995). In accord with this idea, Chris Steele, working in Virginia Penhune's lab, used diffusion imaging to show that the callosal pathway that is enhanced with musical training early in life interconnects the two motor cortices and that it is related to motor synchronization accuracy (Steele et al. 2013); see also Bengtsson et al. (2005) for a related finding.

In addition to these interhemispheric motor pathways, which are relevant for music but are largely symmetrical, there are broader patterns of interhemispheric connectivity that display interesting asymmetries (Iturria-Medina et al. 2010). Interestingly for our purposes, it appears that the way that the auditory cortices in each hemisphere are connected to the rest of the brain differs between the left and the right side. My Montreal Neurological Institute colleague, Bratislav Mišić, applied an interesting model to examine these asymmetries in several large samples of diffusion MRI data (Misic et al. 2018). Rather than simply computing the shortest topological connections in the white-matter pathways between any pair of nodes in a brain network, he developed a spreading diffusion model, which captures the relative amount of time it would take for activity from one spot to spread to another spot, based on the interconnections between them (Mišić et al. 2015). When applied to the auditory cortex, the model shows that activity from the right side spreads more widely and quickly to the rest of the brain, especially to the other hemisphere. In contrast the left side shows a more focal pattern of communication within the left temporal and frontal lobe, as shown in Figure 5.9. The existence of this greater left intrahemispheric auditory connectivity has also been suggested by other diffusion imaging studies (Cammoun et al. 2015). Thus, we may conclude that the cascade of information flow from the right auditory cortex is more globally widespread, reaching more of the brain, than the equivalent process on the left, which is more confined within the left hemisphere.

These differences in the way that the right and left auditory cortices interact with other brain regions are consistent with functional data as well. In fact, the analysis of asymmetries just described was motivated by a curious observation we obtained with a brain stimulation study. My former postdoc Jamila Andoh, now at Mannheim University, was interested in how transcranial magnetic stimulation applied to the auditory cortex influenced other brain areas. She measured functional connectivity patterns with resting-state MRI, before and after delivery of inhibitory transcranial magnetic stimulation to auditory cortex on each side (Andoh et al. 2015). Surprisingly, she observed that stimulation of the right auditory cortex led to much more widespread changes in both auditory and motor functional networks, in both hemispheres, compared to stimulation of the left. Furthermore, this influence of the stimulation was mediated by individual differences in the structural organization of the interhemispheric pathway in the corpus callosum.

Based on the Mišić model, this pattern makes sense: when the right auditory cortex is stimulated, it influences more of the brain, compared to the left auditory cortex, because of its greater integration into the entire connectome. Further hints about these asymmetries also emerge from analysis of resting-state fMRI connectivity patterns. Thus, one study showed that the left superior temporal region has greater exchange of information with other left-hemisphere structures than its homologue on the right does within the right hemisphere (Liu et al. 2009). Another study revealed that left-hemisphere cortical regions interact more among themselves, whereas right-hemisphere regions interact in a more integrative manner across both hemispheres (Gotts et al. 2013). All these functional differences are in close accord with our conclusions based on the structural asymmetries.

The functional implications of these interhemispheric connectivity asymmetries are not yet entirely known. Furthermore, the way in which these large-scale asymmetries at the network level interact with the local asymmetries within the auditory cortices described earlier, also remains to be understood. But if we think of the differential patterns in terms of

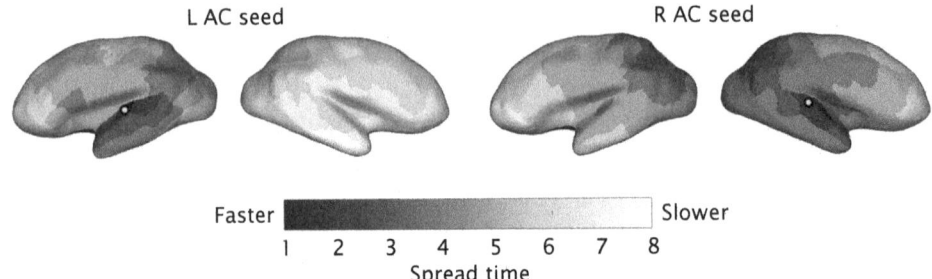

L AC seed   **R AC seed**

Faster ▮▮▮▮▮▮▮▮▮▮▮▮▮▮▮▮▮▮▮▮▮▮▮▮ Slower
1   2   3   4   5   6   7   8
Spread time

FIGURE 5.9 Network-based asymmetries of auditory cortex. Diagram shows simulated spread of information from seed regions in left or right auditory cortex (yellow circles) according to a graph-theory model using diffusion imaging of white-matter pathways. The right auditory cortex facilitates more efficient communication with a broader set of brain regions, both within and across hemispheres, while the left auditory cortex participates in a more local, within-hemisphere network. Reproduced under the terms of a Creative Commons CC BY license from (Misic et al. 2018).

segregation versus integration of information processing, it leads to some useful and testable ideas. The left hemisphere auditory system may be optimized for certain kinds of processes, with its fast, low spectral resolution local organization, and its focal intrahemispheric connectivity pattern in the ventral stream, with more limited interhemispheric interaction. Many aspects of language, including analysis of rapidly changing speech sounds and tight motor control over a specific and dedicated set of vocal effectors could benefit from this organization. The right hemisphere auditory network, with its lower temporal resolution, higher spectral segregation at a local level, and its more widespread interhemispheric connectivity, along with relatively flexible mapping of auditory-motor interactions of different effectors via the dorsal stream, could all prove optimal for musical perception and production, which requires control over a wide range of effectors.

In a sense, our current understanding, although still limited in many ways, comes full circle to some of the earlier ideas that had been proposed about hemispheric specialization. In particular, the different patterns of interhemispheric connectivity detailed above are reminiscent of some proposals from half a century ago, which were meant to explain why lesions of the left hemisphere often entailed much more specific, focal symptoms, than lesions of the right hemisphere (Semmes 1968). The idea then was that cognitive functions are represented in a more diffuse manner on the right than the left. But we may reformulate this concept based on the knowledge recently uncovered: it's not necessarily the case that cognition is more distributed across the right hemisphere, but rather that, because of its heavily interconnected nature, focal lesions on the right have different and maybe less discernible manifestations than lesions of the left, which often result in specific syndromes, due to the more focused connectivity within the left hemisphere.

Reprise

The specialization of structures and networks within and across the two hemispheres manifests itself in many ways, but the lateralization of linguistic and musical processes offers a salient and relevant example of asymmetric processing. An enormous amount of evidence, from both lesion studies and various imaging techniques, points to preferential processing of some elements of language and music to the left and right hemispheres, respectively. But the nature and underlying mechanisms behind these asymmetries continue to be the subject of debate.

Rather than considering spoken language and music as unitary constructs, in this chapter, I argue that what is most useful to understand neural specializations are the acoustical cues that are especially relevant to perceive these communicative signals. Music and speech exploit opposite ends of an acoustical continuum, which can be characterized in terms of temporal and spectral modulations. The different ways in which these cues are used are not absolute: both communicative domains make use of spectral and temporal features to carry information. But experimental manipulations demonstrate that speech is much more dependent on preservation of temporal than spectral cues, compared to music; whereas the melodic aspects of music are dependent on fine-grained spectral information to a greater extent than for speech.

The auditory nervous system jointly encodes spectral and temporal modulation rates that are present in sounds, as shown by many physiological studies in nonhuman species and functional imaging in humans. This neural property provides the basis for proposing that hemispheric asymmetries arise at this level of sensory processing, resulting in enhanced temporal resolution in left auditory cortex and enhanced spectral resolution in right auditory cortex. More generally, there is a close match between the rates of spectrotemporal modulation contained in communicative signals of a given species, and the neural tuning to those modulations, leading to efficient encoding of ethologically salient information. I propose that humans possess two principal auditory communicative systems—speech and music—and that neural networks within each hemisphere are optimized to process the features that are most relevant for each of these systems.

Descending influences from higher levels of the hierarchically organized auditory nervous system can alter the patterns of activity in lower levels that are driven by acoustical features. These top-down processes can lead to either enhancements in lateralization, via iterative processing of long complex stimuli, or can also lead to shifts in hemispheric responses, as a function of cognitive factors such as learning, familiarity, and attention.

Anatomical asymmetries can be observed at both local and global levels, within and across hemispheres. Current evidence suggests that the organization of white matter and the patterns of corticocortical microconnectivity within left and right auditory cortical areas lead to higher temporal resolution on the left, and higher spectral resolution on the right, consistent with the functional evidence. Long-range connectivity between auditory areas and other brain regions also suggests differences in organization such that networks originating in the left auditory cortex are more focally organized within the hemisphere, while networks originating in the right auditory cortex are more globally organized across both hemispheres. These differential patterns of embeddedness may be relevant for understanding the complex auditory-motor interactions that are relevant for both speech and music.

PART II

Pleasure

6

The Reward System

One of the greatest discoveries of 20th-century neuroscience took place just a few blocks away from where my lab is located, for it was at McGill that James Olds and Peter Milner first identified what we now recognize as a key component of the brain's reward system (Olds and Milner 1954). As Peter would later recount, he had been targeting electrodes toward certain structures in the rat brain for his PhD work, to understand the effect of electrical stimulation on arousal. But instead of the intended targets, the electrodes ended up in a different position in a region that, when stimulated, caused the animal to continually return to the part of the enclosure where the stimulation had taken place, as if it were seeking something it had experienced at that specific location. When they closed the loop and had the stimulation be triggered by the rat's own action (pressing on a lever), they found to their astonishment that the rats would self-stimulate for hours, even neglecting food. It seemed that the stimulation produced such a profound experience that it supplanted the need for anything else. They concluded with characteristic modesty: ". . . we have perhaps located a system within the brain whose peculiar function is to produce a rewarding effect on behavior" (Olds and Milner 1954).

And thus was launched the entire field of what today we call "affective neuroscience," that focuses on the role of the reward system on behavior and cognition (Panksepp et al. 2017). It is ironic to consider that Olds and Milner used strictly behaviorist methods, using rats pressing levers to receive reinforcement—even referring to their apparatus as a Skinner box—to show how behavior was strongly influenced by internal signals. This concept was anathema to the classic behaviorists of the time, who would have denied the validity of a construct such as "pleasure."

Yet we cannot understand behavior at all unless we understand what drives it in the first place. In this sense, pleasure may be seen as essential to guiding behavior: it's evolution's way of signaling that certain stimuli and the actions needed to obtain them are necessary for survival or fitness (perhaps in contrast to pain or fear, which are evolution's ways of making us avoid danger). Another way to think of it is that survival depends on achieving a balance within internal states, that is, homeostasis. So, behavior needs to be optimized to

From Perception to Pleasure. Robert Zatorre, Oxford University Press. © Oxford University Press 2024.
DOI: 10.1093/oso/9780197558287.003.0006

satisfy these needs. The internal state can be based on interoceptive neural signals related to metabolism (e.g., hunger or thirst); or they can be sensory in nature coming from the environment (e.g., heat or cold); or, of greatest interest to us of course, they can also be cognitive (e.g., anxiety, curiosity, or boredom). Thus, the reward system guides us toward actions that regulate our internal states to maximize well-being.

The term "pleasure center," which has been widely used to describe the results of Olds and Milner and other later studies, is perhaps not ideal, for reasons that we shall discuss below. Yet, it is abundantly clear that our subjective hedonic experience of pleasure is intimately tied to the reward system (Berridge and Kringelbach 2015) and that its activity is directly related to the subjective experience of pleasure under many different circumstances (Kühn and Gallinat 2012). But the reward system is certainly not confined only to hedonic processes; it is also critical to several other phenomena, in particular learning and motivation/action, both of which enable goal-directed behavior. One of the leading researchers in the domain of reward, Wolfram Schulz, makes the point that reward is composed of two components: sensory and value-related—as shown in Figure 6.1 (Schultz 2016). The sensory aspect is processed by neural systems involved in perception, whereas the value component largely depends on the reward system. According to this model, reward value computations occur after perceptual processing and lead to decision-making and action in relation to the potentially rewarding event.

The neural substrates associated with reward are, for the most part, quite distinct from the auditory pathways and structures described in Part I of this book. Whereas those cortico-cortical auditory loops, as we've seen, are essential for perceptual, mnemonic, and predictive processes, they do not by themselves account for reward-related affective or hedonic responses. In the context of music, Schulz's framework fits with the idea that auditory networks handle perceptual processes, whose outcome is then evaluated by the reward network. This chapter will therefore focus on the structure and function of the reward system, which is of prime importance since our ultimate goal is to understand why music gives us pleasure. Furthermore, another critical goal of this chapter is to understand how these two distinct systems communicate with one another anatomically and functionally.

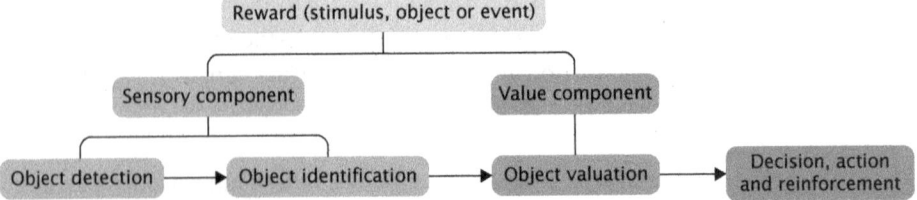

FIGURE 6.1 Reward-related processing components. Diagram illustrates theoretical model for reward processing containing two main components, organized serially. An incoming stimulus (gray box) is first processed in sensory channels (blue boxes). The reward value is assessed within reward channels (orange boxes) after identification is complete, leading finally to decision and action (purple box). According to this model, reward value does not directly reflect the stimulus parameters but rather the utility of the stimulus for survival, fitness, or well-being. Reprinted by permission from Springer Nature: Springer Nature, *Nature Reviews Neuroscience* (Schultz 2016), copyright 2016.

Anatomical Outline of the Reward System

The complexity of the reward system, as shown in Figure 6.2, can seem daunting, as it encompasses core structures in portions of the cortex, basal forebrain, and midbrain, together with thalamus, amygdala, hippocampus, and brainstem nuclei (Haber 2017). As is the case with the anatomy of the auditory pathways, most of our knowledge of the reward system comes from animal studies. And while it is reasonable to assume that these basic survival-related circuits are largely preserved across evolution, we should keep in mind that details of their organization or function may differ in unknown ways in humans. The following sketch of the connections represents a simplification of the complex wiring that is known to exist. In the context of this book, it is meant to provide a framework with which we can interpret the functional data that are of greatest relevance to our needs.

The reward system is very much embedded within several other brain systems, which is one reason why its boundaries are not entirely agreed upon. Nonetheless, there is wide consensus that its most critical components include the midbrain dopaminergic nuclei in the ventral tegmental area and substantia nigra, and a set of structures in the basal forebrain that are collectively known as the striatum (Haber and Knutson 2010). The striatum,

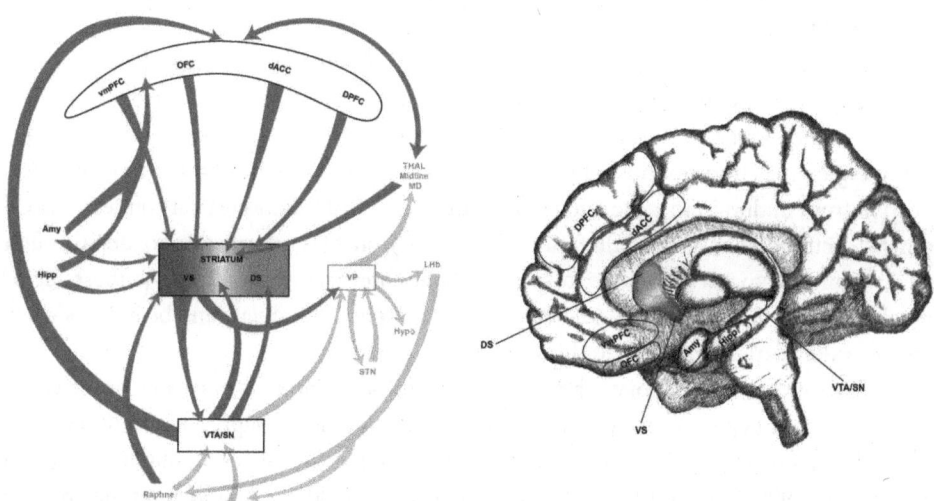

FIGURE 6.2 Anatomy of reward system and its connections. Left: Diagrammatic depiction of some of the principal inputs and outputs to/from the core reward system. The ventral tegmental area and substantia nigra (VTA/SN) provide the main dopaminergic inputs to the dorsal and ventral subdivisions of the striatum (DS, VS). The VTA/SN also innervates large swaths of the cerebral cortex. The striatum receives inputs from many structures in a topographically organized manner, shown by the color shading. Ventral striatum (red–yellow) receives inputs from the amygdala (Amy) and hippocampus (Hipp), together with cortical inputs coming primarily from ventral regions including ventromedial prefrontal cortex (vmPFC) and orbitofrontal cortex (OFC); the dorsal striatum (green–blue) receives cortical inputs primarily from more dorsal cortical regions, including the dorsal anterior cingulate cortex (dACC) and the dorsal prefrontal cortex (DPFC). Other connections of the system not discussed here are shown in light gray. Adapted with permission from (Haber 2017).

Right: Anatomical drawing of the medial aspect of the right cerebral hemisphere indicating the position of the various structures shown in the diagram on the left. Drawing by E.B.J. Coffey, used with permission.

which also forms part of the basal ganglia, can in turn be further subdivided into dorsal and ventral components (although it is probably more accurate to consider it as a continuum rather than as two separate structures because there is no sharp demarcation between them, either anatomically or physiologically). The nucleus accumbens, located within the ventral striatum, is one of the structures stimulated by Olds and Milner and one of the most important in the entire reward circuit, due to its conspicuous recruitment in many reward-related behaviors.

The connectivity between these core reward structures and the rest of the brain is critical to understanding the functional characteristics of the entire system. The striatum receives projections from frontal-lobe cortical areas that are important for affective and cognitive processing (Zald and Kim 1996). As one proceeds from the most ventral part of the striatum to its more dorsal portions, the cortical inputs gradually shift from the ventromedial and orbitofrontal cortex to the anterior cingulate cortex and finally to the dorsolateral prefrontal cortex (see Figure 6.2). The most ventral part of the striatum also receives inputs from the hippocampus and amygdala, structures which play critical roles in many functions, especially memory and emotion (Friedman et al. 2002). These topographically organized patterns of connectivity imply that different portions of the striatum are involved in different aspects of reward processing, progressing from affective/motivational to more cognitive/executive control processes as one proceeds from more ventral to more dorsal portions of the system.

The striatum, in turn, sends projections to the midbrain dopaminergic structures, also in a topographic manner. Conversely, the midbrain structures have a massive reciprocal dopaminergic input to the striatum. In addition, these midbrain regions send diffuse dopaminergic projections to most of the cerebral cortex (Lidow et al. 1991). This direct connection from midbrain to cortex, together with indirect midbrain-cortical connections via other structures (such as the pallidum and thalamus), enables information to loop through all structures from cortex via striatum to midbrain and vice versa. Thus, via these reciprocal pathways, the core reward-related structures can both modulate and be influenced by higher-order cortical-based operations.

The way that the auditory cortex connects with the striatum is of particular relevance to us. Based on evidence from monkey tract-tracing studies, nonprimary auditory areas (especially more anterior fields) project to portions of the caudate and putamen (Yeterian and Pandya 1998, Jung and Hong 2003). Therefore, these dorsal striatal areas may receive some direct auditory input. But the most relevant auditory inputs to the dorsal part of the striatum are indirect and arrive from the dorsolateral prefrontal cortex and other dorsal cortical structures that form part of the auditory dorsal stream, as discussed in Chapter 4 (see Figure 6.3).

Direct inputs from auditory cortex to the more ventral components of the striatum, including the nucleus accumbens, seem to be quite limited or absent, according to the same anatomical studies just cited. The ventral striatum, therefore, most likely receives the majority of its auditory inputs indirectly, via the orbitofrontal and/or ventromedial frontal cortex, which is known to be a multisensory hub (Barbas 1993) and specifically receives inputs from the most anterior portions of the auditory ventral stream (Petrides and Pandya

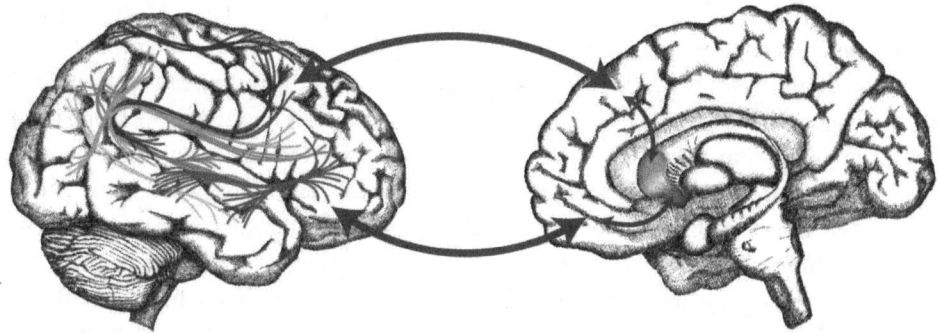

FIGURE 6.3 Connectivity between auditory dorsal and ventral pathways and reward system. The auditory ventral (red–yellow) and dorsal (green–blue) cortical pathways are illustrated on the left; their topographically organized interconnections with the reward system are shown schematically by the large arrows. Red arrows illustrate how the auditory ventral pathway connects via the ventromedial and orbital frontal cortex to the ventral striatum (red–yellow on the midsaggital view on the right). Blue arrows illustrate how the auditory dorsal pathways connct via the dorsolateral prefrontal and cingulate cortex to the dorsal striatum (blue–green on the midsagittal view on the right). Drawings by E.B.J. Coffey, used with permission.

1988, Cavada et al. 2000). The orbitofrontal cortex then directly connects to the ventral striatum (Figure 6.3).

The orderly topographic mapping between the auditory dorsal stream with the dorsal striatum, and the auditory ventral stream with the ventral striatum (Figure 6.3) thus provides the basis for bringing together the perceptual circuits described in Part I of this book with the reward system (the focus of Part II of this book). As such, it is crucial for understanding how reward activity is elicited by music because it underlies the functional interactions between these systems. We turn next to that question.

Functional Interactions Between Frontal Cortical Regions and the Reward System

Each of the various frontal cortical areas that link to the reward system are related to many complex functions. Their interactions with the reward system must be understood in order to gain knowledge about how pleasure and other affective responses emerge. The white-matter connections described above in animal studies are paralleled by findings from functional connectivity studies in humans, thus providing validation of the anatomical model (Di Martino et al. 2008). First, we will consider the interactions between ventral frontal regions and the ventral parts of the striatum, followed by the more dorsal portions of the system.

Traditionally, the ventromedial prefrontal cortex has been associated with aspects of decision-making that require evaluation of reward outcomes. This is the reason why patients with lesions to this region often display enormous difficulty making good decisions and are especially prone to favoring short-term gains despite long-term losses (e.g., Bechara et al. 2000b); yet they usually do not have difficulties with most cognitive processes or language. This reward-evaluation functional feature of the ventromedial frontal cortex is also found throughout the entire orbital region of the frontal cortex. The orbitofrontal cortex is of particular interest to us because its activity tracks the magnitude of the reward value of a given stimulus, in both monkey physiological studies (e.g., Padoa-Schioppa and Assad 2006) and human imaging studies (e.g., O'Doherty et al. 2001). Furthermore, this value is also related to subjective pleasantness (Kringelbach 2005, Kühn and Gallinat 2012).

One of the first demonstrations of this value-tracking phenomenon comes from a neuroimaging study carried out over two decades ago by my colleague Dana Small. For her PhD work, Dana was interested in responses to food reward (Small et al. 2001). In order to dissociate the value assigned to the stimulus from the sensory input associated with that stimulus, she fed people multiple identical pieces of chocolate over the course of the brain imaging experiment, leading to lower value over time as people became satiated (I admit to being an eager pilot participant in this study, though by the end of it I never wanted to see chocolate again). The activity in the medial portion of the orbitofrontal cortex closely tracked subjective pleasantness ratings, decreasing as more chocolate was consumed, even though the stimulus had not changed. This finding not only shows that the value/pleasantness (rather than physical stimulus features) is most important in eliciting the response of this region, but it also shows that interoceptive cues (associated with increasing satiety as more chocolate was consumed) influence the response (see also [Kringelbach et al. 2003] for a related finding).

But the orbitofrontal cortex does not respond only to pleasant tastes nor to any one single stimulus class. Rather, it responds to the reward value of many different stimuli (Kringelbach and Rolls 2004). An idea that has gained support is that neurons in the orbitofrontal cortex are able to represent the value of different kinds of stimuli using a common code, which enables comparisons across different stimulus classes. This capacity was shown, for example, in an fMRI study in which people made judgments about how much they would be willing to pay for various items, including food and trinkets (Chib et al. 2009). The authors reported that activity in the medial part of the orbitofrontal cortex responded to the various items according to their value, regardless of the item category. More recently, based on a meta-analysis of over 200 studies, the idea of a subjective, supramodal valuation system has been confirmed. It has also been extended to include not only the medial orbitofrontal cortex but also the ventral striatum itself as being specifically involved in registering the magnitude of positive reward (Bartra et al. 2013). A contrasting view is that orbitofrontal neurons may only be important for coding value as part of some broader function, such as integrating environmental information to predict behavioral outcomes (Zhou et al. 2021).

These findings provide an idea of the mechanisms that allow us to make everyday sorts of judgments based on subjective reward value (Padoa-Schioppa and Conen 2017). Whether we would prefer to listen to a favorite jazz piece or eat some pizza depends, in part, on how we code the value of these disparate activities; the existence of a common neural code allows us to compare them to one another and decide which activity to select. The decision also depends on many other factors, including prior exposure (Am I a jazz fan? Do I always like pizza?), and especially internal states (like boredom versus hunger, in this example), which the orbitofrontal cortex is well-positioned to integrate because it also receives interoceptive and visceral inputs (Öngür and Price 2000).

The capacity of orbitofrontal neurons to respond in a similar manner to different types of stimuli indicates that this region is a convergence zone for inputs from many modalities (Cavada et al. 2000). Indeed, in monkeys, it receives projections from all the sensory cortical areas, including auditory cortex—and especially the more highly processed auditory information—coming from the anterior portions of the temporal lobe within the ventral stream, as discussed in Chapter 3 (Hackett et al. 1999, Romanski et al. 1999a, 1999b). Therefore, the orbitofrontal cortex stands as a critical interface between auditory cognitive processes and the affective processes carried out within the reward system.

Moving on to the functional role of the dorsal circuitry, based on the anatomy, we would expect that dorsal portions of the frontal cortex would interact most with the dorsal part of the striatum. Direct support in favor of a functional relationship between these structures was provided by Antonio Strafella and colleagues, who showed that transcranial magnetic stimulation (TMS) of the dorsolateral frontal cortex induced dopamine release in the dorsal striatum, as measured with dopamine-specific ligands and positron emission tomography (Strafella et al. 2001); see also (Cho and Strafella 2009). Modulation of the dorsolateral frontal region with TMS also induces an increase in blood oxygenation in the dorsal striatum but not in auditory cortex, as shown with fMRI scanning (Dowdle et al. 2018). These findings will become relevant again in the following chapter, for our discussion of how music-induced pleasure can be modulated by brain stimulation.

These modulations of activity in the reward system, enabled via frontal cortex connections to reward-system structures, can in turn lead to changes in reward-related behaviors. For example, inhibition of the dorsolateral frontal cortex with brain stimulation reduces the craving that smokers experience when exposed to smoking-related cues that signal impending access to cigarettes, and this effect is mediated via a modulation of activity in several structures of the reward system (Hayashi et al. 2013). In general, such findings support the idea that the dorsolateral frontal cortex exerts some level of control over the reward system, particularly in terms of weighing the value of immediate rewards versus delayed rewards (Figner et al. 2010). This conclusion is consistent with the effects of frontal lesions which, as noted earlier, often lead to disruptions in appropriate evaluation of behavior and can even lead to compulsive behaviors and addiction (Bechara et al. 2000a).

Given that our interest in this book is about how music engages the reward system, the next obvious question is: how do auditory signals exert their influence on the reward circuitry? As mentioned earlier in this chapter, there is little or no direct anatomical input from auditory areas to the ventral striatum. Hence, the indirect pathways linking auditory cortical areas to the striatum are especially important. The topography of these inputs suggests different functional properties for the ventral versus dorsal pathways.

As we saw in Chapter 3, highly processed auditory information proceeds along the ventral pathway to the inferior frontal and orbitofrontal cortex. Therefore, the inputs from ventral frontal areas, especially orbitofrontal cortex, to ventral parts of the striatum are well-situated to provide information about the value of a particular auditory event. The ventral stream enables retention of events in working memory, encodes auditory patterns at an abstract level, and generates predictions about the expected content of those patterns. Those features of the ventral pathway suggest that melodic and harmonic structure, and how well they conform to known statistical expectations, would be transmitted to the reward system via this pathway.

In a complementary manner, the inputs from more dorsal frontal areas (especially dorsolateral frontal cortex) to more dorsal parts of the striatum provide a second pathway for information to flow from auditory cortex to the reward system, via the auditory dorsal stream. As reviewed in Chapter 4, the auditory dorsal stream is closely linked to action and to temporal properties of music, including higher-order temporal organization such as metrical structure. Thus, it can generate predictions about when events may occur in relation to other events. Therefore, the inputs to the striatum via the auditory dorsal stream would be expected to influence reward processing most strongly in the temporal domain—for example, anticipating the timing of specific events (cadences, harmonic modulations, entry of new voices, etc.) and rhythmic expectancies (downbeats, syncopations) as the music unfolds over time.

Reward Prediction Error

Next, we turn to distinguishing responses of the reward system to positive events from the more subtle concept of responses to expectations about those events. Early descriptions of the reward system focused on responses to stimuli that were known to act as biological reinforcers for behavior, typically juice or food in a conditioning paradigm, for instance. These studies consistently demonstrated that neurons in the midbrain and striatum responded when such a stimulus was presented and that the neurotransmitter dopamine was involved. But in a critical series of experiments, Wolfram Schultz and his colleagues went beyond this direct stimulus–response relationship to reveal that dopamine neurons changed their responses as a function of learning when the reward appeared (Schultz et al. 1997).

In one of Schultz's typical experimental setups, a monkey would be trained to press a lever to receive fruit juice—a tasty treat that constitutes a primary reward for the animal (i.e., a stimulus which is biologically rewarding). Under these conditions, a response would occur in midbrain dopamine neurons immediately upon receipt of the reward. Subsequently, the monkey would be trained to press the lever when a light goes on which signals the availability of the juice. After a few days of learning, the neurons responded quite differently: instead of discharging to the delivery of the juice itself, as they did before learning, they would now only activate when the light went on. If the juice failed to appear at the expected time, that is, after the lever press, the dopamine neurons would become inhibited, decreasing their response precisely at the moment when the reward should have occurred. On the other hand, if more juice were delivered than usual, dopamine neurons would respond with a greater magnitude than usual. These features of the reward response indicate that dopamine neurons in the midbrain are doing something more sophisticated than merely responding to the presence of reward: they are responding to the *expectation* of reward (for a recent review with additional detail about these responses see [Schultz 2016]).

This idea of expectation can be formalized via the concept of reward prediction error (Schultz 2017). Essentially, this term refers to the difference between what is predicted to occur and what actually occurs. The word "error" here should not be taken in its everyday meaning of "mistake" but rather simply indicates the degree to which the prediction matched the outcome. More importantly, reward prediction errors are not always negatively valenced events, because a positive reward prediction error occurs when an unexpected or better-than-expected reward occurs. Although the details are still debated, dopamine neurons in many parts of the reward system seem to encode reward prediction error. These neurons are inhibited when an expected reward does not occur (negative reward prediction error), they do not respond when a reward occurs as expected (zero reward prediction error), and they are strongly active when an unexpected, or larger than expected, reward occurs (positive reward prediction error), as shown in Figure 6.4. The computation being carried out appears to be simply a subtraction between obtained and expected events (Eshel et al. 2015). However, more complex models with different timing of activity are needed to explain how dopamine neurons respond to a cue signaling later reward, but not to the reward itself, when it is fully predicted (Watabe-Uchida et al. 2017).

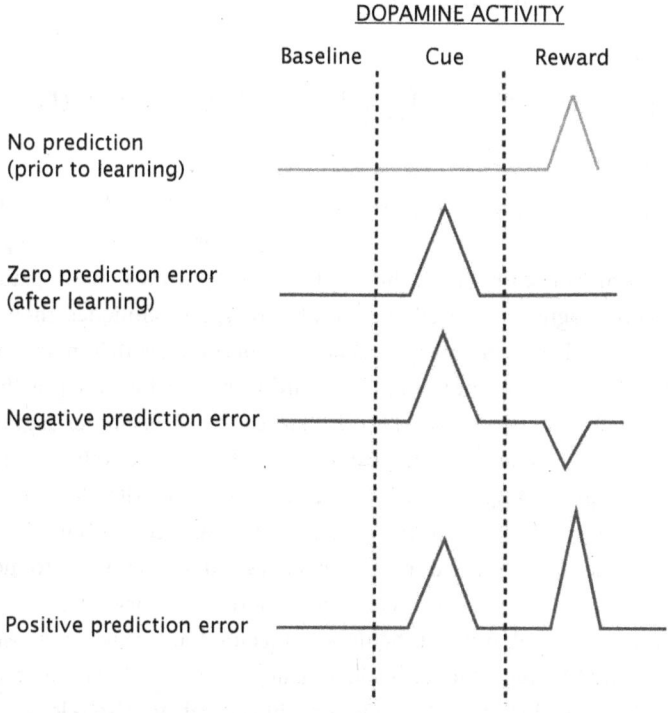

FIGURE 6.4 Schematic depicition of dopamine activity and reward prediction error. Prior to learning (green line), dopamine activity is elicited by a reward but not by a cue preceding it. After learning an association between a cue and the reward that follows it (blue line), dopamine activity occurs when the cue is presented, but not when the reward is presented, if the reward is exactly as expected because the prediction error is zero. If the reward is absent or of lower value than expected (red line), then dopamine activity is inhibited at the time when the reward would be presented because it is a negative prediction error. If the reward is of higher value than expected (purple line), there is a potentiation of the dopamine response at the time the reward is presented because it is a positive prediction error. Adapted from figure by R Clements in SITNBoston. Reproduced under a CC BY-NC-SA 4.0 license.

From a broader perspective, the operations carried out in reward-related dopamine neurons can be seen in a predictive coding context as providing information to update internal models (Friston 2010). In earlier chapters, we saw how prediction signals flow in a feedback manner, from higher-order structures down to earlier stages of processing, and conversely, how information coming from hierarchically lower neural structures is sent via feedforward pathways to higher structures when there is some discrepancy between the model expectancy and the input, leading to mismatch responses and updating of the internal model. Reward prediction errors can be considered as a special case of signals that update internal models, where information across multiple sensory modalities is integrated and compared to optimal goal states, to compute the subjective value of the input in terms of survival or fitness. If the reward prediction error is either small or zero, then no change, or learning, needs to take place, because the input agrees with what the model predicts. But if the error is larger—whether positive or negative—it can guide changes to behavior, in order to minimize negative rewards and maximize positive ones (via approach or avoidance of a stimulus, for instance). But, unlike neural interactions within sensory systems, the processes involving dopamine neurons in the reward system are closely linked to affective responses, including pleasure and motivation.

Human Neuroimaging Studies of Reward and Reward Prediction

All the evidence on reward prediction errors reviewed up to this point comes from neurophysiological studies in animals. To what extent do neuroimaging studies support the conclusions drawn from such studies in the human brain? Despite the limited spatial resolution of functional neuroimaging compared to neurophysiological techniques, there is considerable agreement across these different approaches. Furthermore, with human neuroimaging, one can begin to study much more complex reward responses than are possible in animal models, especially in terms of abstract rewards that require higher-order cognition. Hence, such studies are especially relevant to reward responses for music or other aesthetic stimuli.

Numerous brain imaging studies have been carried out with different types of rewarding stimuli. One may distinguish between biologically rewarding stimuli typically, used in animal studies (also referred to as primary rewards), as opposed to more complex abstract stimuli (often referred to as secondary rewards). Primary rewards are usually of direct relevance for survival and well-being and include food or water, as well as sexual stimuli, which would be relevant for reproduction and hence survival of the species, even if not necessarily the individual. Social interactions, in general, are also relevant for survival, and social reward reaches its apex when we form strong attachments to others, whether it be parents with their children, lovers with one another, or attachments formed in the context of close friendships. These relationships are also mediated, at least in part, by dopaminergic activity in the reward system (Feldman 2017).

Psychoactive drugs constitute another class of stimuli that activate the reward system in a biologically direct way, although they are by no means needed for survival—quite the opposite since drug abuse often leads to very negative health consequences. Nonetheless, certain drugs, such as amphetamine (Leyton et al. 2002), alcohol (Boileau et al. 2003, Aalto et al. 2015), nicotine (Brody et al. 2004), and cocaine (Cox et al. 2009), all result in significant dopamine release within the striatum when administered to healthy individuals. In some cases, subjective ratings of pleasure or euphoria elicited by the drug correlate with the degree of dopamine activity (Drevets et al. 2001, Aalto et al. 2015). Thus, dopaminergic responses in the striatum seem to play an important role in the powerful action of these substances, although it is also known that complex interactions across other neurotransmitter systems are involved (Koob and Volkow 2010).

The processing of secondary rewards is most often investigated using money, which is considered rewarding, based on learned associations. That is money has no inherent value in and of itself (it doesn't taste good, you can't use it to shelter from the elements, etc.). Instead, its value is derived from learning and a shared understanding that it is fungible (i.e., you can exchange it for things that are good to eat or that protect you from the cold). Therefore, money is clearly more abstract than food, drugs, or erotic stimuli. However, I would draw a distinction between secondary rewards and aesthetic rewards, such as music and other art, which are neither biologically necessary for survival, strictly speaking, nor have any value based on how fungible they are (the way money's value is based on how well it can be exchanged for certain goods). Music seems to be intrinsically rewarding. So, we cannot simply extrapolate from secondary to aesthetic rewards; they need to be studied in their own right. We will have much more to say about that issue in Chapters 7 and 8.

One relevant question that can be asked is whether the processing of primary and secondary rewards depends on the same reward structures, given that primary rewards are, obviously, phylogenetically ancient compared to secondary rewards and hence might depend on phylogenetically older neural substrates (Knutson and Bossaerts 2007). To address this question, Guillaume Sescousse and colleagues carried out a meta-analysis of a large number of neuroimaging studies, comparing primary and secondary rewards (Sescousse et al. 2013). The primary rewards involved either presentation of juice, food, or other desirable, edible substances or presentation of erotic films or pictures. The secondary rewards all involved some kind of gambling or other game where money could be won or lost. The main results, as shown in Figure 6.5, indicated that all three classes of rewards recruited most of the core reward regions, including the striatum and ventromedial frontal cortex, as well as insula, amygdala, and portions of the thalamus.

The significant overlap across these very different stimuli strongly supports the idea that neural activity in these structures underlies the hedonic value associated with both primary and secondary rewarding stimuli. However, there were some relevant differences in activation patterns between the different stimuli, especially at the cortical level. Monetary rewards uniquely recruited a region in the most anterior portion of the orbitofrontal cortex. Assuming that this response is related to the more abstract nature of monetary stimuli, as compared to food or sexual stimuli, this finding would be consistent with the idea that more cognitively complex functions are represented in the most phylogenetically recent, anterior portions of the frontal cortex (Badre and D'Esposito 2009), which is a principle we also

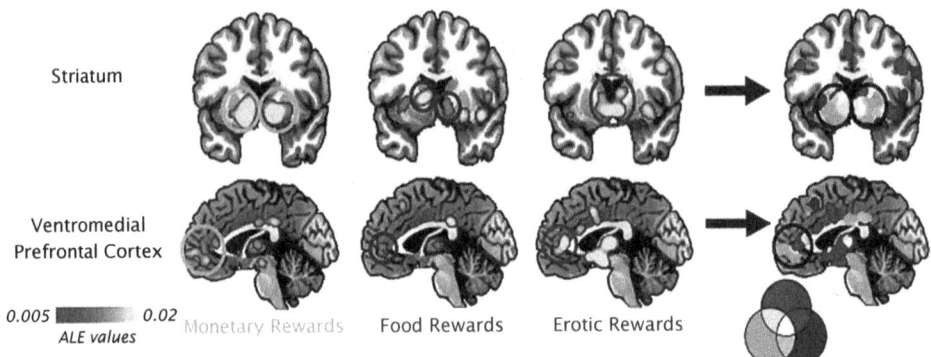

Striatum

Ventromedial
Prefrontal Cortex

0.005 ▮▮▮▮▮ 0.02
ALE values

Monetary Rewards   Food Rewards   Erotic Rewards

FIGURE 6.5 Meta-analysis of human functional imaging reward studies. Brain images show strength of response across many studies to three kinds of rewarding situations: monetary tasks (green), food stimuli (blue), and erotic images (red). The top row (coronal slices) shows the response within the striatum and the bottom row (midsagittal slices) shows the response in the ventromedial prefrontal cortex. On the far right is an overlap map color-coded according to the type of rewarding input. There is significant overlap across all three classes in the two structures. Adapted from (Sescousse et al. 2013) with permission from Elsevier.

encountered in Chapter 4, with regard to the organization of dorsolateral prefrontal cortex (Petrides 2005).

In addition to the common recruitment of reward regions, the analysis also revealed some modality-specific effects. Specifically, food rewards elicited activity in primary gustatory cortex, which makes sense since food has a taste. In contrast, erotic stimuli elicited activity in an inferior visual cortical region known as the extrastriate body area since the images contained, well, quite a few body parts. These differential responses in the relevant sensory regions suggest that although the pleasurable aspects of the different stimuli are mediated via the common reward network, each type of stimulus is initially processed via its relevant sensory pathway. Thus, these findings support the conclusion that the reward system acts in concert with different modality-specific input pathways, that are responsible for sensory analysis, as well as modality-independent memory processes and higher-order cognitive processes. Interactions between reward structures and the rest of the brain provide a good way to think about why different rewarding experiences are so experientially different from one another, and yet they all provide pleasure: it's because they share engagement of the same reward structures, yet access it via distinct pathways. We shall return to this principle in subsequent chapters, when we discuss how music engages the reward system.

The engagement of reward structures for many types of pleasurable stimuli has been consistently reported in many studies (e.g., Lindquist et al. 2016). But several questions remain, in particular with regard to the reward prediction error model. Were the reward areas activated because of the receipt of the reward or because of cues leading to the expectation of the reward? Are the same structures involved during different time points of the tasks? Many of the studies reviewed above were not attempting to distinguish between different possible phases of the prediction/reward cycle; but there are several ways to separate them. One of the most productive ways to do so experimentally is via games in which one makes a judgment in order to obtain a monetary payout. An interesting factor in such games is that, subjectively, the amount of money one wins is not the only determinant of the pleasure experienced; rather, the amount that one expects to win is very important. Thus, it feels more exciting and pleasant to win $50, if one only expected to win $10, than to win $50 if one had expected to win $100 even if, as in this example, the amount won is identical (Mellers et al. 1999). These emotional behavioral phenomena fit well with a reward prediction error account. Furthermore, these sorts of games allow the experimental manipulation of the amount expected and received, as well as the time at which the reward is delivered, making them valuable for understanding the neural responses to more complex reward situations.

One of the most common approaches in neuroimaging to study complex reward situations is the monetary incentive delay task (Knutson et al. 2000). In a typical experiment, a visual cue first appears informing the volunteer about how much money is at stake on that trial. This phase corresponds to the anticipation of reward since the individual is primed that they may be able to win (or lose) a small or large amount. The cue is then followed by a speeded response, which is adjusted such that neither wins nor losses occur on every trial, allowing the experimenter to evaluate trials containing correct or incorrect predictions. There is some debate in the literature about whether the striatum responds only to positively valenced reward (Bartra et al. 2013) or also to negative outcomes (Oldham et al. 2018). Nonetheless, studies using this monetary task consistently show that neural activity in the striatum (Knutson et al. 2001) and the midbrain reward areas (Schott et al. 2008) is elicited during the anticipation phase, before any reward is delivered, indicating that the system responds to the cue signaling that a reward will arrive, as in animal studies. Typically, such studies also show recruitment of the orbitofrontal cortex along with the striatum (Oldham et al. 2018) upon receipt of the reward, particularly if the reward is larger than expected, which would correspond to a positive reward prediction error (Knutson et al. 2003). This pattern is consistent with the evidence reviewed earlier that the orbitofrontal cortex encodes reward value upon receipt (O'Doherty et al. 2001).

It's also interesting to recognize that although recruitment of the reward system is typically accompanied by subjective reports of pleasure in human studies, such activity can also happen for subliminal stimuli that do not enter awareness or elicit any overt behavior (Childress et al. 2008). Thus, the absence of a behavioral response does not always indicate the absence of a neural response in the reward system. The precise role of such unconscious reward activation in overt behavior is not yet very well understood. But it raises interesting questions for the appreciation of complex stimuli, such as music, where one may sometimes experience certain feelings yet not be fully aware of their source.

Precision Weighting

How does the nervous system evaluate the reliability of a predictive cue? So far, we have discussed the expectation of a reward that is based on a predictive cue, such as the light in Schultz's monkey experiments that indicated that food will be available, or the symbol in the monetary tasks, indicating that a certain sum of money may be won. However, those studies did not all consider how precisely the cue actually predicts the target (that is, the variance associated with the prediction). In the real world, as opposed to in a nicely controlled experiment, it is often the case that a reward is only imperfectly predicted by some antecedent—there's some noise in the signal. If I want to eat a banana, its color provides a high-precision cue about how good it will be: green means it will be hard and inedible, yellow means it will be creamy and sweet, black spots mean it will be overripe and nasty. So, you always know what to expect before you eat it. But color provides a low-precision cue to the reward of eating a melon: it's hard to tell by just looking at it whether it will be ripe or too hard; its color may provide some information, but it's not very reliable. So, if the melon turns out to be juicy and sweet, it's a pleasant surprise. Thus, the precision with which one can make a prediction about an outcome might also impact its hedonic value.

This concept of precision weighting has important implications for our understanding of how the reward system responds. Essentially, the precision with which a cue predicts that a reward will occur represents uncertainty, or in some contexts (economics for instance) can be thought of as risk. This variable is distinct from the reward prediction error itself (which represents the difference between expected and obtained reward), yet is related to it because the precision, or degree to which a prediction can be generated, influences the error. A very precise prediction is easier to disconfirm and hence generates a larger error (positive or negative) when violated. In contrast, a fairly uncertain prediction, where various outcomes are possible, generates a smaller error. On the other hand, when the outcome is close or identical to the predicted value, there is more prediction error from the low-precision prediction, because it did not predict the outcome as well as the higher precision prediction. As it turns out, the nervous system represents both these features—uncertainty and prediction error—once the contingencies between a cue and its associated reward have been learned.

The neural firing patterns in midbrain dopamine neurons demonstrate this phenomenon quite clearly. In a landmark study, Christopher Fiorillo and colleagues presented monkeys with different visual cues which signaled how probable it was that they would receive a juice reward two seconds later (Fiorillo et al. 2003). Once the monkey had learned the relationship between cues and rewards, the experimenters varied the reliability of the cue, as shown in Figure 6.6. When the cue was 100% reliable, neural responses were elicited by the cue and not the juice, in keeping with the typical reward-prediction error idea (the reward was fully predicted, so the difference between expected and obtained reward is zero). On the other hand, when the visual cue signaled that no juice would be forthcoming there was no response either, since no reward was expected or delivered. All these results come from the phasic response of the neurons, meaning their rapid, transient activity, and conform to previous understanding of reward prediction error.

The most interesting finding in this experiment, however, came from the sustained neural responses which extended over time. The maximum sustained activity was elicited to the eventual reward at intermediate probabilities—when the cue provided some information but was not perfectly predictive—compared to lower or higher probabilities. Thus, under these conditions of maximum uncertainty, there was some response to the cue followed by a slow ramping up of activity over time as the potential reward moment approached, reaching its maximum at the point when the reward was actually delivered (or

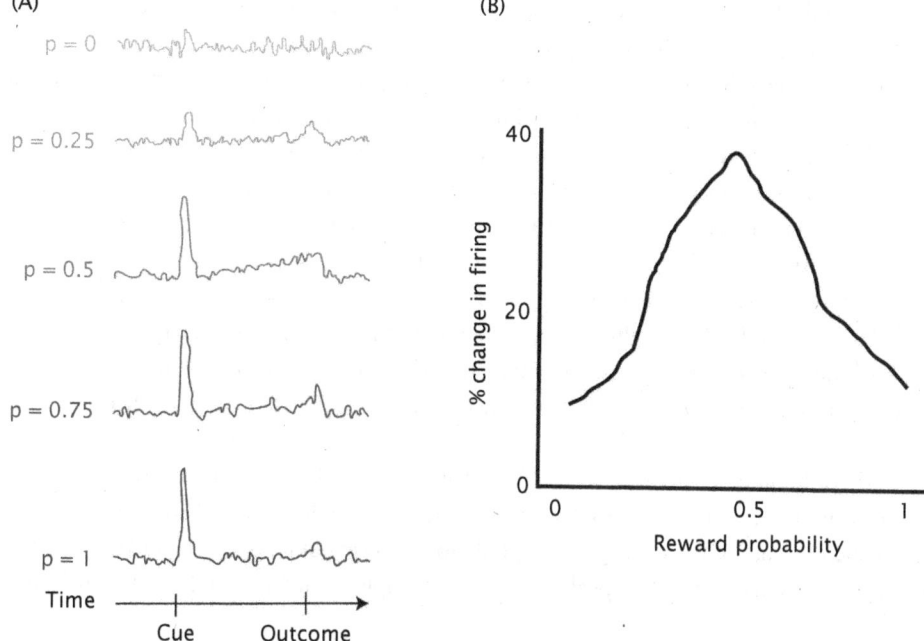

FIGURE 6.6 Activity in monkey dopamine neurons. A: Depiction of activity in a population of neurons as a function of presentation of cue or outcome, with different probabilities of reward being obtained from zero (top) to 1.0 (bottom). When the cue signals that a reward may appear, it elicits a phasic neural response whose magnitude increases linearly as the probability of reward increases. Following cue presentation there is a ramping up of activity at intermediate probabilities, peaking when the outcome occurs. Sustained response does not occur for zero or 100% probability, because the outcome was fully predicted in those cases. B: Change in neuronal firing rate for the tonic (ramping up) component, showing that peak activity is produced for maximal uncertainty (50% chance of reward), with little response at the two extremes. Adapted with permission from (Schultz et al. 1997).

not). This condition elicited a greater response than in other conditions where the predictive certainty was either higher or lower (Figure 6.6, right). In other words, this population of neurons codes for uncertainty, displaying a classic inverted-U shape function. When the animal knows for sure that a treat is either coming or not, then there is no uncertainty; but when the probability is somewhere between zero and one, then the occurrence of a reward is more notable, just as the ripe melon is all the better because it wasn't entirely certain beforehand whether it would be good or not.

This kind of encoding of uncertainty in the reward system can also be seen in human neuroimaging studies. To test the difference between expected probability of a reward versus the uncertainty of receiving it, Kerstin Preuschoff and colleagues devised a gambling game in which the player makes a bet at the start of each trial on whether a playing card drawn second will have a higher or lower number than a card drawn first (Preuschoff et al. 2006). In this situation, using only cards 1–10, if the player first draws a 1, then the probability that the second card will be higher is 100%, so that if the player made that bet, then a win is assured. But, as the number of the first card increases, the probability of winning that bet goes down linearly. On the other hand, the uncertainty of winning follows a different function: if the first card is a 1, a win is certain; if the first card is a 10, a loss is certain; and if the first card is in the middle, a 5 or a 6, a win or a loss are about equally possible. Therefore, this second value follows an inverted-U curve, rather than a straight line, because different cards give different amounts of information about what can happen.

Remarkably, functional imaging data from this study showed that both of these functions exist in the brain response, just as in the neurophysiology study described in the previous paragraph (Fiorillo et al. 2003): the early neural activity in reward structures followed a linear function for probability of reward, whereas a slightly later neural response followed the curvilinear shape that characterizes uncertainty. As shown in Figure 6.7, this latter result was particularly clear in the ventral striatum and midbrain regions, where the activity peaked when the first card was an intermediate value (representing maximum uncertainty), and activity was very low when the outcome (win or loss) was completely determined after the first card.

These findings imply that the activity in the reward system not only represents the expectation of receiving a reward, based on some prior predictor, but that it also incorporates some estimate of the quality of the predictor, in terms of its precision or the extent to which it can be trusted, so to speak (recall the example from Chapter 1, of the unreliable

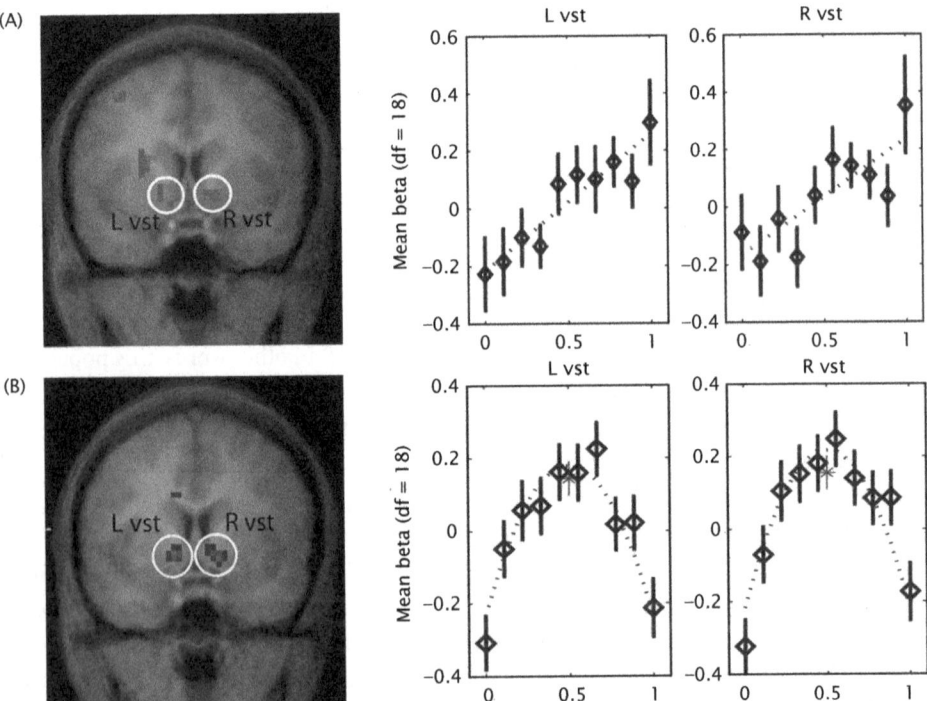

FIGURE 6.7 Reward probability and uncertainty in functional imaging. Functional MRI data collected during a card game in which reward probability and uncertainty were manipulated. A: Activity in ventral striatum (L vst and R vst on brain image, orange) followed a largely linear trend as a function of reward probability (greatest at 100% probability of winning). B: Activity in the same structures (blue) showed an inverted-U shape as a function of uncertainty (greatest at 50% probability of winning). Adapted from (Preuschoff et al. 2006) with permission from Elsevier.

information provided by the drunken sailor). Another way to think about it is that the eventual outcome in these ambiguous cases provides more information than in the case of complete determinacy, when the outcome is a foregone conclusion. Or, in the parlance of predictive coding, it provides a way to update the internal model: so the reward, if it happens in a low-precision situation, is not only rewarding because of its intrinsic properties (e.g., taste in the case of food) but also because it provides potentially useful information, such that even the absence of a primary reward can still serve as a useful signal, to enhance one's understanding of the environment and its contingencies. This relationship, between learning and the reward associated with obtaining information that promotes learning, is very powerful, and we turn our attention to that topic next.

Learning, Information-Gathering, and Curiosity

From the very beginning, research in this domain made it clear that rewards and learning were intricately intertwined. Recall that Olds and Milner's animals spontaneously returned to the location where they had last received stimulation, showing that they had learned some association between that spot and the reward experienced (even though the link between them was largely coincidental). More generally, as the behaviorist school emphasized, rewards act as reinforcers of behavior, which is a direct way that organisms can learn about how to act on their environment to survive: by trial and error, they either receive or miss out on reward and adapt accordingly. This basic idea of reinforcement learning fits well with both the reward prediction error idea and the predictive coding idea since the difference between what is expected and what actually happens upon performing an action serves as the signal that updates the model, results in learning, and changes behavior. An extensive body of evidence links dopamine activity in the reward system to this basic form of reinforcement learning (Daniel and Pollmann 2014).

This type of stimulus–response learning is important for understanding the functionality of the reward system. But it is harder to see how such a straightforward mechanism might explain the rewarding properties of more abstract types of stimuli, including music, which also engage the dopaminergic reward system, as we will detail in the next chapter. But if we think of these more complex stimuli, in the context of the information they might provide to inform our understanding of the environment and promote learning thereof, then it becomes clearer why they might engage the reward system. Put simply, the reduction of uncertainty provided by stimuli that carry information is itself rewarding.

As an example of this phenomenon, we can look at one class of stimuli, often termed social rewards. These stimuli, such as attractive faces (Aharon et al. 2001), are known to elicit activity in the ventral striatum and other reward-system structures, even if they are unfamiliar, and especially if they are gazing directly at us (Kampe et al. 2001). This response can be well accounted for by the importance of other people in our everyday lives, so that seeing their expressions reveals something about their state of mind and allows us to infer their attitudes toward us. Such information is valuable as it adds to our stockpile of knowledge, so that we can plan future actions toward those individuals (Baumeister 2005). Similarly, acquiring secondhand knowledge about other people's actions, intentions, or relationships that you were not aware of (otherwise known as gossip) can also generate reward-related brain activity (Alicart et al. 2020). Thus, simply viewing faces or hearing about others is intrinsically rewarding because of the importance of social information for navigating our complex social environment—it reduces the uncertainty about the context for our social actions.

The rewarding value of social information is a specific example of a much more general phenomenon, sometimes termed "epistemic curiosity" (Kidd and Hayden 2015). It refers to the constant seeking of information that characterizes a lot of human (and also animal) activity. Importantly, the drive to obtain information need not be linked to the delivery of any kind of primary reward—the knowledge itself is rewarding, even if it never leads to direct reward. This kind of curiosity was described by William James over a century ago as a knowledge gap that needs to be filled (James 1890). Behavioral evidence shows that people indeed value information in relation to such a gap: the difference between what they know and what they want to know or expect to learn (a kind of information prediction error, if you will). In turn, acquisition of knowledge drives learning (Marvin and Shohamy 2016) and generates pleasure (Litman 2005).

Uncertainty also plays a role here. People show a strong preference to acquire information about an uncertain upcoming event, even if the information has no effect on the eventual outcome itself. A personal anecdote can illustrate this phenomenon. In Montreal when you're waiting for a bus it's possible to consult an online app that announces updated times at which the next bus will arrive at a given stop. When the bus seems to be late, I somehow find it satisfying to keep checking the expected time of arrival, even though it does not make the bus move any more quickly and even if I still have to wait in the freezing cold just as long. This behavior is especially prominent during a snowstorm (a situation with greater environmental uncertainty and correspondingly low-precision predictions).

Several experiments have demonstrated and quantified these phenomena. For instance, people will pay to find out the outcome of a lottery, information that they would otherwise have to wait for, thereby reducing the amount of time that they have to live with the uncertainty (Bennett et al. 2016). Novelty itself seems to be a kind of shortcut that the reward system uses to determine whether a given stimulus needs to be investigated (Wittmann et al. 2008). Information-seeking also becomes more relevant as a function of curiosity—the informational gap, once again. Thus, people are willing to spend more resources to find out answers to questions that they are especially curious about, even for trivia questions that have no particular significance to their everyday lives (Kang et al. 2009). Curiosity can indeed be a potent driver of behavior. I wouldn't have written this book—and you, dear reader, probably would not be reading it—were it not for excessive epistemic curiosity about music on both our parts.

Ample evidence implicates the reward system in the acquisition of information, irrespective of its usefulness in making a decision for a given task. Even monkeys demonstrate the value of obtaining information: when a variable amount of water is provided to them, they seek advance information about how much water they will get each time. Most critically, their dopamine midbrain neurons show activity that parallels their behavior (Bromberg-Martin and Hikosaka 2009). These are the same neurons that respond to primary rewards, or to cues signaling their presence, thus showing that the reward system is engaged for these more complex cognitive functions, even in monkeys.

Human studies have also shown the interaction between the reward system and curiosity. In experiments using a trivia game format, brain activity in the striatum was detected after the presentation of the question (Kang et al. 2009)—a kind of anticipatory response, reminiscent of what we have already seen in many other experimental contexts. The degree to which striatum and midbrain dopamine regions are activated also scales as a function of the degree of curiosity expressed by the individual, so that the questions one is more curious about engage those areas to a greater extent (Gruber et al. 2014), as seen in Figure 6.8. In this way, curiosity acts as a kind of multiplier, raising the reward value of a given stimulus. In addition, uncertainty also plays a big role in these circumstances, in accord with the predictive coding idea and the fact that responses to the rewarding stimulus are maximal under uncertainty, but minimal when outcomes are determined, as discussed earlier (Preuschoff et al. 2006). Thus, even in a trivia game, if the answers to only some questions are to be revealed, but the player does not know which ones those are, there is much greater striatal response when the answer is revealed compared to a situation where the player knows that their curiosity will always be satisfied for every question (Ligneul et al. 2018).

In the context of the reward system's involvement in learning, it is important to note that information which triggers reward activity is better remembered later, an effect which emerges from interactions between reward and hippocampal systems (Adcock et al. 2006). This phenomenon was nicely shown for the abstract reward of knowledge gained by my colleague Pablo Ripollés and the team in Barcelona, who measured brain activity while people were reading a passage that contained a new, unknown word (Ripollés et al. 2014). When the reader eventually figured out the meaning of the word from the context, activation was seen in the ventral striatum; and the degree of retention of the new word was linked to the functional interactions between the striatum and the hippocampus (Ripollés et al. 2016). A similar effect was demonstrated in the trivia game mentioned above (Gruber et al. 2014).

Therefore, just like any other reward that fulfills a need (such as water when one is thirsty), receiving information that one is curious about is intrinsically pleasurable and facilitates learning, especially under uncertain conditions. It's no wonder that people talk about a "thirst for knowledge"! It's important to emphasize that in many of these studies, no primary, or even secondary, reward is ever received. This fact provides us with a critical clue for understanding even more complex situations, such as aesthetic rewards, as we will discuss in subsequent chapters, where the link to actual physical rewards is essentially absent. Furthermore, inciting curiosity by enhancing reward value also leads to deeper encoding and better memory. This effect on memory is consistent with the observation that the gap between information expected and information received is a better predictor of learning than the absolute value of the information itself (Marvin and Shohamy 2016).

In these experiments, curiosity varies for every individual on an item-by-item basis (e.g., I might be more curious about the sound of an obscure organ than about how much money a hockey player earns, while other people would have the reverse tendency). But one can also characterize curiosity as a global trait rather than being linked to a specific type of content. Indeed, personality theories often include curiosity-related factors that have to do

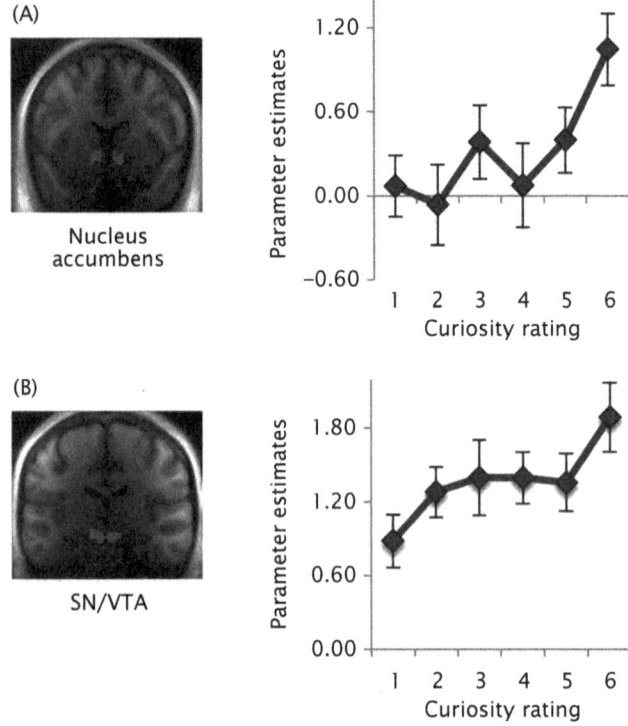

FIGURE 6.8 Brain activity in reward regions modulated by curiosity. Functional MRI data collected during a trivia question and answer game showed that activity in the nucleus accumbens (A) and substantia nigra/ventral tegmental region (B) increased as volunteers indicated greater curiosity about the answer to the question. Reprinted from (Gruber et al., 2014) with permission from Elsevier.

with features such as novelty-seeking, openness to experience, or extraversion. There are several inter-related constructs that describe stable individual differences in the degree to which people are open to exploration and to take risks in order to obtain rewards (Segarra et al. 2014). But most relevant for us is that these variations have also been linked to dopaminergic function (Depue and Collins 1999, Smillie 2008).

People with higher scores on sensitivity to reward scales (that is, those who respond more to rewarding events than average) show greater responses in the striatum and orbitofrontal region in the monetary incentive delay task described above (Simon et al. 2010). Greater striatal responses have also been reported in various reward tasks as a function of novelty-seeking (Wittmann et al. 2008) and openness (Passamonti et al. 2015). In the drug studies mentioned earlier, the degree of response to amphetamine in the reward system was also linked to personality traits such as novelty seeking (Leyton et al. 2002). Even in a resting state scan in which no task is presented, functional connectivity between reward-related regions is greater in people with higher sensitivity to reward (Adrián-Ventura et al. 2019). So, these individual personality differences play an important role in how any given individual's reward system will respond under different circumstances. These findings will be revisited in Chapter 8, when we discuss how sensitivity to musical reward can differ across individuals.

Liking Versus Wanting

As we have seen throughout this chapter, our experience of pleasure from different sorts of events or stimuli is strongly tied to the functionality of the reward circuitry. However, a distinction can be drawn between motivational as opposed to hedonic components of reward. This concept has been most developed by Kent Berridge and his associates at the University of Michigan, who have done extensive work on the neurobiology of the reward system, using a rodent model of response to sweet liquid (Berridge 2019). In these sorts of experiments, "wanting" is characterized by approach behaviors or actions that are needed to obtain the reward, whereas "liking" is indexed by stereotyped facial expressions that are characteristic of sweet taste, such as lip-smacking and relaxed facial muscles (expressions also made by human babies when they taste something sweet). These two behaviors usually occur in order, with wanting preceding liking, as the animal must first show motivated behavior in order to subsequently obtain the reward. In a more human cognitive context, we might equate wanting with desire (or even craving), as well as seeking out rewards or paying money to obtain them; whereas liking would be more related to the hedonic experience of pleasure, or even happiness, that results from obtaining the desired item. This conceptual framework has some important consequences for our interpretation of the findings from the literature.

Perhaps the most important impact of the dissociation between wanting and liking for our purposes is that it challenges the assumption that dopamine is directly involved in the hedonic experience of pleasure. For a number of years, it was thought that dopamine was directly involved in generating positive affective states (Wise 1980). And, as a corollary to that idea, it was also thought that people who experienced a generalized lack of pleasure (anhedonia) suffered from a dysfunction of the dopaminergic reward system (Smith 1995). However, according to more recent models, dopamine does not directly cause pleasure. Instead, it modulates wanting (that is, motivational responses) as seen, for instance, in the type of anhedonia that is characteristic of certain disorders, such as Parkinson's disease or depression (Treadway and Zald 2011). In many such cases, there is evidence that even though their dopamine system is depleted, Parkinson's patients are still capable of experiencing pleasure when a rewarding stimulus, such as a sweet taste, is given to them (Sienkiewicz-Jarosz et al. 2013); but they lack the interest or incentive required to obtain the reward.

The clearest evidence that dissociates the two components comes from a series of carefully conducted experimental studies in rodents, showing that depletion of dopamine via chemical lesion had no effect on the facial expressions associated with liking sweetness, but that it did eliminate wanting to eat (Berridge et al. 1989). Conversely, electrical stimulation of dopaminergic reward structures led to a four-fold increase in eating, but no change in licking behavior (Berridge and Valenstein 1991). Similarly, in human studies, blocking dopamine did not consistently alter the euphoric effect of amphetamine (Brauer and De Wit 1997), nor did pharmacologically increasing dopamine result in elevated mood (Liggins et al. 2012). Instead, manipulating dopamine seems to change the desire to obtain drugs, or other rewarding substances, without changing the hedonic experience elicited by the outcome, in line with the idea that dopamine primarily affects the wanting aspect of reward. It should be noted, however, that some studies have reported an association between the degree of dopamine release and subjective ratings of euphoria to amphetamine (Drevets et al. 2001) or to the pleasantness of food (Small et al. 2003). These findings leave open the question of how dopamine affects subjective pleasure and whether this relationship may be mediated by interactions with other neurochemicals in addition to dopamine.

Even the findings from stimulation of so-called pleasure centers may need to be reinterpreted as perhaps more reflective of changes to wanting than to the experience of pleasure, per se, as indicated by verbal reports from a few rare human patients with implanted electrodes in the ventral striatum. Rather than experiencing a hedonic or pleasurable state per se upon stimulation, they often report a desire or motivation to act in order to obtain a reward. For example, in one study, when the electrodes were switched on, patients reported specific desires, such as wanting to visit a famous cathedral, or a wish to go bowling (Schlaepfer et al. 2008). On the other hand, in other individuals, stimulation of regions close to the ventral striatum does sometimes result in signs of pleasurable emotions, such as smiling or even laughter, as well as verbal reports of euphoria or giddiness (Haq et al. 2011). All these reports must be interpreted cautiously, as the regions being stimulated are not always precisely documented, nor is the spread of activation from the stimulated site well understood. Additionally, since the patients tested were suffering from psychiatric or neurological disorders, their responses may not be representative of a normal, functioning system. Furthermore, the complex cascade of neurotransmitters that may be released by the stimulation is largely unknown. Nonetheless, these observations do suggest that directly stimulating the reward system may, under some but not all circumstances, result in an experience of pleasure whether it is mediated by dopamine or not.

One outcome about which there is more agreement is that striatal stimulation often alleviates the symptoms of depression (Bewernick et al. 2010), including anhedonia (Schlaepfer et al. 2008). These sorts of studies are important not only for their potential clinical benefit to patients, but also on a theoretical level. It's not entirely clear to what extent liking and/or wanting responses are modulated under such circumstances, nor what the precise role of dopamine might be. But these long-term beneficial effects of brain stimulation lend support for the role of the striatum in regulation and expression of pleasurable experiences and positive mood. This topic will be considered further in the context of music in Chapter 9.

Dopamine Versus Opioids

If it is indeed the case that dopamine does not necessarily cause pleasure directly, what is the chemical pathway by which pleasure is experienced? One line of evidence indicates that structures within the reward system contain hedonic "hot spots" and "cold spots," which are responsible for hedonic reactions and respond to opioids. Berridge has described them as "anatomically small pleasure-generating islands of brain tissue, tucked within larger limbic structures" (Berridge 2018). They are distributed throughout the reward system, including the ventral striatum, brainstem, and orbitofrontal regions. When these hot spots are stimulated with opioids, or other compounds such as endocannabinoids, they double or triple the rates of liking responses to sweet taste, as defined, again, by the animal's facial expression. Yet, when the same tiny structures are stimulated with dopamine, they do not elicit such responses (Berridge and Kringelbach 2015). Although these hot spots are distributed across several brain structures, they appear to act in concert to generate pleasure responses because when one spot is stimulated, the others respond as well (Castro and Berridge 2017). Interestingly, the so-called cold spots, which are distributed in topographically distinct locations, respond to opioids in exactly the opposite manner—suppressing liking responses or even generating avoidance, fear, and disgust responses to the same stimuli. Thus, although these small regions may be spatially adjacent, they show very different outcomes when stimulated.

The behavioral outcomes associated with opioid stimulation of these hot spots would appear, at first glance, to represent a fixed, mechanical kind of affective response. However, a series of experiments shows that, rather than being strictly determined according to their anatomical location, these responses depend on context. For example, in a calm and safe environment, such as the animal's home nest, which is dark and quiet, the responses of the different hotspots to stimulation tend more toward positive affect. However, when the animal is placed in a stressful environment with bright lights and loud rock music (which would stress me out too), then the same regions that previously generated positive responses when stimulated will produce outward signs of fear and loathing (Reynolds and Berridge 2008). These contextual modulations of the affective responses are most likely mediated by top-down control signals coming from frontal cortex. As we reviewed above, the striatum receives inputs from orbitofrontal and other frontal cortices. When these areas are excited, they modulate both the valence (positive versus negative) as well as the intensity of the response from the hotpots in the striatum (Richard and Berridge 2013). These context effects may also be thought of in terms of uncertainty, as discussed earlier: stressful environments are more uncertain than calm ones, where events are more predictable.

Human studies also show evidence for opioid-mediated pleasure. For example, after administration of the opioid antagonist naltrexone, liking ratings for a sweet-tasting liquid decreased; conversely, morphine, an opioid agonist, increased liking ratings; these effects were seen only for the most desired items however (Eikemo et al. 2016). This finding parallels what has been observed with rats (Cooper and Turkish 1989). Similarly, pharmacological manipulation of the opioid systems results in changes to ratings of attractive faces, but again, for the most highly attractive stimuli (Chelnokova et al. 2014). A study combining opioid antagonists with functional imaging reported that blocking opioid receptors reduced the subjective pleasure associated with viewing erotic pictures and that this occurred to a greater extent than for the reward associated with the monetary incentive delay task (Büchel et al. 2018). Of greatest interest in that study, a reduction of activation in the ventral striatum, orbitofrontal cortex and medial prefrontal cortex was observed after the drug was given. This effect correlated with the individual degree of pleasure modulation, showing that the reward system mediated the interaction between task and opioid manipulation.

All these findings, from humans and rodents, implicate the opioid system, rather than dopamine neurons, as being responsible for hedonic responses that we call "pleasure." However, the situation may be more complicated than it appears because these neurochemicals interact with one another in complex ways. For example, administration of drugs such as amphetamine not only results in dopamine release in the striatum but also promotes the release of endogenous opioids, such as endorphins, both in rats (Olive et al. 2001) and humans (Colasanti et al. 2012). It's possible that the euphoric response associated with these drugs is secondarily mediated by opioids in such situations. But even if that is the case, dopamine would still play an important role in promoting their release. In practice, it is very difficult to determine whether a pleasurable response is exclusively due to the action of one or another chemical system or due to some interplay between them.

There is another particularly important caveat in interpreting the findings about dopamine versus opioids in relation to the hedonic experience of pleasure. Almost the entire literature that has so elegantly succeeded in dissociating these processes from each other is based upon the use of primary rewards (or drugs) in relatively simple experimental contexts. These are typically stimuli that elicit innate responses—such as food, water, or sexual stimuli—that do not require much, if any, learning. There is a good reason for this choice of stimuli and paradigms in those studies: they afford excellent control over all the necessary features that enable the distinction to be made, based on well-defined behavioral responses. But it is not certain whether the conclusions drawn from this well-controlled, but relatively narrow, experimental approach apply in exactly the same way to more complex rewards, of the type that interest us in the context of music and aesthetic responses. And of course, it is not possible to study these abstract aesthetic rewards in an animal model.

A critical distinction in this respect is that abstract rewards do not depend solely on the activity profiles within the reward system but rather emerge from interactions between the reward system and higher-order cognitive, mnemonic, and perceptual systems. The thirsty mouse is motivated to obtain water because of a homeostatic imbalance (i.e., it's thirsty). Its hedonic experience of sipping water does not particularly depend on statistical learning of past experiences with the precise pattern of drops and their timing as the water is delivered from the feeding spout. Aesthetic rewards instead depend to a great extent on prior statistical learning of abstract patterns and the contingencies and expectancies that emerge from such learning. All these responses are largely mediated via the cortico-cortical loops in the ventral and dorsal streams, described in Part I of this book. And much of this auditory/cognitive cortical hardware is phylogenetically recent and hence not well-developed in rodents. Therefore, these considerations need to be kept in mind as we attempt to understand both the basic functionality of the reward system and also how it may respond to abstract, nonprimary rewards in the human brain. We will pick up this thread again in the next chapter, where we directly address the role of dopamine and the reward system in the experience of musical pleasure.

Reprise

The reward system is of prime importance for understanding cognition and behavior, as it is responsible for many aspects of motivation, emotion, and memory. Furthermore, its activity is directly tied to subjective feelings of pleasure. Anatomically, it consists of a set of interconnected structures spanning midbrain, striatum, amygdala, and ventromedial/orbit-ofrontal cortex as well as several other structures (see Figure 6.2). The striatum specifically receives highly processed and topographically organized inputs from the rest of the brain, including via connections from the frontal lobe. These frontal cortical regions are also the termination points for the auditory ventral and dorsal streams that were discussed in Chapters 3 and 4. As such, they constitute the critical link between auditory processing networks and the reward system. More specifically, the auditory ventral stream provides inputs to the ventral striatum related to abstract patterns of sounds and their contingencies (*what* event to expect), whereas the auditory dorsal stream feeds information related to temporal patterns and expectancies in time (*when* to expect the event) to the dorsal striatum.

The subjective value of a stimulus is coded within the orbitofrontal cortex in a modality-independent manner. Thus, very different types of stimuli or events generate a common response, allowing comparisons and decisions to be made about behaviors that lead to favorable outcomes. More generally, the entire reward circuit responds to highly valued stimuli of many types, ranging from innate, survival-related primary rewards (such as food) to secondary rewards that acquire value through learning or association (such as money) and then to more abstract stimuli or events (such as information or aesthetic stimuli). Depending on the nature and modality of the rewarding stimulus, there is co-activation of the reward system with higher-order modality-specific cortical areas. These interactions across systems serve to distinguish dissimilar events that have similar reward value from one another.

The reward system is intimately tied to learning since positive outcomes from behaviors generate rewards, which act as learning signals or reinforcers. Activity of dopamine neurons in the striatum and other reward regions is linked to the expectation of future delivery of reward. This phenomenon can be formalized as reward prediction error—if the reward value is exactly as predicted, then there is zero error and no response to the reward; but if the reward is better or worse than predicted, then there is a positive or negative error, respectively, leading to an enhancement or inhibition of the response to the rewarding item. Reward prediction error, therefore, serves to update internal models of goal states, leading, in principle, to optimization of behavior.

Responses within the reward system are also weighted according to how precisely a cue predicts the reward. Imprecise predictors yield greater response to the eventual reward because the outcome in such cases provides more information than when it is more uncertain. Information reduces uncertainty and engages reward structures, even in the absence of any primary or secondary rewarding stimulus. Curiosity—the gap between what one knows and what one wants to know—amplifies reward responses. Individual personality differences in information seeking also modulate recruitment of the reward system.

A distinction may be drawn between the motivation to seek out or act in order to obtain a reward, termed "wanting," and the hedonic experience associated with receiving or consuming the reward, termed "liking." Animal studies indicate that the modulation of dopamine neurons in the reward system is primarily associated with the wanting phase, while opioids are largely responsible for the liking component. It remains to be seen if this model applies in the same way to abstract or aesthetic rewards that do not depend directly on innate, biologically determined drives and that, instead, require interactions with perceptual, mnemonic, and cognitive systems.

7

Music Recruits
the Reward System

A few years ago, I was on a plane, absent-mindedly thumbing through the in-flight magazine (they still had those then), when I came across a full-page ad for some headphones. I was astounded to see that the ad mentioned how listening to music would result in great pleasure because of the release of dopamine and that this was a good reason to buy their product. It made me realize that the idea of music and dopamine had become part of popular culture, for better or for worse. But it was not always so.

Indeed, for many years, studying music-induced pleasure or emotion was not a mainstream scientific pursuit. Our lab, and those of several other colleagues, had spent decades working on the more basic aspects of music processing in the brain, especially related to perception and production (as presented in Part I of this book). But I, for one, had tended to shy away from anything to do with emotion induced by music to avoid veering too far into terrain that might be considered esoteric. Yet, the persistent question remained: "Why do we love music?" And none of our work on the functional properties of the auditory cortex could answer it.

This situation changed in the 1990s, due to several factors. First, there were earlier advances in the psychology of music that not only broke ground on important aspects of music cognition but also developed powerful behavioral methodologies that could be applied to study it (Deutsch 1982). Second, the extensive knowledge generated from research by many groups, including ours, about the neural basis of music perception and production provided a platform upon which to expand into the affective realm of music. Finally, during this decade, great empirical and theoretical strides were being made in the neuroscience of emotion in general (Damasio 1994, Ekman and Davidson 1994, LeDoux 1998), making it more tractable to study and also lending it scientific respectability. All of these developments led to the emergence of a new science of music and emotion (Juslin and Sloboda 2001). These advances also nicely dovetailed with the many physiological explorations of the reward system, as discussed in the previous chapter. Of course, most of

From Perception to Pleasure. Robert Zatorre, Oxford University Press. © Oxford University Press 2024.
DOI: 10.1093/oso/9780197558287.003.0007

this latter research dealt with primary or basic rewards. As such, we could not take it for granted that abstract patterns of sounds would yield similar responses to those elicited by food or money. Indeed, certain philosophical traditions argued strongly against it on the grounds that aesthetic emotions were somehow different from quotidian "utilitarian" emotions (Skov and Nadal 2020). Nonetheless, all this prior research set the stage for us to tackle the potentially more complex situation exemplified by music.

Neuroimaging Studies on Music and the Reward System

Early Studies

It was fortuitous timing then that around this time, Anne Blood, now a researcher at Harvard, joined my lab as a postdoc. Anne had a strong interest in music and emotion and insisted that we should take the plunge into that area. But when we initially discussed how to approach the problem, we were faced with two difficult issues. First, we realized that it would be tricky to elicit consistent affective responses to a given musical stimulus for the simple reason that people's tastes in music vary a great deal. And so, if we wanted to be able to examine responses that were in common across individuals, we would have to find a way to deal with that problem. Second, we did not want to rely solely on verbal self-report about emotional experiences, because they are entirely subjective and could prove to be unreliable. We needed to convince potential skeptics that we were doing rigorous science! Therefore, we needed some objective index that could serve to validate what people might say about their musical experience.

To tackle the first of these issues, we initially came up with the idea that it might be easier to elicit consistent displeasure by introducing dissonance into an otherwise conventional tonal melody, rather than try to elicit agreement on pleasurable responses (Blood et al. 1999). The processing of dissonance is a complex topic, both in terms of its psychoacoustics (McDermott et al. 2010) as well as in music theory. However, we applied it in a very simple context. With the help of our musician colleague Christine Beckett, we parametrically varied the dissonance of chords accompanying a simple tonal melody. Our listeners gave us systematic ratings, in accordance with our expectations, so that as progressively more dissonant tones were added, they judged the music to become more unpleasant. As such, their responses validated the approach, allowing us to move the experiment to the scanner.

The brain imaging data showed a clear principal result: as the stimuli changed from more consonant to more dissonant, brain activity in the parahippocampal gyrus increased, thus implicating this structure in the processing of negative affect. This conclusion was supported by subsequent functional imaging studies (Koelsch et al. 2006) and by studies that showed that patients with excision in this region did not find highly dissonant music particularly unpleasant (Gosselin et al. 2006). But the more intriguing finding in our brain imaging data came from examination of the opposite pattern of activity: the response associated with increasing consonance, which showed a correlated increase in orbitofrontal cortex and frontal pole. Therefore, more activity in those regions was associated with greater consonance and greater pleasure. This finding is well in keeping with the literature reviewed in Chapter 6 regarding the role of the orbitofrontal cortex in valuation of diverse stimuli with greater activity being associated with preferred items (Small et al. 2001, Kringelbach 2005, Kühn and Gallinat 2012). Thus, for the first time, this experiment linked at least one component of the reward system to musical pleasure and therefore encouraged us to go further.

We realized, however, that we had only probed one side of the pleasure continuum—the negative side—since the pleasant music was indeed reasonably pleasant but only in relation to the dissonant version. Even the most consonant version of our stimulus was a bit bland and hardly amounted to anything like real music that one would enjoy and want to listen to again. So, what we really needed to do was to go to the other extreme and look at the very positive, intense pleasure that we all know to be a central aspect of the musical experience. We reasoned that, if we could find a way to elicit the near-ecstatic feeling that music can sometimes produce, we would really be able to test the role of the reward system in mediating such experiences.

Chills and Liking

To tackle this question, we needed to solve the problem that different people may experience pleasure from very different music. Anne was a drummer in a rock band, whereas I, trained as a baroque organist, had quite different musical preferences. At first, we were discouraged that it would prove impossible to figure out how to design the right experiment. And yet as we thought it through, at some point we realized that (despite our opposite favorite genres) we both experienced a similar and tremendously powerful pleasure from our own preferred music and that this experience was accompanied by the presence of highly pleasurable "chills". This was one of those "aha" moments, leading to a simultaneous solution to the two problems mentioned above: how to get both consistent and objective responses associated with pleasure. Rather than focusing on finding the music that would elicit pleasure in everyone, we turned it around, focusing on the spontaneous individual pleasure response elicited by letting people choose their own music. This approach also has the advantage of permitting a much more naturalistic experience because we could use real music instead of laboratory-synthesized sound patterns.

There already existed some valuable behavioral research indicating that peak pleasure is associated with musical chills (also sometimes referred to as "frisson," "shivers," or "goose bumps"). Questionnaire-based studies (Sloboda 1991, Panksepp 1995) showed that although not experienced by everyone, chills do occur with some regularity in most people and are most common among those who have some degree of musical sophistication. These studies also found that chills are almost always experienced as highly pleasurable, that they often occur at consistent time points in the music, and that they are often linked to surprising moments, such as an unexpected harmony, sudden change in musical texture or loudness, or the entrance of a new voice (Sloboda 1991). These sorts of reports are entirely subjective, and hence don't entirely fulfill our need for objective, quantifiable indices of the pleasure experience. However, they do provide a good rationale for the use of chills as an index of pleasure.

In order to address the need for objective measures, we took advantage of the fact that chills are accompanied by psychophysiological responses. These include changes in heart rate, respiration, skin conductance, skin temperature, and other measures that are associated with engagement of the autonomic nervous system. We used these measures because they cannot be easily faked, thus ensuring objectivity. They can also be plotted as a function of time, allowing us to see the temporal evolution of the experience in a quantitative manner. Such measures had already been shown to correlate with emotional responses to music in several other studies (Krumhansl 1997, Grewe et al. 2007). In our lab, we were able to show that peaks in these physiological variables were quite well-correlated with the peak self-reported pleasure associated with chills (Salimpoor et al. 2009): heart rate, skin conductance, and respiration all go up, while skin temperature goes down (due to constriction of blood vessels on the skin surface, and that's also why they are called "chills," after all), as shown in Figure 7.1. Finally, because the chill events tend to occur at specific moments in the music, it enabled us to select the right musical excerpts to present in the scanner during the most likely time that volunteers would experience high pleasure. This way, we ensured that we could capture the real peak experience during the limited timeframe of the brain imaging protocol.

There remained one problem: if we used different music for different listeners, how could we ensure that any neural responses that we identified as representing chills were not actually due to some other, simpler factor instead, such as the acoustical features of the music that often accompany, or even generate, chills in the first place (a new instrumental timbre, a chord change, etc.)? To solve this issue, we capitalized on our insight about differing musical tastes: I could listen to heavy metal all day and not have a chill, even though Anne would likely have many of them. Therefore, the sound content of the music can be dissociated from its effects on the brain by comparing across different listeners. Thus, we instituted a pairing procedure, whereby each listener in the study would bring in their favorite music and would also choose a piece of music from another listener's list that they found more or less neutral to serve as a control. In this way, we could ensure that the acoustical features would be perfectly controlled across the entire pool of people. In addition, we only used instrumental pieces, to make sure that any emotion induced was not due to the words of a song (with pieces ranging from a Beethoven piano sonata to jazz, tango, and even bagpipes!).

The outcome of this neuroimaging study was pretty clear (Blood and Zatorre 2001). First, listeners reported experiencing chills to their favorite pieces 77% of the time; this is not a trivial outcome, because listening to music while being immobile inside a positron emission tomography (PET) device with an IV in your arm, is not exactly as conducive to pleasure as compared to a concert hall or sitting in your favorite easy chair. Second, the psychophysiology validated individual subjective reports—skin conductance, muscle tension, and respiration—were all significantly higher during the chills-inducing music than the control condition. Third, and most important, when we correlated the magnitude of brain activity as a function of degree of subjective pleasure (see Figure 7.2), we saw significant brain activity in the major components of the reward system (striatum, orbitofrontal cortex, midbrain) and a few other related structures (anterior cingulate, insula). Thus, for the first time, we could confidently conclude that music, despite its abstract nature, does indeed

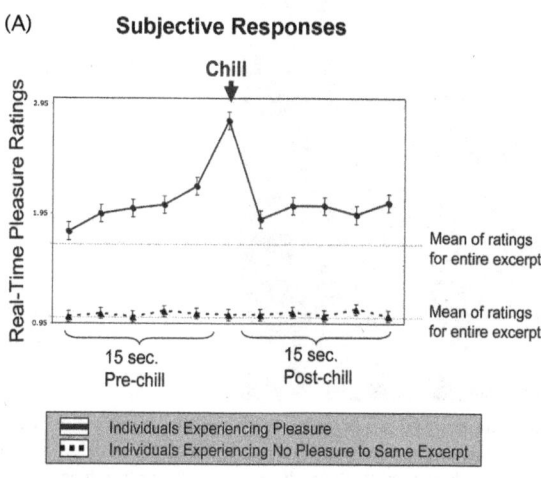

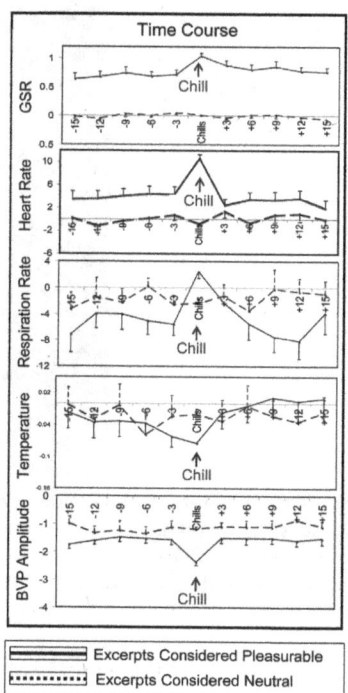

FIGURE 7.1 **Psychophysiological responses to musical chills.** A: Real-time subjective pleasure ratings as a function of the time at which chills response occurred reveal that chills correspond to the peak of pleasure ratings (solid line). Individuals who did not experience pleasure to the same excerpts showed no increases in pleasure during the same epochs (dotted line). B: Real-time physiological recordings as a function of the time at which the subjective chills response occurred reveal that chills are associated with peaks in autonomic nervous system activity (solid lines). Individuals who did not experience pleasure to the same excerpts did not show significant changes in psychophysiological responses during the same epochs (dashed lines). Abbreviations: GSR: galvanic skin response (electrodermal activity); BVP: blood volume pulse. Adapted from (Salimpoor et al. 2009). Reproduced under the terms of a Creative Commons Attribution license.

recruit the classic reward system, a finding which really opened the door to an entire new set of questions.

Several subsequent studies over the following years confirmed and expanded on the engagement of the core reward structures by music. Many of these studies used functional magnetic resonance imaging (fMRI), which has higher spatial resolution than our initial studies using PET (which yields blurrier images), and adopted several different paradigms, allowing to test the generality of the phenomenon. Thus, activity in the striatum and other reward structures was reported in studies comparing liked versus disliked music (Montag et al. 2011, Brattico et al. 2016); comparing pleasant music versus music rendered unpleasant by randomizing some of the tones to generate dissonance (Menon and Levitin 2005, Koelsch et al. 2006, Mueller et al. 2015); measuring correlations with emotional valence ratings (Mitterschiffthaler et al. 2007); and comparing intact music to musical segments scrambled at different timescales (Farbood et al. 2015).

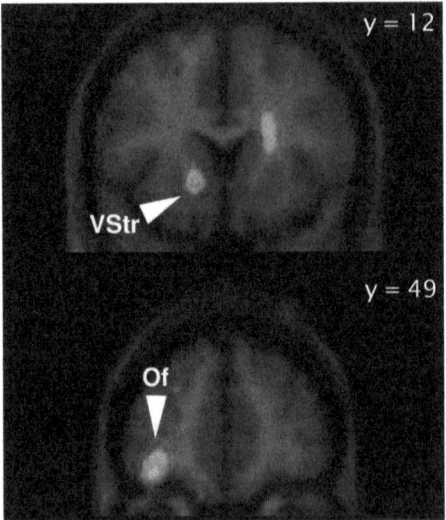

FIGURE 7.2 Cerebral blood flow increases in reward structures with pleasant music. Coronal sections showing positive correlation of blood flow in the ventral striatum (top) and in the orbitofrontal cortex (bottom) with increasing intensity of music-induced chills. Reproduced with permission from (Blood and Zatorre 2001). Copyright (2001) National Academy of Sciences.

These numerous studies varied in terms of the musical training of the persons tested, indicating that this variable was not critical to striatal recruitment. They also demonstrated that chills per se are not necessary for the engagement of the reward system, since the activation pattern was observed even in their absence, during milder pleasure experiences. This latter finding addressed the concern that perhaps our earlier results were limited only to the very particular experience of chills. Had that been the case, then the findings would not be especially representative of the everyday experience of enjoying music nor be typical of all listeners, since some of them never get chills even though they enjoy music. As we had shown, chills represent a peak pleasure moment and hence are associated with particularly strong activity in the reward system (Blood and Zatorre 2001). But these additional findings show that they are by no means required for the reward structures to be recruited.

This conclusion was further bolstered by another experiment carried out by my former PhD student Valorie Salimpoor (Salimpoor et al. 2013). Valorie was interested in addressing several outstanding questions about music-induced pleasure. The first one was whether we could move away from the idea of peak pleasure/chills and instead deal with the more normal experience of simple enjoyment of music. Second was the issue of whether familiarity played an essential role in recruitment of the striatum and associated structures. It is well-established that as music becomes more familiar, it also generally becomes better liked (Peretz et al. 1998). But using familiar music as many prior studies had done, including our own, raised questions about how memory may interact with hedonic responses (Pereira et al. 2011). To overcome both these issues while retaining ecological validity, Valorie came up with the idea of using a purchasing paradigm, modeled after the way in which iTunes sold music clips at the time. We adopted a well-validated behavioral economics protocol (Becker et al. 1964) frequently used to estimate value or utility. In this task people are given a fixed budget and then make bids for each item, as at an auction, revealing how much they are willing to pay to receive a given item. This method also got around individual differences in preference for any given musical excerpt since we modeled the dollar amount to each stimulus and to each listener separately.

So, in this study, you would listen to a brief musical excerpt and decide how much money you would be prepared to spend to obtain the full recording and take it home with you; the dollar amount represents the assigned value (including zero, for disliked music). To limit the music-selection search space, we sought out listeners who favored indie and electronica music genres, which were popular at the time on the McGill campus. We then used music recommendation software to select music clips within these genres from very recently posted music web sites, making it very unlikely that anyone would be familiar with them. As predicted by previous data, an analysis using multivariate regression showed that only the ventral striatum correlated significantly with the reward value, as estimated from the monetary bid. Thus, as shown in Figure 7.3(A and B), even with music not previously

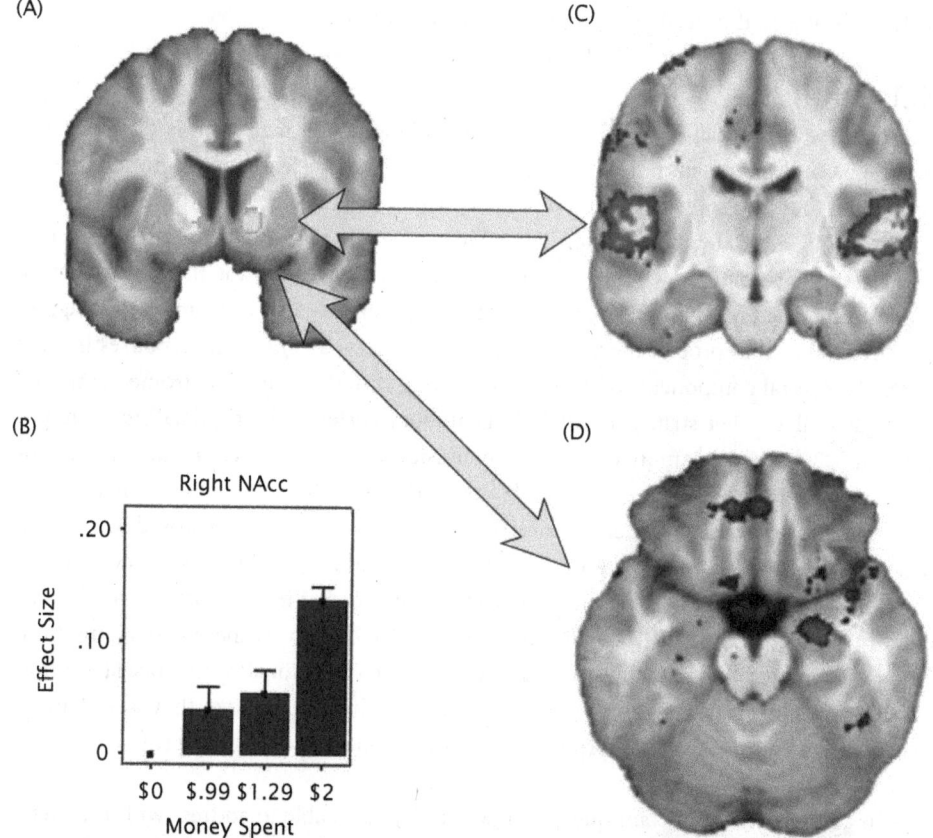

FIGURE 7.3 Modulation of functional connectivity with increasing music monetary value. A: functional MRI signal in the right ventral striatum increases as a function of value assigned to music, as indexed by monetary bids made to each excerpt (shown in B). C and D: Brain regions whose functional connectivity with the ventral striatum increases as a function of increasing value (yellow arrows), including the auditory cortex (C) and the orbitofrontal cortex and amygdala (D). Abbreviation: Nacc Nucleus Accumbens. Adapted with permission from (Salimpoor et al. 2013).

heard and without necessarily evoking chills, the reward-system response is present for music that one wants to hear again and scales with the value assigned to it by the listener.

These findings align very well with research coming from the domain of economics and consumer studies. Gregory Berns and Sara Moore asked adolescents to listen to new, unfamiliar pieces of music and rate how much they liked them; at the same time, they also took measures of brain activity with fMRI while listening to the music. Three years later, they looked at sales data from these songs to see which ones had become best sellers (Berns and Moore 2012). The song ratings provided by the original listeners did not predict which songs went on to become hits. But, remarkably, the activity in the ventral striatum and ventromedial prefrontal cortex of the teenagers who first heard those songs three years earlier correlated significantly with subsequent sales. In other words, their brains were able to predict which songs would become hits better than their own subjective liking, as shown in Figure 7.4. This result not only validates all of the experimental findings about the role of the ventral striatum in musical enjoyment reviewed in this section, but it also demonstrates that the activity of the reward system really does reflect real-world phenomena and is not solely applicable to the sterile environment of the laboratory.

Meta-analysis

Each of these experiments from different labs looking at the neural basis of music and pleasure has advantages and disadvantages. Collectively, they clearly point in the same direction. But they also show some differences in the precise pattern of activity detected. How do we formalize the degree to which the findings actually overlap? Meta-analyses across multiple studies are a very useful way to test more systematically for the consistency of responses reported from different experiments. In one such analysis, Stefan Koelsch aggregated 47 studies that probed musical emotion and observed consistent responses in both dorsal and ventral components of the striatum (along with the orbital/ventromedial frontal cortex) as well as other structures, including auditory cortex and amygdala/hippocampus (Koelsch 2020). Focusing more exclusively on musical pleasure, a recent meta-analysis carried out by Ernest Mas-Herrero in my lab analyzed 17 neuroimaging studies that specifically reported on positively valenced affective responses to music (as opposed to emotion in general). Here, as shown in Figure 7.5A, we also found strong consistency in responses from the dorsal and ventral striatum and orbital/ventromedial frontal cortex as well as the auditory cortex and insula (Mas-Herrero et al. 2021b). Therefore, these converging findings, across many systematically selected studies, give us additional confidence about how pleasure conveyed by music engages the reward structures. In passing, they also demonstrate that the findings from the prior studies, using modest sample sizes, prove to be quite reliable.

In addition, an important question that we were also able to address with the meta-analysis approach is to what extent music-induced activity patterns are similar or different from those associated with more primary rewards. In Chapter 6, we saw that the core structures of striatum, ventromedial/orbitofrontal cortex, and midbrain nuclei were commonly active for many kinds of stimuli—primary ones, like food, opioid drugs, or sexual images; secondary ones, such as money; and more abstract ones, such as games or information. But,

(A)

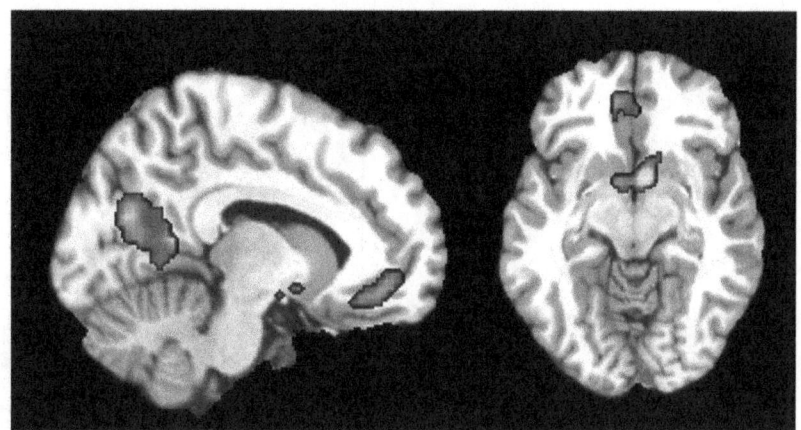

(B)

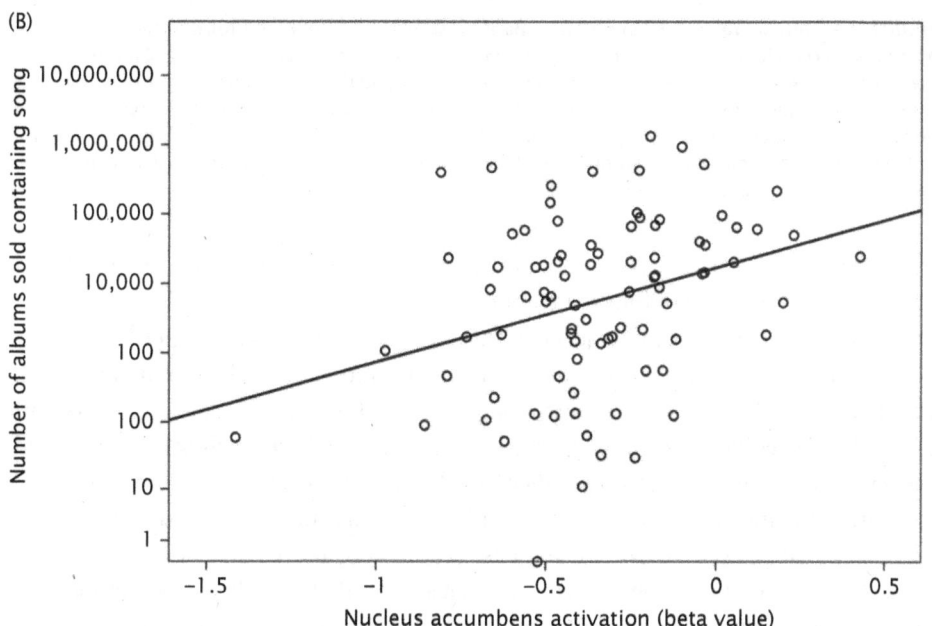

FIGURE 7.4 Reward system activity and music sales. A: Brain maps showing correlations between functional MRI activity and likability of new songs in ventromedial prefrontal cortex and ventral striatum. B: Correlation between ventral striatum activity elicited by songs and subsequent sales of albums containing those songs. Adapted with permission from (Berns and Moore 2012).

as I argued earlier, music and other aesthetic stimuli are dissimilar from both primary and secondary rewards on several levels. Therefore, the outcome of this comparison was not a foregone conclusion. Ernest was able to compare the findings from the musical pleasure literature with similar studies done on food-induced pleasure (Mas-Herrero et al. 2021b). In accord with the idea of a common neural pathway for pleasure, most of the relevant regions (ventral striatum, ventromedial frontal cortex, and insula) were recruited for both music and food. Interestingly, music elicited more activity in the striatum than food did. But this

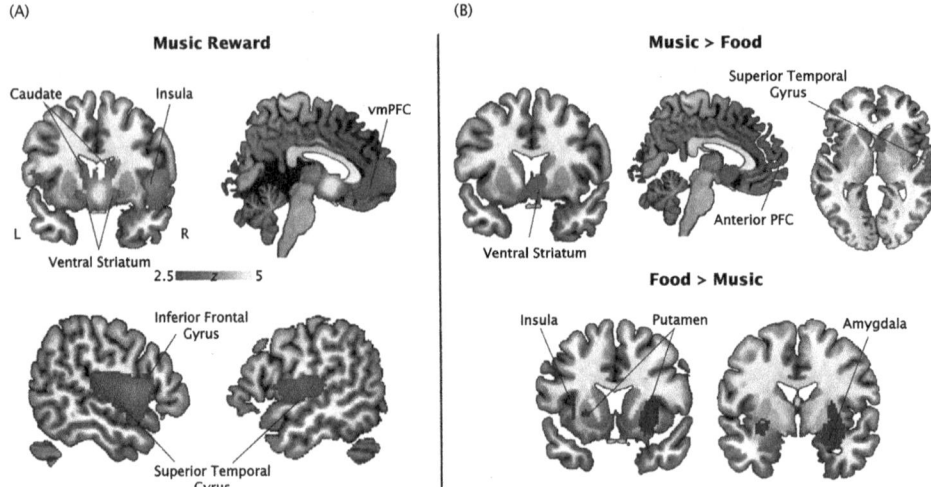

FIGURE 7.5 Meta-analysis of functional imaging studies of music- or food-induced pleasure.
A: Music-induced pleasure showed consistent activity in ventral and dorsal striatum, insula, ventromedial prefroncal cortex (vmPFC), superior temporal gyrus bilaterally, and right inferior frontal gyrus. B: Top: areas with greater response to music compared to food-induced pleasure include the ventral striatum, anterior prefrontal cortex (part of vmPFC), and right superior temporal gyrus. Bottom: areas with greater response to food compared to music-induced pleasure include the insula, putamen, and amygdala. Adapted from (Mas-Herrero et al. 2021b) with permission from Elsevier.

effect could be because, whereas many of the music studies used real, beautiful music that people enjoyed, the food studies were limited to delivering a few squirts of juice or the like (it's rather difficult to consume a delicious three-course meal inside a scanner!).

The most important result of the music–food comparison, as shown in Figure 7.5B, was that activity in auditory cortex was only associated with pleasurable music, not with food; whereas recruitment of the gustatory cortex was only associated with food reward, not with music. This finding of specificity in the sensory cortex for each modality mirrors earlier findings examining gambling tasks and erotic stimuli (Sescousse et al. 2013), suggesting that what drives reward responses within any given stimulus category greatly depends on the cortical input systems associated with that stimulus. More importantly, Ernest's meta-analysis specifically focused on responses associated with pleasure by including only studies that used certain conditions, such as comparing pleasurable to unpleasurable music or liked to disliked music. Therefore, the recruitment of the auditory cortex cannot be ascribed to the mere presence of sound in one condition over another. We turn next to the question of how auditory cortex may influence reward activity.

Interactions Between the Reward System and Auditory Pathways

In one sense, it might seem obvious, maybe even trivial, that the auditory system would be active (along with the reward system) during the experience of pleasure from music. After all, music consists of sounds, and sounds are processed in the auditory networks. But what's more interesting to consider is that the contribution of the auditory system during pleasurable music listening may reflect a particular kind of processing that is intricately tied to the emotional experience. In other words, the various perceptual and cognitive processes enabled by the auditory pathways, described in Part I of this book, may contribute in some essential manner that results in musical pleasure. This idea also plays an important role in moving us away from the simplistic interpretation of the early studies, that a single region or set of regions is responsible for a complex function, and instead guides us toward a more contemporary network-oriented understanding of neural activity associated with the pleasure of music.

One of the first and clearest indications that the auditory cortical pathways were specifically involved with musical hedonic experiences came from the same music-purchasing experiment described above (Salimpoor et al. 2013). The key goal of that study was to try to understand, at a brain network level, how the increased activation of the striatum in response to more highly valued music was related to what was going on in the rest of the brain. With the help of our colleague Randy McIntosh from the University of Toronto, we applied a multivariate technique to investigate this matter, measuring how the temporal correlation between different structures changes as a function of different conditions. Using the right ventral striatum (the structure with the highest sensitivity to musical reward) as a seed region, we determined that a series of cortical and subcortical areas all increased their temporal coupling with the ventral striatum as the reward value of the music increased. In other words, these regions interacted with each other to a great extent when processing highly valued music, but very little, or not at all when processing low-value music (see Figure 7.3 C and D).

The most fascinating aspect of this result is that large swaths of the auditory cortex bilaterally were among the brain regions that showed the highest increase in functional coupling with the striatum. This finding means that when the music was most highly valued, there was a great deal of crosstalk between auditory perceptual mechanisms and the reward-based mechanisms that assign hedonic value. In addition to the auditory areas, increased functional connectivity was also noted in the right inferior frontal gyrus, part of the auditory ventral stream, and some parietal and superior frontal regions that are most likely part of the auditory dorsal stream. Furthermore, other regions belonging to the reward network were also revealed by this analysis, most notably the ventromedial/orbitofrontal region and the right amygdala.

For us, this finding represented a real breakthrough because it finally linked the two principal sets of networks we had been studying for so long in the context of music: the auditory and the reward systems. Notably, many of the structures revealed by the connectivity analysis were also active during music listening. Yet, the magnitude of their activity was similar across the different dollar amounts representing value assigned. This pattern means that what changes as a function of increased value is not necessarily the processing load on any of these regions, or not as indexed by the hemodynamic response at least; but rather, it indicates that there is a change in the interactions between the two main networks. These findings fit well with the anatomical networks described in the previous chapter: the auditory cortex and striatum, while not directly connected anatomically, still interact—on the one hand via inputs going from the inferior frontal and ventromedial frontal cortex to the ventral striatum, and on the other hand via dorsolateral frontal regions to the dorsal striatum. These cortical structures in the ventral and dorsal pathways were precisely the ones that showed up in the analysis of the functional connectivity modulation, strongly suggesting that they mediated the communication between auditory and reward systems.

Subsequent studies have also explored the idea that auditory and reward systems interact during pleasurable music listening. Using a paradigm in which people were scanned while listening to longish sections of real music (including pieces by Phillip Glass and Györgi Ligeti), Shany et al (2019) showed that online ratings of arousal and valence of the music correlated with moments of musical surprise, as independently determined by musically expert raters. As expected, activation in the striatum was more pronounced among those individuals who found the music most pleasant. But a very interesting finding came from the analysis of functional connectivity between the right auditory cortex and the striatum: those same people also exhibited stronger connectivity between these structures as a function of surprise ratings during the most pleasant pieces relative to those who found the music less pleasant.

This finding is in accord with our prior results (Salimpoor et al. 2013) but extends them to more naturalistic music listening and implicates musical surprise as one of the contributing factors to the neural interaction between the two systems. Another functional imaging study specifically looked at the degree to which a final chord in a cadence conformed to conventional musical rules or not (Seger et al. 2013). This study reported that dorsal striatum, as well as superior temporal and inferior frontal regions, were recruited as a function of increasing unexpectedness, and that these regions were functionally interconnected. Yet again, this result supports the contention that cortical auditory loops interact with reward regions, and that these connections are modulated by degree of surprise.

Further evidence that pleasure is related to functional interactions with auditory cortical networks comes from an EEG experiment carried out by Alberto Ara and Josep Marco (Ara and Marco-Pallarés 2020), who observed that theta-band phase synchronization between right frontal and temporal regions increased with degree of pleasure experienced to realistic music, thus showing that network interactions in the right auditory ventral stream, possibly related to working memory and predictive mechanisms, play a key role in music-evoked pleasantness. This synchronization is strongest for unfamiliar music, implying that schematic expectancies require greater engagement of predictive mechanisms (Ara and Marco-Pallarés 2021). The connectivity between reward and auditory systems may even be relevant to the effects of psychotropic drugs on responses to music, as one study found an enhancement of functional interactions between right ventral striatum and auditory cortex after administration of cannabis containing cannabidiol, as compared to cannabis without this compound (Freeman et al. 2017).

The auditory-reward interactions in these studies suggest a potential mechanism by which music gains reward value. The cognitive processes carried out in ventral and dorsal streams not only include perceptual analysis and working memory but also extensively relate to predictive processes at various levels, as we have seen in Chapters 2, 3, and 4. As a result, it seems like a very interesting possibility that these perceptual and predictive functions would engage the reward system, given the latter's important role in anticipating and evaluating outcomes, as reviewed in Chapter 6. This idea will be developed much further in the next chapter, when we discuss the emergence of pleasure responses as a function of musical structure. But first, we need to evaluate further the implications of the interactions between auditory and reward networks. If this model—that that the communication across these networks is responsible for the experience of pleasure is in any way correct—then we can make a rather bold prediction: if musical pleasure is absent, then the interaction should also be absent, and vice-versa. But what would it mean for musical pleasure to be absent? Are there people who do not experience pleasure from music?

Does Everyone Love Music?

The Case of Specific Musical Anhedonia

During a sabbatical visit to Barcelona in 2011, in addition to benefitting from the exceptional culture, gastronomy, climate, and natural beauty of Catalonia, I was fortunate enough to be able to work with the team of Antoni Rodríguez-Fornells and Josep Marco-Pallarés, who have considerable expertise in the domain of individual differences of affective responses. We quickly became friends and, over many a glass of Cava, began to think about how we could combine my interest in music cognitive neuroscience with their knowledge. A natural topic of discussion was, of course, musical pleasure.

At some point, we thought of asking what now seems a rather obvious question: does *everyone* actually love music? We realized that we had taken for granted the idea that musical pleasure was universal. It may well be so at a societal level (we know that all human cultures seem to have some sort of music). And, as previously noted in this book, according to surveys, most people report that music is one of the greatest sources of pleasure (Dubé and Le Bel 2003). Yet, anhedonia is a well-established psychological disturbance in which pleasure and motivation to obtain stimuli such as food, touch, or sex are greatly blunted, and that reward-system dysfunction has been implicated (Der-Avakian and Markou 2012, Ritsner 2014, Rømer Thomsen et al. 2015). It has also been shown that people with generalized anhedonia respond less strongly to music, showing reduced activity in the reward system (Keller et al. 2013). But we were more interested in whether musical anhedonia could also exist independently of generalized anhedonia. Ernest Mas-Herrero, a PhD student at the time, enthusiastically took on the challenge of exploring this question.

In order to characterize hedonic aspects of individual musical experience, the first step was to develop a behavioral scale that could be used to quantify the degree to which a person experiences music as rewarding. No such scale existed at the time, so to achieve this goal, Ernest went through several iterations of a set of questions focused on musical pleasure and administered them along with established questionnaires on mood states and general anhedonia as controls (Mas-Herrero et al. 2013). The music-oriented questions included items such as "I get emotional listening to certain pieces of music" or "music calms and relaxes me" or "music makes me bond with other people," and so forth. The results, based on over a thousand online responses, indicated that music-based reward scores could be ascribed to five separable factors: mood regulation, emotion evocation, musical seeking (desire to learn more about music or seek new music), sensory-motor aspects (wanting to move or dance), and social-related reward. Most of these factors map well onto descriptions from prior behavioral studies of how people use music in real-life situations (Juslin and Laukka 2004). They also give us clues about musical emotion more generally, a topic that will be discussed in Chapter 9.

But undoubtedly, the most interesting outcome of this study came from the comparison of this new scale (that we christened the Barcelona Music Reward Questionnaire, or BMRQ) with standard scales used for assessment of global anhedonia (Chapman et al. 1976): although the vast majority of respondents fell within the healthy normal range on both scales, a subset of people (around five percent in several samples) reported no symptoms of generalized anhedonia or mood disorder yet scored relatively low on the BMRQ, as seen in Figure 7.6. In other words, these individuals indicated that they derived little or no pleasure from music, despite not being globally depressed or otherwise emotionally impaired. Could these people then represent a specific form of anhedonia confined to music?

To find out, Ernest followed up with a second study, in which he specifically sought out a new set of individuals with a low BMRQ score but no evidence of generalized anhedonia or low sensitivity to reward (Mas-Herrero et al. 2014). In addition, he also found two control groups: one with average scores and another with higher-than-average BMRQ scores, the latter of whom we termed hyperhedonics (I admit to being in that group!). As a control, we further verified that the persons with music anhedonia scored normally on the amusia test developed by Isabelle Peretz's group (Peretz et al. 2003), thus ruling out the possibility that lack of pleasure to music could be secondary to a perceptual deficit. Nor could their indifference to music be explained by social deprivation, extreme poverty, or other similar socioeconomic factors, which were all carefully evaluated. Yet, despite all this, when given pieces of music that their peer group generally enjoyed, those with low BMRQ scores reported little or no pleasure from it; and most convincingly, objective psychophysiological measures of skin conductance or heart rate, which are usually very sensitive to musical pleasure as discussed earlier (Salimpoor et al. 2009), did not show any significant modulation either (see Figure 7.6).

These findings show that what these folks were telling was true: they simply don't derive pleasure, or any real emotional or physiological response, to music. But how specific is this situation? Could it be that, despite normal scores on anhedonia questionnaires, they were suffering from some sort of more generalized disorder? The answer seems to be no. When asked to rate the pleasure they experience to non-musical rewarding events (food, money, sex, exercise, drugs, etc.) their ratings were normal. Most importantly, when tested with a monetary incentive delay task (see Chapter 6), they showed completely normal behavioral and psychophysiological responses, indicative of arousal, on trials in which they win money, unlike their responses to music. Finally, one may ask, to what extent musical

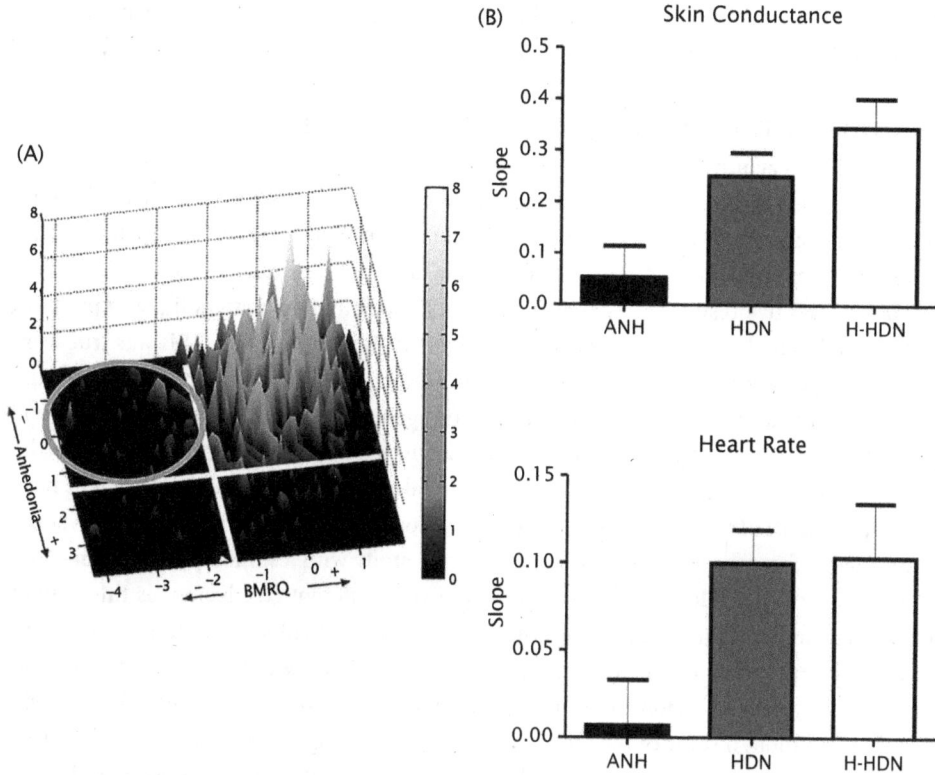

FIGURE 7.6 Specific musical anhedonia. A: Distribution of scores on the Barcelona Music Reward Questionnaire (BMRQ) and of global anhedonia scores. Most individuals fall in the top right quadrant, indicating no anhedonia on either scale; but some individuals (gray circle) score low on the BMRQ, indicating little or no pleasure to music, despite normal scores on the global anhedonia scale. Adapted with permission of University of California Press—Journals, from (Mas-Herrero et al. 2013); permission conveyed through Copyright Clearance Center, Inc.

B: Psychophysiological responses to music in three groups of persons: specific musical anhedonia (ANH), average musical hedonia (HDN), and hyper-hedonics for music (H-HDN). Adapted from (Mas-Herrero et al. 2014) with permission from Elsevier.

anhedonia is specific to music, as opposed to other aesthetic pleasures. While we don't know the full extent of the answer, when tested with visual art at least, musical anhedonics showed no particular disturbance (Mas-Herrero et al. 2018b), once more supporting the absence of response to music in particular.

Although all these results were somewhat surprising to us, the people we identified as having specific music anhedonia were generally pleased that we had scientifically validated what they had always known to be true. As a group, they are healthy and normal, but simply don't enjoy music. As such we have always been careful not to label the condition as a disorder per se. It's better to consider it as part of natural cognitive/affective variation. Indeed, there is no clear demarcation in the BMRQ distribution corresponding to anhedonia, meaning that sensitivity to musical reward is a continuum in the population, ranging from very high (hyperhedonics) to high/medium (most people) to low (anhedonics). Some of the individuals with music anhedonia even thanked us for allowing them to "come out of the closet," as one of them put it, so that they could get on with their lives without their friends and family constantly trying to make them enjoy music!

Neural Correlates of Musical Anhedonia

A few studies of acquired (as opposed to congenital) musical anhedonia as a result of brain damage exist in the literature, but they provide few clues to possible neural substrates. One of the first such cases was reported by Tim Griffiths and his colleagues. They described a patient who, after suffering a stroke to the left insula and amygdala, reported a complete loss of any affective response to his favorite pieces, espcially Rachmaninoff piano preludes that he used to find wonderful prior to the stroke (Griffiths et al. 2004). This was true even though his perception of the piece seemed to be intact. This finding may implicate the insula in musical pleasure, and/or emotion more generally. Indeed, that structure is one that is reliably activated in music reward studies in healthy volunteers, as shown in the meta-analysis (Mas-Herrero et al. 2021b). But it's also possible that the lesion might have interrupted some of the pathways linking the auditory system to the reward system via the ventromedial frontal cortex. However, another case study with a similar presentation—loss of musical pleasure following stroke—came to a very different conclusion, as this patient had a lesion in the right parietal lobe (Satoh et al. 2011). Individual case studies are always tricky to interpret; but a more systematic study of 78 patients with focal brain damage did not generate any clear lesion location associated with acquired musical anhedonia (Belfi et al. 2017), although some of the areas identified included right temporal lobe and ventro-medial frontal cortex, regions associated with music-induced pleasure in prior studies, as we've seen.

Our model predicted that specific musical anhedonia, or at least the congenital va-riety not caused by an acquired lesion, would be associated with some alteration in the reward-auditory interaction, based on the neuroimaging data reviewed above. To test this idea directly, Noelia Martínez-Molina and the rest of the Barcelona team carried out two additional studies, one examining functional connectivity and another examining struc-tural connectivity.

In the first experiment, we selected people in the three subgroups (low, medium, and high scores on the BMRQ) and scanned them with fMRI while they listened to music care-fully chosen to span a range of liking for this population (Martinez-Molina et al. 2016). As a control, we also administered the monetary incentive delay task. As expected, those with musical anhedonia did not recruit the striatum or other reward-related structures during music listening, whereas the other two groups did (Figure 7.7A). But on the monetary task, all three groups showed clear striatal responses, supporting the specificity of the music an-hedonic response. Most importantly, as shown in Figure 7.7B, the functional connectivity between the right auditory cortex and the striatum was reduced or absent in the musical anhedonic group, intermediate in the average hedonia group, and somewhat enhanced in the musical hyperhedonics. These findings strongly support the conclusion drawn from the prior functional imaging studies in persons with average music reward sensitivity, that au-ditory and reward networks interact in support of musical pleasure: those people who don't feel pleasure don't have that interaction.

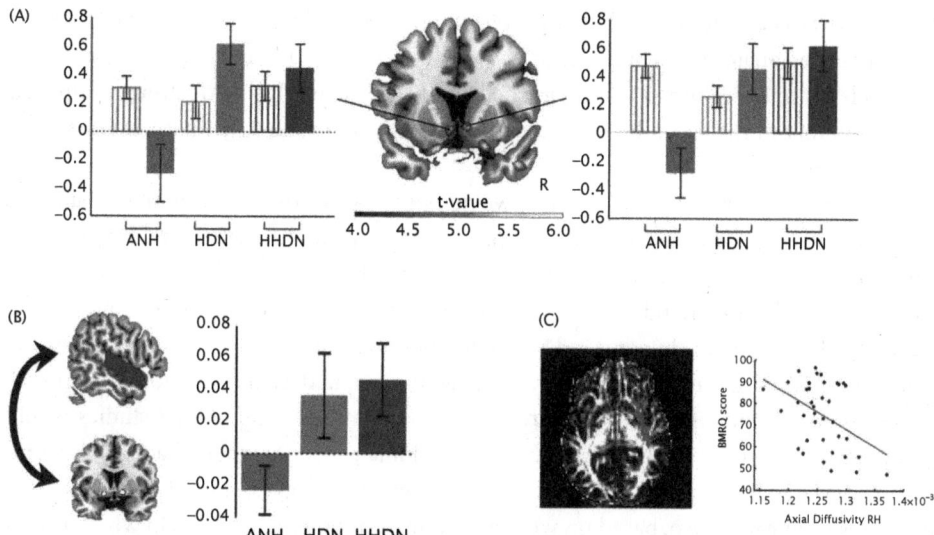

FIGURE 7.7 Neuroimaging in specific musical anhedonia. A: functional MRI activity in the left and right ventral striatum for three groups of persons: specific musical anhedonia (ANH—red), average musical hedonia (HDN—green), and hyper-hedonics for music (H-HDN—blue) during a music listening task (solid bars) and a control monetary task (striped bars). Only the ANH group failed to show positive activity to music, but they did not differ from the other groups in the control task. B: Functional connectivity between the right auditory cortex (blue area on the saggital MRI slice) and the ventral striatum in the three groups during music listening. Only the ANH group failed to show any coupling between the two structures. C: Diffusion weighted imaging data showing diffusivity as a function of BMRQ score in the white-matter tract interconnecting auditory and orbitofrontal regions (shown in red). More diffusivity (suggesting less well-organized pathway) was associated with lower BMRQ scores. Adapted with permission of the Society for Neuroscience, from (Martinez-Molina et al. 2019); permission conveyed through Copyright Clearance Center, Inc.; figures 7.7A and 7.7B adapted with permission from (Martinez-Molina et al. 2016).

Is the relative decrease in functional interaction due to a different anatomical organization in this population? To address this question, we carried out a second study to examine patterns of white-matter structure using diffusion MRI scanning, in a sample of individuals drawn from the previous fMRI study (Martinez-Molina et al. 2019). As we saw in the previous chapter, the anatomical connections between auditory regions and the ventral striatum are indirect, involving waystations in ventromedial and orbitofrontal cortex. Therefore, these tracts were analyzed separately. More diffuse white-matter organization was observed as a function of lower BMRQ scores in the connection between auditory and orbitofrontal areas (in the right hemisphere) (see Figure 7.7C) and in several other related tracts. Furthermore, this study also showed that the degree of functional connectivity between these structures, as identified in the prior study (Martinez-Molina et al. 2016), was related to the degree of anatomical connectivity, thereby directly linking function to structure.

An interesting study from California showed a similar relationship between the experience of musically induced chills in a large healthy population and white-matter organization of the tracts connecting auditory areas with medial frontal cortex, as well as insula (Sachs et al. 2016). Therefore, taken together, these findings indicate that, at a population level, there is a continuum, such that people with lower music reward sensitivity have less well-organized anatomical connections between the auditory and reward networks, which in turn leads to lower functional interactions between them. This work strongly supports the model that musically induced pleasurable feelings are enabled by interaction between these systems.

One caveat to these findings is that we do not yet know the direction of causality. It's a chicken-and-egg problem: it could be that the anatomical differences are innate or the result of a congenital developmental process and that this factor is what drives the lack of experience of pleasure in individuals with low music reward sensitivity. Conversely, it could be that such individuals do not develop much functional connectivity in these systems because of disuse of the relevant pathways, leading, in turn, to decreased anatomical connectivity. To answer this question definitively, developmental or longitudinal studies would be necessary. However, people with musical anhedonia generally report that they don't remember ever enjoying music as children or in adolescence, suggesting a phenomenon that expresses itself early. Also, based on what we know from other conditions in which white-matter structural variability is associated with atypical behavioral outcomes (e.g., amusia—see Chapter 3), the congenital explanation seems most likely. However, the two possibilities are not necessarily mutually exclusive: a small congenital effect could predispose people to not be interested in music, leading to a reduced development of the interactions due to lack of experience over time.

Modulating Music Reward Directly: Now You Like It, Now You Don't

The research carried out on specific musical anhedonia is in line with the model that interactions between reward and auditory circuitry underlies musical pleasure. But in a sense, this evidence could be construed as providing support for the hypothesis in only one direction. We've shown that when pleasure is absent, the operation of the reward system is altered. But we have not shown the converse: that by altering its operation, pleasure will, consequently, also be altered. Moreover, there is still the potential objection that we have not yet demonstrated directly that there is a true causal relationship between the engagement of these neural systems and musical pleasure because in musical anhedonia there was only selection of individuals—not active intervention.

To address the issue of causality head-on, it becomes necessary to find a way to manipulate brain activity directly. Noninvasive brain stimulation techniques have been increasingly adopted in cognitive neuroscience, exactly because they are especially well-suited to this purpose (Parkin et al. 2015, Bergmann and Hartwigsen 2021). Our goal then was to use such techniques to modulate the reward system and then observe if musical pleasure would change as a consequence.

The first difficulty we faced was that noninvasive brain stimulation can only be applied at the surface of the head, whereas the reward structures of interest, notably the striatum, are located quite deep in the brain. However, it is possible to modulate deep structures, by targeting cortical areas that are heavily interconnected to them. We know from the anatomy reviewed in the previous chapter that the dorsal striatum receives direct projections from the dorsolateral frontal cortex, providing a good target for our first transcranial magnetic stimulation experiment (Mas-Herrero et al. 2018a).

Prior studies had already demonstrated a direct link between dorsolateral frontal cortex stimulation and changes in dopamine in the dorsal striatum (Strafella et al. 2001), validating the use of this cortical target to modulate reward function. More importantly, depending on the parameters of stimulation, it is possible to both upregulate dopamine activity in the striatum (Pogarell et al. 2007) or downregulate it (Ko et al. 2008). We wanted to do more than merely interfere with musical pleasure—that would provide only a weak test of the hypothesis because brain stimulation might have any number of nonspecific negative effects (e.g., it might make people anxious, or give them a headache), which might, in turn, reduce musical pleasure. Therefore, we applied both inhibitory and excitatory stimulation to see if we could demonstrate a bidirectional modulation effect.

We asked our volunteers to bring in a selection of their favorite music, and we also provided them with some top 40 pop music that they were likely to enjoy, based on their expressed preferences. Excitatory, inhibitory, or sham stimulation was applied on different days (listeners did not know which condition was which). We then had them judge how much they liked the music (using a subjective scale), how much money they would be willing to spend to buy a recording (as in previous studies measuring "wanting" responses [Salimpoor et al. 2013]), and we also measured skin conductance, as an objective index of physiological arousal (Salimpoor et al. 2009). All three measures significantly increased following excitation and decreased following inhibition, indicating that liking, wanting, and arousal were all similarly modulated bidirectionally by the brain stimulation. The effect was seen for both self-selected music and the pop songs we had chosen, showing that neither familiarity nor novelty play a major role. An important detail is that the skin conductance measures taken during silent moments prior to music listening were not affected by the stimulation, which indicates that the modulation of brain responses did not lead to a generalized modification of arousal but rather specifically modified the arousal induced by the music.

This finding is rather remarkable: the degree of pleasure we feel can be modulated in either direction by transiently changing the excitability of certain brain structures. But which brain structures are they? Although prior studies had shown that dorsolateral frontal stimulation altered dopamine activity in the dorsal striatum, they also showed changes in orbitofrontal and cingulate areas (Cho and Strafella 2009). Furthermore, changes to any one region are also likely to induce changes in various networks of remote regions. Thus, to really provide a strong test of our hypothesis—that pleasure emerges from interactions between reward and auditory circuits—requires us to actually demonstrate that the stimulation produces the predicted changes in those networks.

Armed with these positive behavioral results, then, the next step was obvious: perform the same experiment but this time scan people with fMRI immediately after stimulation, while they listen to the music, to determine the pattern of brain responses (Mas-Herrero et al. 2021a). We recruited a new sample of listeners and used different musical materials; after applying the same stimulation protocol as before, the behavioral results were essentially

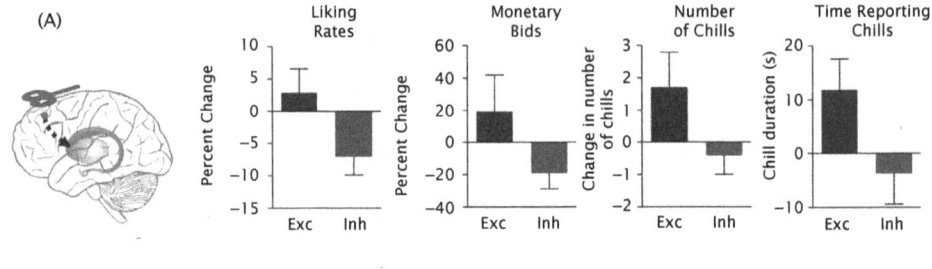

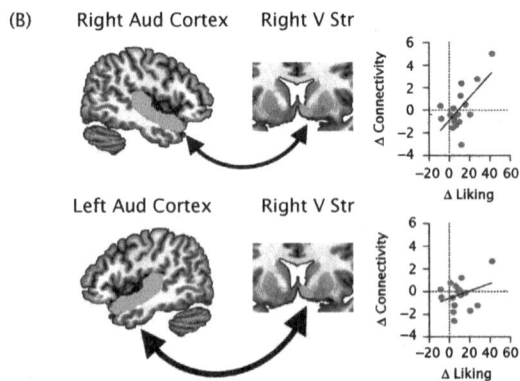

FIGURE 7.8 Modulating musical pleasure via brain stimulation. A: Transcranial magnetic stimulation applied to the dorsolateral frontal cortex modulates striatal activity (diagram). Bar graphs show bidirctional changes to msuic liking rates, monetary bids, number of chills, and time reporting chills as a function of excitatory or inhibitory stimulation (blue and red, respectively). Adapted with permission from (Mas-Herrero et al. 2018a).

B: functional MRI data collected during music listening after brain stimulation shows that changes in subjective liking are related to changes in functional connectivity between the right auditory cortex (pink) and the right ventral striatum (blue); similar but not significant trend is seen for the left auditory cortex. Adapted with permission from (Mas-Herrero et al. 2021a). Reproduced under the terms of a CC BY 4.0 license.

identical to the prior study. This outcome provided independent replication of the bidirectional modulation of liking and wanting responses, as shown in Figure 7.8A, demonstrating that these effects are robust and reliable even without huge sample sizes. For the fMRI data, we analyzed several regions of interest within the reward circuitry (including dorsal and ventral striatum, and ventromedial frontal cortex). We observed that brain activity was indeed different in all these structures when listening to music after excitatory stimulation, which significantly enhanced the responsiveness of the circuitry to music, compared with inhibitory stimulation in which neural responses were blunted. This finding shows that the stimulation did indeed modulate neural responses within the reward system (but not in control regions) as predicted.

But the most important finding of this study was that the functional connectivity between the right auditory cortex and the right ventral striatum was enhanced in direct relationship with the degree to which the stimulation enhanced pleasure ratings (see Figure 7.8B). That is, the difference in subjective pleasure experienced by each person when comparing excitatory to inhibitory stimulation was predicted by the degree to which communication between auditory and reward systems was modulated by the stimulation: those people whose pleasure was most boosted were also the ones who showed the greatest increase in connectivity and vice-versa. This finding provides definitive causal evidence in favor of the hypothesis that this interaction underlies the experience of musical pleasure. And the fact that the link was stronger from the right auditory cortex fits nicely with both our earlier functional data with anhedonics as well as the broader hemispheric differences we discussed in Chapter 5.

The Role of Dopamine in Music-Induced Reward Responses

In Chapter 6, we reviewed the evidence for the important role of dopamine in reward-related processes. Dopamine is the primary neurotransmitter involved in the midbrain-striatal-cortical loop (see Figure 6.2), and its role in signaling anticipatory responses and reward prediction error is well-established. Yet none of the research carried out on music-related reward reviewed up until this point tackled this issue. Most of these studies used functional neuroimaging methods that rely on hemodynamic signals and therefore have no specificity for any given neurotransmitter. There is evidence that when dopamine is released in the striatum, there is a correlated fMRI signal (Knutson and Gibbs 2007). However, the converse need not be true. As such, establishing whether music-induced reward activity involves the dopaminergic pathway becomes a critical issue if we are to link the abstract, cognitive type of pleasure elicited by music to the more basic biological mechanisms that have been intensively studied in the neurophysiological studies described in the previous chapter.

In order to address this issue, Valorie Salimpoor carried out a study with a different imaging method than the ones we have discussed until this point (Salimpoor et al. 2011). In our first study of music-induced pleasure, we used PET to measure blood flow (Blood and Zatorre 2001). But to test for the role of dopamine, we took advantage of another PET scan technique using Carbon-11 labeling of the synthetic molecule Raclopride, which binds to dopamine receptors with high specificity and has an affinity for the striatum (Gunn et al. 1997). However, the technique has very low temporal resolution (since it takes about an hour to get a single scan), and because of radiation safety considerations, only two scans are possible per person. Nonetheless, it provides one of the only ways to measure the level of striatal dopamine neurotransmission associated with a given stimulus condition in-vivo in healthy volunteers. This technique had already been extensively used in humans to show dopamine release in relation to primary rewards, such as food (Small et al. 2003) as well as certain euphorigenic drugs, such as cocaine (Cox et al. 2009), but had not been used for music.

The experimental design we applied was similar to our prior study of the musical chills phenomenon (Blood and Zatorre 2001), as this approach provides a convenient index of maximal pleasure that is objectively verifiable via psychophysiology (Grewe et al. 2007). Furthermore, it was straightforward to assign people to listen to two music selections, during each of the two possible scans: their own favorite, preselected music and that of

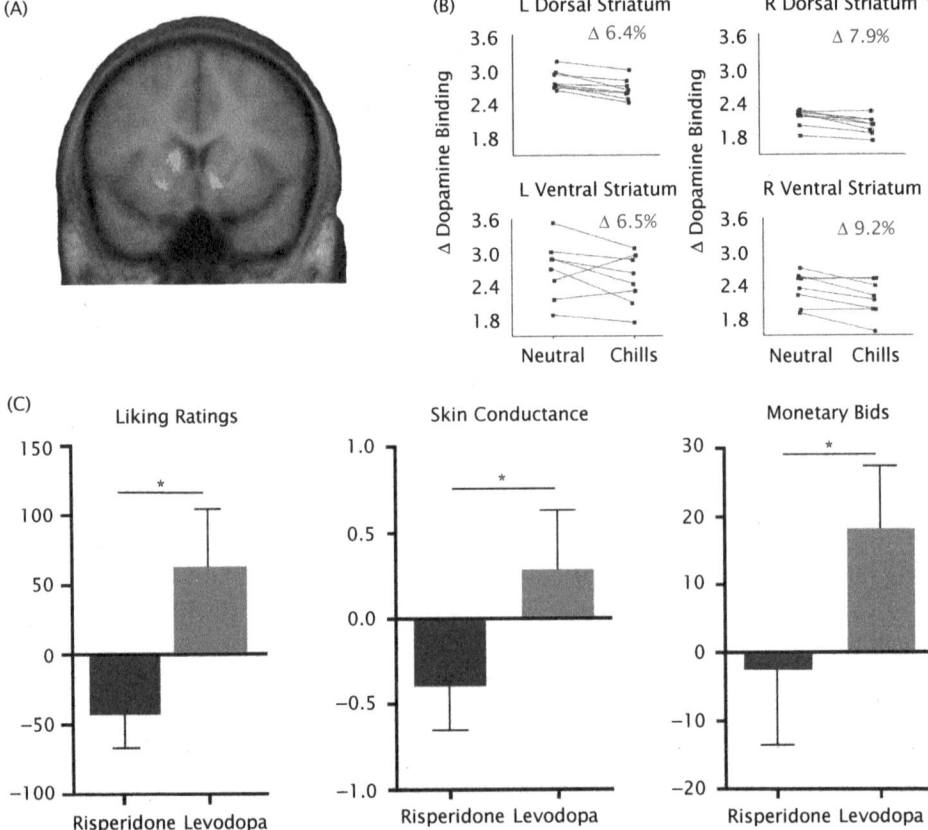

FIGURE 7.9 Dopamine and musical pleasure. A: Positron emission tomography scan using the dopamine-sensitive radioligand C¹¹ Raclopride shows dopamine release during listening to favored music in dorsal and ventral striatum. B: Change in dopamine binding for each individual and average percent change (lower binding indicates greater dopamine release) for the image shown in A. Reproduced with permission from (Salimpoor et al. 2011).

C: Modulation of liking ratings, skin conductance, and monetary bids to music after pharmacological manipulation, relative to placebo control. All three measures are significantly altered in the negative direction by Risperidone (dopamine antagonist) and in the positive direction by Levodopa (dopamine precursor). Adapted with permission from (Ferreri et al. 2019).

another listener from the same experiment that they judged to be neutral, thereby counterbalancing the diverse stimuli used. As predicted, comparing the two conditions showed significant change in both reported pleasure (number and intensity of chills), as well as psychophysiological variables (skin conductance, heart rate, respiration, etc.). Most importantly, there was significant modulation of the uptake of the tracer, in both dorsal and ventral striatum, indicating dopamine release (on the order of 6–9% magnitude) during the pleasurable music, as shown in Figure 7.9 A and B. This finding, therefore, brings music in line with other known natural rewards in demonstrating that they all depend on the same dopaminergic circuitry.

The two main responses, in dorsal and ventral striatum, correspond to both the specificity of the Raclopride technique (which is largely insensitive to cortical dopamine) and the existing literature that shows that both components of the striatum are often active during reward tasks (see Chapter 6). But there was an intriguing difference in the brain-behavior correlations between the two structures. Whereas dopamine expression in the dorsal striatum was significantly correlated with the number of chills experienced during the music listening period, that was not the case for the ventral striatum, whose dopamine response instead was correlated with the intensity and degree of pleasure, as subjectively reported by the listeners. This distinction has important implications because it implies that the ventral striatum is more related to the subjective experience of pleasure itself, while the dorsal striatum is more linked to the presence or absence of a peak response. We will return to this dissociation between dorsal and ventral striatal contributions in the next chapter, when we discuss the different temporal sequence in which the two structures are engaged.

The demonstration that dopaminergic activity is elicited by music was very exciting. But in order to achieve even stronger support for the proposition that musical pleasure actually depends upon dopamine, we needed to go further for several reasons. First, neuroimaging—even a neurochemically specific technique such as PET raclopride—is always correlational. We cannot show with imaging results alone that dopamine is actually *required* for the experience of pleasure, only that it correlates with it. We know from the brain stimulation experiments described above that pleasure can be causally modulated by manipulating reward system activity (Mas-Herrero et al. 2018a, Mas-Herrero et al. 2021a), but those techniques are not chemically specific. Finally, although we saw a clear increase in dopamine in the PET study, it cannot be excluded that the experience of pleasure was actually mediated by another neurochemical pathway (the obvious candidate being the opioid system, based on the evidence presented in Chapter 6).

To address this issue of nailing down dopamine's role, we return to beautiful Barcelona, this time to the *Modernista* site of the Hospital de la Santa Creu i Sant Pau. Here, Laura Ferreri and Ernest Mas-Herrero were able to design a study with a team of researchers who had extensive experience in pharmacological studies. (This was the first—and probably the last—time that I was involved in a study carried out at a UNESCO World Heritage Site!) The basic idea was to provide a definitive test of the causal influence of dopamine on musical pleasure by manipulating the levels of dopamine in healthy volunteers with pharmacological agents that would either increase or decrease dopaminergic neurotransmission (Ferreri et al. 2019).

This pharmacological approach has previously been used by several labs to demonstrate, among other things, the role of dopamine in enhancing verbal learning (Knecht et al. 2004) as well as the association between learning and pleasure (Ripollés et al. 2018). More importantly, a similar manipulation of dopamine in a monetary task not only demonstrated the expected modulation of fMRI activity in the ventral striatum with the drug, but also that administering a dopamine agonist led to a greater propensity to choose the most rewarding option in a way that could be directly explained by alterations in reward prediction error computations (Pessiglione et al. 2006). Thus, these studies show that dopamine is directly involved in learning and the pleasure resulting from learning, via reward prediction error mechanisms, making the technique well-suited for our research questions.

Our music experiment (Ferreri et al. 2019) involved three conditions on three days, during which either a dopamine precursor (levodopa) or a dopamine antagonist (risperidone) or a neutral placebo (lactose) were administered in counterbalanced order. The design was a classic double-blind study, such that neither the volunteers nor the experimenter was aware of which drug was administered. The music tasks were very similar to those used in our prior studies (Salimpoor et al. 2013, Mas-Herrero et al. 2018a): listeners were presented with musical excerpts, both self- and experimenter-selected after drug administration; we then measured their subjective liking ratings, monetary valuation, and skin conductance. As a control, we also administered the monetary incentive delay task, described in the previous chapter, which serves as a known probe of dopaminergic reward-system function.

The findings showed that after being given the dopamine precursor levodopa, listeners reported feeling high pleasure, including chills, for longer time periods than under the dopamine antagonist risperidone, and showed higher pleasure ratings (see Figure 7.9C). Similarly, the skin conductance ratings showed a significant change in the predicted direction (higher or lower, after levodopa or risperidone, respectively). But more importantly, this modulation was specific to the self-rated pleasurable moments of the music and did not occur throughout the entire excerpt. These effects were similar for the familiar music selected by each listener, as well as for the less-familiar music selected by us. Also, as expected, monetary valuation of the music was modulated by the drug: people were more willing to spend money to purchase music after levodopa, than they were after the antagonist. Finally, the control monetary incentive delay task demonstrated the predicted modulation in skin conductance during trials in which a large monetary reward was obtained.

These results confirm the causal role of dopamine in music-induced subjective pleasure (as shown by the rating scale), as well as arousal (skin conductance), and even motivation (monetary valuation). Since the control task (monetary incentive delay) had already been shown to be dopamine-dependent (Schott et al. 2008), its modulation in our music study validates that the pharmacological manipulation successfully changed dopamine neurotransmission. But critically, there was no effect of the drug during neutral trials of this task, when no money was won or lost. Similarly, the drugs amplified or dampened the moments of greatest pleasure generated by the music rather than more neutral moments. In addition, the skin conductance values were not modulated by the drug during baseline conditions, but only during music listening. These aspects of the results indicate that the drugs did not globally alter mood in one direction or another but rather caused a differential response to the pleasure-inducing stimulus—music or money. This means that the dopamine was still generated by the music, but its effect was changed depending on which drug had been administered.

Dopamine also influences memory for music (Ferreri et al. 2019), as well as verbal learning (Ripollés et al. 2018). This effect is consistent with the very significant role that dopamine plays in many aspects of learning, as reviewed in Chapter 6. It is well-established that highly pleasant and arousing music—the kind that would activate the reward system—is better recognized after a delay compared to less pleasant music (Eschrich et al. 2008), as is also true for other kinds of stimuli, such as pictures (Bradley et al. 1992). We may also recall the important role of the dopaminergic reward system in mediating information-seeking and curiosity. All these memory-related effects offer an additional clue to the potential mechanisms by which dopamine has such effects on hedonic response, suggesting that various dopamine-dependent cognitive processes, including learning and memory, likely play a role in the pleasure generated by abstract stimuli. Although it is not our focus here, we should also point out, for completeness, that memory encoding and consolidation effects for highly arousing and emotional stimuli also involve various other pathways, especially the amygdala and hippocampus (Phelps and LeDoux 2005). However, their role in memory, in relation to rewarding musical stimuli, is not as clear (Frühholz et al. 2014). We shall return to the question of how learning mechanisms may influence rewarding aspects of music in the next chapter.

Dopamine Versus Opioids?

The findings from all these experiments conclusively show that dopamine is not only released in response to pleasurable music (PET study) but that it plays a causal role in mediating the rewarding experience (pharmacological manipulation study). Yet, they do not necessarily prove that no other neurochemical signaling may be involved. Strong evidence from the literature using primary rewards in both humans and animals, indicates that opioid mechanisms most likely mediate the hedonic response, as discussed in the prior chapter. Yet, as we also saw, almost all of these studies focus on basic rewards, and none of them tested aesthetic ones. What evidence do we have then for a role of opioids in music-induced pleasure?

Only a handful of studies have explored this question. An early attempt to measure the effect of opioid blockage on music reported that out of 10 individuals, three subjectively reported an attenuation of music-induced chills after injection of naloxone (an opiod antagonist) compared to saline (Goldstein 1980). Although this study is frequently cited as evidence for the role of opioids in musical pleasure, one could just as easily point out that, for most people tested, blocking opioids did not have any apparent effect. Furthermore, this study did not report any quantification of the presumed effect, so we don't know how much attenuation occurred in the three individuals in question, nor can we tell if it was related to side-effects of the drug. Most importantly, only the presence and/or number of chills was recorded, not how much subjective pleasure was experienced. Therefore, even if lower arousal was elicited in a few individuals leading to fewer chills, this study does not indicate whether or how much the actual hedonic enjoyment was modified.

A more recent study used a similar pharmacological blockade approach (with naltrexone rather than naloxone) and did report a decrease in subjective pleasure ratings to familiar music as well as a decrease in objective physiological measures, including electromyogram measures of facial muscles (Mallik et al. 2017). However, listeners continued to report substantial pleasure with the opioid block (mean ratings decreased only about 9%), which is notable considering that the dosage of naltrexone used is thought to block 80–90% of opioid receptors (Lee et al. 1988). On the other hand, the magnitude of decrease in physiological measures was much larger in that study, implying that the drug affected arousal more than pleasure. Another recent pharmacological study is of interest because it used the largest sample size (49) of all other studies published to date (Laeng et al. 2021). In this experiment, listeners rated subjective pleasure to self-selected music after receiving naltrexone, but there was no significant change in subjective pleasure ratings; instead, the study did demonstrate a significant attenuation of pupil dilation—yet another physiological marker of arousal—in response to the most pleasant moments in the music. As such, this result suggests a dissociation between subjective measures of pleasure and objective measures of arousal, with only the latter being blocked by the opioid antagonist.

In any pharmacological manipulation, including the three studies mentioned in this section, it is difficult to establish how specifically the agent used impacts the function of interest, because of uncontrolled side effects that may cause generalized negative responses. This is especially true for opioid blockers, which can sometimes result in unpleasant physical effects (e.g., nausea), and which could lead to decreases in pleasure to music. Since these effects are only indirectly related to the drug, they don't inform us about the neurobiological consequences of the medication. To provide a greater test of specificity, it is better to use a bidirectional modulation approach, with both agonists and antagonists, as discussed above in the case of dopamine. It's also important to use a control, non-musical task for the same reason: it adds specificity to the conclusion.

To address these points, our Barcelona team got to work yet again, carrying out a study with very similar procedures to the one with dopaminergic modulation described above (Ferreri et al. 2019), but this time using opioid agonists and antagonists, oxycodone and naltrexone, respectively (Mas-Herrero et al. 2023). Looking at the subjective pleasure ratings first, unlike the dopamine-mediated effects, we did not observe any significant modulation in either direction with the two drugs, for either familiar self-selected music or experimenter-selected music—nor was there any effect on monetary valuation. However, the drugs clearly influenced arousal, as indicated by significant modulation of skin conductance during music listening (increased for the agonist and decreased for the antagonist). A similar pattern was observed on skin conductance alterations for the monetary gambling task, showing that the effect is not confined to music.

The results across these various studies of opioid function and music are not perfectly consistent. But overall, there seems to be little or weak evidence for an effect of opioid modulation on subjective musical pleasure. In contrast, all four of the studies reported robust and consistent changes in several objective variables related to arousal, including number of chills (Goldstein 1980), electromyography of facial muscles (Mallik et al. 2017), pupil dilation (Laeng et al. 2021), and skin conductance (Mas-Herrero et al. 2023). This pattern of results is surprising, given the vast literature on opioid-mediated hedonic responses in animals mentioned in Chapter 6 (Peciña et al. 2006) and the fact that in humans there is substantial evidence, including from similar pharmacological manipulations, that supports a role for opioid signaling in hedonic responses to food

(Nummenmaa et al. 2018), sexual stimuli (Büchel et al. 2018), or facial attractiveness (Chelnokova et al. 2014).

Therefore, we return to the possibility that, whereas hedonic responses to primary or more basic rewards primarily depend on opioid systems in the hedonic hotspots of the reward system, the pleasure experienced from music (and possibly other more abstract rewards too) may be mediated by partially dissociable systems, at least some of which are regulated via dopamine. This interpretation is in line with the point made earlier in this chapter, that musical pleasure emerges from the interactions between the reward system and cortical pathways responsible for perceptual, mnemonic, and cognitive processing. Those higher-order mechanisms may not be dependent on opioid neurotransmission and could influence the reward system primarily via dopaminergic action. Specifically, the proposal I am defending in this book—that musical pleasure arises to a significant extent from the important role of reward prediction mechanisms—would be in line with a prominent role for dopaminergic neurotransmission, given that those prediction signals are largely dependent on that neurotransmitter system. I will develop this point further in the next chapter.

Despite the critical role of dopamine, however, the fact that physiological indices of arousal were consistently modified by opioid manipulation in the studies just reviewed suggests that opioids likely play a role in the expression of pleasure via the autonomic nervous system and could thus influence hedonic responses indirectly, perhaps by modifying the intensity of the experience (Laeng et al. 2021). The concept could also tie into models in which valence and arousal are considered separate, dissociable dimensions of emotional response, a topic we shall revisit in Chapter 9.

An intriguing possibility that emerges from this discussion is that modulation of the intensity, if not valence, of affective responses might be mediated via proprioceptive and interoceptive feedback from somatic signals (including, for example, skin sensations, muscle tension, heart rate, and other autonomic physiological variables that are linked to arousal). Such feedback would be targeting ventromedial frontal cortex, insula, and amygdala, which, of course, are also involved in reward processing. This idea would need much further development to be viable, but it would be compatible with the so-called somatic marker hypothesis, developed by Antonio Damasio and colleagues (Damasio 1991, 1994; Poppa and Bechara 2018). The hypothesis stipulates that visceral and somatosensory afferent inputs influence emotions and behavior via their inputs to various structures, in particular the insula. The meta-analysis of musical pleasure referred to earlier (Mas-Herrero et al. 2021b) did show consistent responses in the insula, which has been independently highlighted as an integration region for somatosensory and interoceptive inputs with other sensory modalities, and with awareness of emotion and bodily states (Paulus and Stein 2006). It also plays an important role in Damasio's model of emotion (Damasio et al. 2000). More work on the role of the insula, opiods, and music-induced pleasure would seem to be in order.

Reprise

Around 20 years ago, neuroimaging studies began to document the consistent recruitment of the reward system for positively valenced music. This result has been replicated under many conditions and is directly linked to the subjective experience of musical pleasure. Musical pleasure can be characterized both via subjective ratings as well as through psychophysiological measures of arousal, such as skin conductance, heart rate, and respiration, among others. All these responses peak during moments of music-induced chills, which many people experience during certain musical passages. Activity in the striatum, orbitofrontal cortex, and midbrain regions is elicited in relation to the degree of subjective pleasure; and although chills often occur at the moment of peak pleasure, they are not necessary to engage the system.

Some of the auditory cortical processing pathways for music perception reviewed in earlier chapters show enhanced interactions with the reward system as a function of the value assigned to the music by the listener. Specifically, there is greater neural coupling between the striatum and auditory, frontal, and parietal cortical regions, for the most preferred music. This finding indicates that the reward value of a piece of music as it unfolds over time depends on the exchange of information between cortical systems involved in auditory perceptual, mnemonic, and other cognitive processes and the reward-related areas that compute value and generate hedonic pleasure.

A critical implication of this model is that lack of musical pleasure should be reflected specifically by a disruption of the interaction between auditory and reward systems. To test this hypothesis, persons with specific musical anhedonia were identified: they show no or decreased pleasure specifically to music, but not to other stimuli. Functional imaging of this population showed normal responses in the reward system to a control monetary task, but music failed to engage their reward system. As predicted, functional connectivity between the striatum and the auditory cortex was specifically disrupted during music listening in the music anhedonics, whereas it was elevated in those with especially high affinity for music. Structural imaging suggests that in specific musical anhedonia, there is a relative disorganization of the pathways linking auditory cortical areas to the reward system, which communicate via waystations in the ventromedial and orbitofrontal regions.

Modulating the reward system directly via brain stimulation results in enhanced or reduced subjective musical pleasure following excitatory or inhibitory stimulation, respectively, together with similar changes in objective psychophysiological measures of arousal and monetary valuation. When excitatory stimulation is applied, there is a concomitant increase in the response of the striatum to music, whereas with inhibitory stimulation, activity is reduced. Most critically, the functional coupling between right auditory cortex and striatum is similarly modulated by brain stimulation and predicts behavioral outcomes. These findings provide direct causal evidence that auditory-reward system communication underlies the experience of musical pleasure.

The role of dopamine has been documented via ligand-specific imaging, which demonstrated release of dopamine in dorsal and ventral striatum, while listening to pleasant compared to neutral music. To prove causality, dopamine neurotransmission can be blocked or enhanced pharmacologically, which results in upregulation or downregulation of both subjective and objective indices of pleasure and arousal. Pharmacological manipulation of the opioid system shows little or no modulation of subjective pleasure, in contrast to measures of arousal which are consistently modulated. This dissociation suggests that dopamine may play a more prominent role than opioids in the subjective experience of hedonic pleasures to aesthetic rewards, such as music.

8

Why Does Music Engage the Reward System?

We saw in the last chapter that pleasure from music is mediated by brain mechanisms that are partially shared with mechanisms which mediate responses to biologically relevant stimuli and behaviors. But none of that really explains *why* music should do so. What exactly are the characteristics of music that lead to these responses? And how is it that those characteristics lead to engagement of some of the brain's most basic circuitry for fitness and survival? This is the key question, and although we may not have all the answers yet, I think we are now in a position where we can reach successive approximations to good answers. And in fact, we already have several clues about plausible explanations from the findings reviewed so far.

One major clue to music's power to engage the reward system is provided by the huge body of evidence touched upon in many previous chapters indicating that predictive mechanisms play a critical role in many aspects of perception, action, and reward. We've specifically seen how predictive mechanisms enable dopaminergic responses, which emerge from computations that compare expected and received rewards, that is, reward prediction errors. A closely related concept is that surprise is important in giving rise to musical pleasure. This idea was already mentioned in the context of chills, which are often generated by novel or surprising moments in a musical passage (Sloboda 1991). We also know that surprises, or at least certain kinds of surprises, not only engage the reward system in the context of music (Shany et al. 2019) but more generally as well (Fouragnan et al. 2018). Surprise cannot happen if there is no prediction.

Musical expectations of upcoming events based on past events, and the surprise generated when those expectations are not entirely met (or, especially when exceeded), are the basis for some fundamental insights coming from music theory. Ideas originated by Leonard Meyer, a musicologist who worked in the mid-20th century (Meyer 1956), and recently developed more fully by David Huron, emphasize the importance of expectancies in generating musical emotions (Huron 2006). These and other scholars have developed the

From Perception to Pleasure. Robert Zatorre, Oxford University Press. © Oxford University Press 2024.
DOI: 10.1093/oso/9780197558287.003.0008

idea that musical emotions arise from the creation of musical tension and the anticipation of its eventual resolution. There are many musical devices that performers and composers use to manipulate tension. In tonal music, for instance, modulating away from the tonal center and returning to it later is a common way to create and resolve tension (as depicted, for example, in the few brief, but ravishing, bars by Brahms shown in Figure 8.1). Other familiar approaches include introduction of delays before the resolution to heighten the sense of tension; or changing the most commonly expected tonal sequences prior to a cadence via addition or prolongation of sounds (these devices are known to music theorists as ritardandi, appoggiaturas, suspensions, etc). We humans seem to be highly sensitive to the very structure of music, as it pushes and pulls us along by generating waves of tension and resolution.

One big sticking point—or to put it more positively, fascination—to explain why music is rewarding is its abstractness. Nonetheless, the review of the functional properties of the reward system in the foregoing chapters gives us clues about how and why abstract stimuli engage this system. The reward system responds to a wide variety of stimuli ranging along a continuum from basic primary rewards, like food, to secondary rewards, like money (which have a learned association to primary rewards), to more abstract entities like information and its attendant curiosity. In each case, it is possible to link the reward back to some fundamental biological imperative. Primary rewards are required for survival or fitness; secondary rewards can be used to obtain primary rewards (else, they lose all their value); and knowledge—even when not immediately useful—can enhance fitness, by guiding adaptive behavior in the future, especially when the environment is uncertain. Yet, it is possible to imagine that a brain that has evolved to value information for good biological reasons would also value it even in the absence of such reasons. So, this capacity to value information and to develop curiosity to obtain it might provide a kind of pivot to go from more real-world knowledge acquisition to entirely abstract information-seeking. Music may thus represent a kind of apex of this epistemic curiosity—a reward for learning.

Finally, we have seen the critical role played by the anatomical and functional networks that link the largely subcortical reward system to the perceptual, mnemonic, motor, and cognitive systems, chiefly instantiated via the cortico-cortical loops described in Part I of this book (Figure 6.3). The importance of these interactions for the ability to experience

FIGURE 8.1 Musical tension and resolution. The opening bars of Johaness Brahms' Intermezzo in A Minor, Op 116 No. 2, illustrate the buildup of harmonic tension and its resolution. Maximum tension occurs at bar 6 (red arrow), a relatively dissonant diminished seventh chord with a bass note of D#, harmonically distant from the tonic of A minor; this moment also corresponds to the highest pitch in the passage, adding more tension. Subsequent bars return to the original A minor key. Public Domain.

pleasure also provides us with a critical clue to understanding the source of musical hedonic responses: they represent the substrate by which learning and information-seeking, perception and action, prediction and surprise, all contribute to pleasure, reward, and emotion. In the rest of this chapter, I will evaluate these distinct pathways to pleasure, incorporating the role of surprise and information, and attempt to tie them together into a coherent, if still preliminary, model of why music has such power over us.

Predictive Processes and Their role in Musical Reward

Theoretical Considerations

First let us review the concepts of prediction and prediction error and how they might specifically apply to music. As we saw in Chapter 2, even at the earliest level of the auditory neural pathway, sounds are processed not merely based on their acoustical properties but also on how well they do or do not fit in relation to surrounding acoustical events. In other words, sounds are not analyzed in isolation but rather according to the context they are embedded in, and that is used to generate predictions of upcoming events. These predictions are compared to the actual input, and an error signal is generated in proportion to how unexpected the event is. We refer to these as sensory prediction errors because they pertain to changes in sensory input and occur within sensory pathways. Based on a lot of the physiological evidence reviewed in Chapter 2, we know that prediction signals propagate via descending (non-lemniscal) pathways, from higher to lower levels of the system, while sensory-based prediction error signals are propagated via ascending (mostly lemniscal) pathways, from lower to higher levels of the system.

At early levels of the neuraxis—in brainstem or midbrain nuclei, for example—context tends to be local, confined only to a few moments before the current event (see Chapter 2). Whereas at hierarchically higher levels, in and beyond the auditory cortex, context can involve working memory mechanisms in the auditory ventral stream (see Chapter 3) that allow integration across progressively longer timescales as one ascends the hierarchy (Farbood et al. 2015). Within the ventral stream there is also access to long-term representations, which form the basis for predictions based on learning of regularities and contingencies acquired over a lifetime. Long-term knowledge is essential, both for predictions based on statistical learning, which generally apply to regularities within a known musical system, as well as for predictions based on veridical knowledge of a specific piece of music that a person has learned (see Chapter 1). These predictions are largely pattern-based, that is, they are about predicting the content of the event that will happen in keeping with the concept that the ventral stream represents invariant object properties (see Chapter 3).

In parallel, the auditory dorsal stream (see Chapter 4) is also involved in predictive processing, especially in relation to evaluating the sensory outcomes of actions, via the perception–action cycle. These processes involve interactions between motor systems and the auditory cortex, with descending influences from motor cortical areas most likely providing the predictions. Unlike the ventral stream, however, these predictions are largely temporal in nature because the dorsal stream tracks events as they unfold in time. This makes sense because actions must be executed in real time; therefore time, just like space, must necessarily be encoded within the dorsal stream—unlike the ventral stream, which accumulates information over time and hence has to discard it in order to form stable long-term representations.

In all these cases, error signals that occur when the event does not match the prediction serve as attention and/or learning signals. Thus, prediction errors are critical to updating internal models (i.e., memory representations), which, in turn, means devoting cognitive resources (i.e., attention) in order to incorporate the new information. This account of perception has been very influential and applied to many different domains (Summerfield and de Lange 2014, Friston 2018), as discussed in Chapter 1.

The nature of prediction just described primarily pertains to perception and thus occurs at various levels of the perceptual pathways. But as we saw in Chapter 6, there is also a parallel prediction mechanism at play within the reward system. Reward prediction errors, mediated via dopamine signals, are valenced because they relate to whether a given stimulus has a higher or lower reward value than expected. As such, reward prediction errors need not necessarily be directly linked to the physical properties of a stimulus, unlike sensory prediction errors (for instance, identical physical stimuli may be more or less rewarding depending on internal states or environmental context). Yet reward prediction errors are nonetheless generated by the presence of a not-fully-predicted event in the environment, which means that the reward system receives inputs from sensory processing systems (see Figure 6.1), coursing along specialized sensory-specific pathways (auditory, visual, gustatory, etc.), as we saw in Chapter 7.

Classic reward prediction errors usually arise from a greater or lesser magnitude of reward (say, more or less food than expected); but it is also possible for them to arise from a sensory event that provides information, as discussed in Chapter 6. Therefore, the valence of the reward prediction error associated with such an event would depend upon its informational value: information relevant to learning is itself rewarding, and hence the degree to which a stimulus provides information that can be incorporated into one's internal model (or not) would result in a positive (or negative) value weighting. According to this view, the two types of prediction error (one related to perception and the other related to reward) would co-occur in most cases because the former causes the latter. Thus, although there are clear differences in the nature of prediction errors in the two domains, involving different neuroanatomical substrates and whether they are valenced or not, they can both be viewed as providing learning signals. In the context of music, I propose that they may often be elicited by the same acoustical events, such that a sensory prediction error generated in auditory pathways is thence propagated to the reward system.

How do these concepts of prediction play out, specifically in terms of musical structure? There are several ways in which music affords opportunities for predictive mechanisms to be engaged. Rohrmeier and Koelsch (2012) make the useful distinction that musical predictions can manifest, at a minimum, in terms of the nature of the events and of their timing; that is, a listener can generate expectations about *what* will occur as well as about *when* it will occur. This distinction maps nicely onto the ventral and dorsal auditory processing pathways, respectively. Both processes may occur at local levels (that is, focusing only on how the current event relates to its immediate predecessor) or also at more global levels (taking into account hierarchical musical structures that may have unfolded over longer stretches of time). This distinction maps nicely onto earlier versus later hierarchical stages along neural processing pathways.

Let's take an example involving prediction at a local level: a listener having heard a set of sounds with certain characteristics (timbre, loudness, pitch range, timing, texture, etc.) would expect subsequent sounds to be rather similar to past sounds, all else being equal; so, a sudden change in any of these variables would generate an error signal. Since they represent a local change in incoming acoustical information, such events would result in a cascading series of sensory prediction errors (such as those measured via the mismatch response, as described in Chapters 2 and 3) all along the ascending auditory pathway and within the ventral stream. Perceptual updating would also occur, together perhaps with orienting or attentional responses. But these sorts of changes might also generate a reward prediction error. However, reward prediction errors are, by definition, valenced. In the case of money or food, it's easy to see how getting more of it than expected is better, while getting less is worse. But how could a sound be considered better or worse than expected?

A critical distinction to make here is that, for a musically positive valence, the event should be relevant or informative in some way. Informative stimuli lower uncertainty, and as we've seen in Chapter 6, that by itself can engage the reward system. A discontinuity in the sound stream could have many different musical functions, some of which can play this informative role. One such prominent function would be to signal the start or the end of a structural section of the music. For instance, an unexpected harmony might signal a modulation to a new key, acting as a pivot from the old to the new key; in that case, the surprising chord serves to mark the shift. Another example, on a slightly more global level, is provided by the change from the verse to the chorus of a pop song, which is often accompanied by changes in instrumentation, tempo, vocals, etc.—any of which would signal that the next section has started; whereas a loud roll of the tympani, often heard at the end of a classical symphony, for example, tells us that the finale has arrived.

Thus, these types of partially unexpected sonic cues give us information to update or confirm our internal model of how the music is structured, as opposed to, say, the completely unexpected sound of someone dropping the crash cymbals in the middle of a performance (which happened to me once, as a hapless music student). Instead of providing useful information to the structure of the music, such a noise is merely distracting and annoying and will not be met with a pleasurable response from the listener. It would therefore constitute a negatively valenced reward prediction error, which inhibits dopamine neurons in the reward system (Figure 6.4). In more general terms, for any unexpected event, what matters is how well it can be integrated, either retrospectively by incorporating it into an ongoing working memory representation, or prospectively by incorporating it into a further prediction about what's coming next. Events that provide appropriate information to update the internal model would generate positive reward prediction errors. On the contrary, events that are useless and cannot be incorporated would generate negative reward prediction errors. As we shall see below, "better-than-expected" events are also related to intermediate levels of complexity, where prediction and surprise are in balance.

Beyond mere discontinuity in acoustical features, there are also characteristic musical patterns that can convey information. A good example is provided by David Huron, who has documented that stereotypical sound features (melodic figures, trills, repeated notes, etc.) frequently precede important cadences at the ends of musical phrases or sections. These are not only common in different genres of Western music but also occur in music coming from such diverse cultures as Chinese, Moroccan, and Pawnee (Huron 2006). These compositional devices serve to signal to the listener that a structural boundary is approaching (even in the absence of a major discontinuity in acoustical features) and hence provide information of high predictability, which in turn lowers uncertainty. Such musical elements are good examples of information facilitating prediction of upcoming events and, thereby, also serve an anticipatory function. And as we know, cues that signal an upcoming reward can themselves become highly rewarding after some learning (Schultz 2016).

Predictions need not be confined only to local elements (what is happening now compared to what just happened or what's just about to happen); they can also occur at more abstract structural levels of the music and over longer time windows. These operations are handled at correspondingly higher levels of the neural pathway (Farbood et al. 2015) and can generate prediction errors when a global feature is violated: for example, when the contour of a melody changes independently of its constituent tones (Tervaniemi et al. 2001) and in many other cases as well (Paavilainen 2013).

The existence of these different levels of abstraction for progressively more complex combinations of features implies a hierarchical organization. The hierarchical structure of Western music, at least, is the province of music theorists who have developed models of how different styles of music typically display complex embedded levels of melodic, harmonic, and metrical structure (e.g., Narmour 2000), all of which can be used to generate predictions at higher levels of abstraction. Furthermore, even relatively simple music would have the capacity to generate predictions not only in one dimension but simultaneously in several (pitch continuations, harmonies, rhythmic and metrical structures, and loudness). Under most musical listening situations, then, there would likely be multiple, parallel, and overlapping predictions being formed continuously and either fulfilled or not, as the music unfolds in time (Gebauer et al. 2012).

The relationship between musical surprise/expectedness and pleasure/preference is not straightforwardly linear. The first person to suggest that it follows an inverted-U shape was probably the German scientist Wilhelm Wundt, widely viewed as the founder of experimental psychology. He proposed that stimuli can vary along some dimension, such as intensity, for instance, and that intermediate values along that dimension would be preferred (Wundt 1904), as shown in Figure 8.2. This concept was taken further by Daniel Berlyne, who more than 50 years ago developed the idea that "arousal potential" (a construct that included complexity and familiarity) was related to hedonic value, such that it would be maximal at intermediate levels, where the stimulus is optimally engaging, without being either too simple, leading to boredom, nor too complex, leading to confusion (Berlyne 1971).

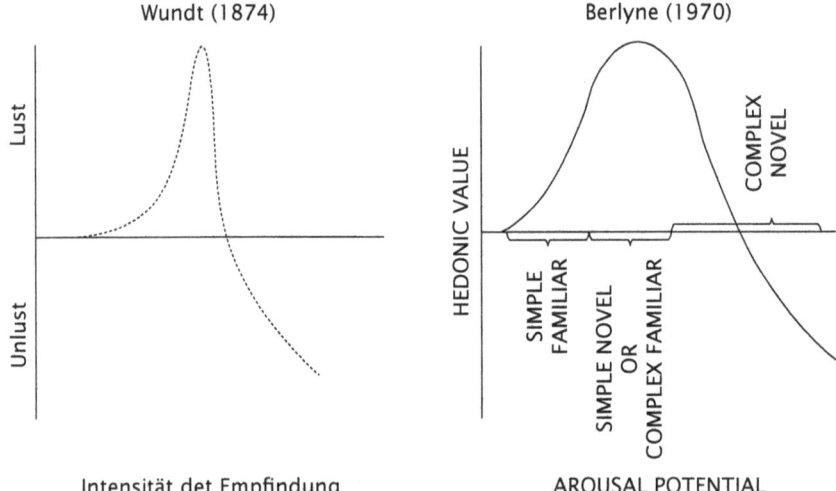

FIGURE 8.2 Inverted U-shaped preference functions. A: Original curve described by Wilhelm Wundt in 1874 indicates a peak in pleasure/desire ("Lust") at intermediate values of the intensity of the sensation ("Intensität det Empfindung"). B: Hedonic value peaks at intermediate values of arousal potential, a combination of complexity and similarity, according to Daniel Berlyne. Adapted from (Cutting 2022). Reproduced under the terms of a CC BY 4.0 license.

This concept has been adopted in various other contexts. For example, the literature on motivation for goal-directed activities (which includes anything from rock climbing to chess playing) posits that, for optimized enjoyment, a medium challenge is preferred (Abuhamdeh and Csikszentmihalyi 2012). The task involved should present difficulties but not be too easy (like climbing a tiny hill or playing chess with a child); nor should it be insurmountable (like climbing a sheer vertical wall or playing a grandmaster). Note that expertise plays a role here too since expert climbers or gifted chess players would indeed look forward to those extreme challenges. Comparable ideas have been developed in fields as diverse as robotics and education, where the basic insight is that learning, whether by a child, an animal, or a machine, proceeds best when there is an appropriate balance of complexity and surprise. In other words, learning is most effective when a model has enough to go on that it can generate predictions, but the outcomes of those predictions are not always perfect (leading to prediction error) and hence are informative for model updating and learning (Oudeyer et al. 2016).

The idea of an inverted U-shaped relationship between complexity and hedonic judgments has been specifically applied to music by several researchers. But although many studies have indeed demonstrated the expected nonlinear relationship, others found evidence of preferences for simpler, more prototypical patterns (Hargreaves and North 2010, Chmiel and Schubert 2017). Some of the inconsistency may be attributed to a lack of objective, quantifiable indices of complexity. But it may also be the case that preference for complexity is modulated by other factors that are not always considered. As we shall see in detail below, recent studies have clarified this issue, supporting the preference for intermediate complexity in musical structures (Cheung et al. 2019, Gold et al. 2019b) and attributing this pattern to predictive processes linked to the reward system.

Experimental Evidence

We already saw in Part I of this book that neurons throughout the entire auditory pathway respond to sounds as a function of the degree to which they are expected, at both lower and higher levels of the processing hierarchy. Within the ventral stream, for example, when musical chord sequences violate the most common statistical expectation, they elicit neural responses from auditory cortex and inferior frontal regions (Figure 3.13), especially on the right (Koelsch and Friederici 2003), in keeping with the right hemisphere's enhanced role in processing spectral information (see Chapter 5). These right anterior responses are not typically sensitive to expressive factors in performance even though they do modulate perceived pleasantness, showing a dissociation between sensitivity to some features compared to others (Koelsch et al. 2008). Therefore it makes sense to interpret those right frontal neural signals as indexing an aspect of sensory prediction error, based on a comparison of the current sound to stable (schematic) syntactic representations of harmonic chord progressions stored in long-term memory.

However, under many circumstances, unexpected chords can also elicit changes in subjective ratings of valence and emotional arousal along with modifications of skin conductance, all of which are modulated by the degree of expectedness (Steinbeis et al. 2006, Koelsch et al. 2008). This result supports the idea that expectancy violation also lead to a reward-related signal. Note that these reward responses might be positive (excitatory) or negative (inhibitory), depending on the nature of the unexpected event; although most of those earlier studies did not attempt to manipulate the valence of the outcome.

We already saw that arousal and peak pleasure moments (chills) in real music are often associated with partly unexpected events (Sloboda 1991, Grewe et al. 2007); and conversely that unexpected musical moments often lead to subjective increases in arousal together with enhanced psychophysiological and facial muscle responses (Egermann et al. 2013). Functional MRI (fMRI) studies also showed greater activity in auditory and reward areas as a function of unexpectedness of chords (Seger et al. 2013) and have demonstrated greater interaction between auditory and reward systems during surprising moments in real music (Shany et al. 2019). Those findings are also in accord with direct electrophysiological recordings from the brains of human listeners, which also show that auditory cortex activity is modulated by the degree of predictability of melodies (Omigie et al. 2019, Di Liberto et al. 2020). These findings, taken together, would therefore support the idea that the predictive processes that give rise to expectations play a role both in perceptual updating as well as in generating arousal and affective responses. In other words, surprising events can generate both sensory and reward prediction errors.

But what is the evidence that true reward prediction errors are, in fact, generated in response to musical events? All the work reviewed so far provides indirect support for the idea. But direct evidence favoring this specific reward prediction proposition has been largely lacking in the literature. To plug this obvious gap, it's necessary to show that the same formal criteria for the existence of reward prediction error that have been applied in other studies using traditional rewards (such as money) would work for music as well. Without such a demonstration, one could criticize the idea that reward prediction error really applies to music at all.

My former PhD student, Ben Gold, decided to tackle this issue head-on. Following closely on prior studies in which reward prediction errors were formally modeled in a reinforcement learning paradigm and were shown to be directly linked to activity in the ventral striatum (Gläscher et al. 2010, Daw et al. 2011). Ben devised a parallel experiment but using musical stimuli (Gold et al. 2019a). In a nutshell, in those prior nonmusical studies, volunteers are shown arbitrary visual symbols that are probabilisitically associated with winning or losing money. One has to learn, by trial and error, which symbols to choose in order to win the greatest amount of money. Successful learning will generally lead to optimal outcomes of winning the most money. But because the relationship between choices made and outcomes received is probabilistic, the reward will sometimes be less than expected (despite having made the right choices), while other times the reward will exceed expectations (despite having made some mistakes along the way). These outcomes would therefore correspond to negative and positive reward predictions errors, respectively. (This arrangement has always struck me as a kind of metaphor for life itself, where choices and rewards, alas, don't always align)

In Ben's experiment, he introduced a twist: instead of using an extrinsic reward (money) as the reinforcer, as in prior studies, he substituted consonant or dissonant versions of Bach chorales as the outcome to test if preferred (consonant) music is really intrinsically reinforcing and whether it can lead to learning. Listeners were not provided with explicit instructions but were only told to make the choices that led to whatever outcome they preferred, which they had to learn by trial and error (even though the probabilistic relation between choice and outcome meant that they could never be entirely sure of what would happen). Although the task was difficult, the group as a whole was able to learn the contingencies, showing that pleasant music can serve as a reinforcer. Most critically, because some trials led to positive and others to negative reward prediction errors, we could model these using the same techniques formally developed in monetary studies (Gläscher et al. 2010, Daw et al. 2011), enabling us to show that the expression of these reward prediction errors was clearly and prominently associated with brain activity in the right ventral striatum. Thus, this experiment goes well beyond the demonstration that pleasant music generates more activity in the ventral striatum (which is well-established by now) but specifically shows that the modeled responses to reward prediction error are directly linked to activity in this region.

Finally, it was also interesting to note that those individuals who learned well were the ones with the strongest reward prediction error-related response in the ventral striatum, compared to those who learned poorly or not at all. This latter finding, therefore, ties the engagement of this system by music into reinforcement learning, just as happens with more conventional rewards, which serve to motivate learning of optimal responses to maximize the most valued outcomes. Except that, in this case, the outcome was nothing more than pleasant music with its intrinsic reward value. And yet, the experiment shows that these music-related phenomena emerge from the same neural circuitry, based on the same reward prediction principles that govern responses to more basic rewards.

Experimental evidence thus indicates that both sensory and reward prediction errors can occur due to musical events, especially if they are not entirely expected. But a critical

aspect of the concept of prediction is that there should be distinct phases of neural responses occuring at different times: at a minimum, there should be a response associated
with the anticipation of an event and another associated with the processing and experience
of the anticipated event (Gebauer et al. 2012). As we saw in Chapter 6, in the context of both
animal studies and monetary tasks with humans, dopamine neurons respond both in anticipation of an upcoming reward (Knutson et al. 2001) as well as upon receipt of the reward
itself (Oldham et al. 2018). Does something similar apply to music too?

In the study described in Chapter 7, Valorie Salimpoor (Salimpoor et al. 2011) used
PET radioligands to show how the experience of musical pleasure was linked to dopamine
activity in both dorsal and ventral parts of the striatum (Figure 7.9A and B). But Valorie
was also interested in the potentially distinct roles that these two components of the striatum might play and whether they were temporally dissociable. So, she designed a second
experiment in which the same volunteers who had been scanned with PET listened to the
same music, but this time, while being scanned with fMRI. By selecting only the regions
that had shown dopamine release to music (the dorsal and ventral striatum), we could
infer that the blood oxygenation response from fMRI in these regions was also linked to
dopamine (Schott et al. 2008). And because we also knew when the moments of greatest
pleasure occurred (the chills moments, indicated by the listeners with a button press), we
could break down the time course of the fMRI signal (which, although slow compared to
EEG, is still two orders of magnitude faster than PET) according to whether it happened
before or after the peak pleasure. The result, as shown in Figure 8.3, was a remarkable dissociation: whereas the dorsal striatum was maximally active prior to the chills moment, its
activity declined at the moment of peak pleasure, at which time the ventral striatum showed
its greatest response.

The anatomical dissociation is especially interesting, given the different connectivity
of the dorsal and ventral striatum that we examined in Chapter 6. The dorsal striatum preferentially receives inputs from dorsal cortical regions, including dorsolateral prefrontal
cortex and the anterior cingulate. These areas are important for higher-order cognition, including temporal processes. The ventral striatum receives inputs from the ventromedial and
orbitofrontal cortex, areas important for establishing reward value. Given these differences
in connectivity, the mapping between dorsal versus ventral striatal activity and anticipatory
versus experiential (chills) aspects of music listening, respectively, fit rather well. According
to our interpretation, the anticipatory phase relates to the expectations associated with the
upcoming pleasurable musical event. These expectations would be computed in the dorsal
cortical circuitry, which is more involved in timing-related computations, and then communicated to the dorsal striatum as part of a reward prediction signal.

The experience of pleasure itself would be related to dopaminergic activity (maybe
along with opioids) in the ventral striatum and related structures, which are known to respond positively to the receipt of reward, especially if it is not fully anticipated (Schultz
2016). Inputs to ventral striatum from ventromedial and orbitofrontal cortex would signal
reward value. This reward value, in turn, would be computed via inputs coming from the
auditory ventral cortical pathway, based on how well the musical events in question fit into
the predicted pattern (i.e., if it was "better than expected" or not). As such, the rewarding

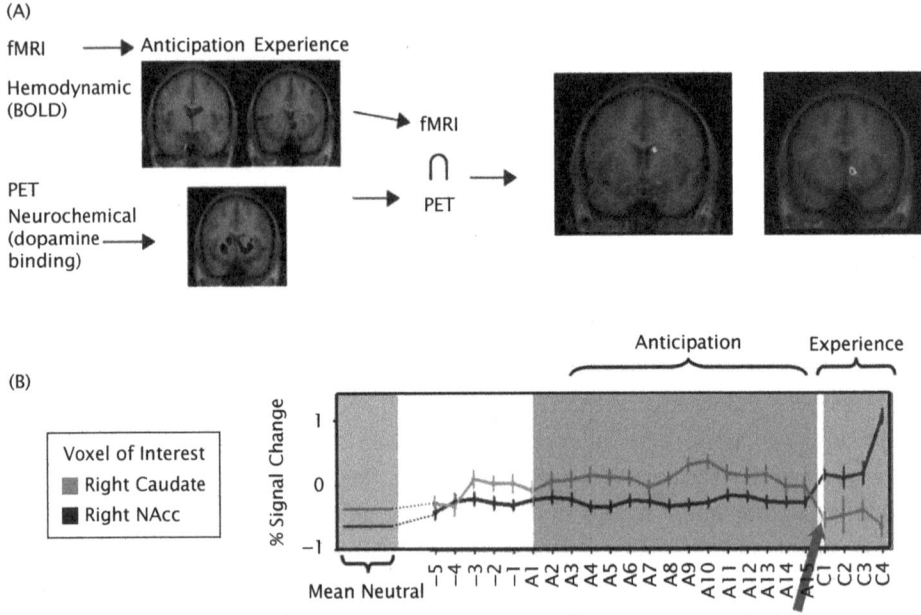

FIGURE 8.3 Anticipatory and experiential neural activity in relation to peak pleasure. A: (left) Combining functional MRI data from periods prior to (anticipation) and during (experience) peak pleasure with functional imaging data from dopamine-specific tracers yields two distinct anatomical regions (right): the dorsal striatum (caudate) during the anticipatory phase and the ventral striatum (nucleus accumbens) during the experiential phase. B: Time-series of functional MRI response corresponding to the two structures in A. Both regions show an increase during the anticipation period relative to baseline, but the dorsal striatum shows a higher response; at and after the peak pleasure, the ventral striatum activity increases while the dorsal striatum activity decreases. Adapted with permission from (Salimpoor et al. 2011).

experience in the ventral portions of the reward circuitry is generated by *what* event occurs, based on comparison with expectations. But *when* the event will happen is predicted in advance, from antecedent cues via a partially dissociable system—the dorsal reward circuitry—which itself generates a dopaminergic reward signal, as we shall also see below when we discuss musical groove.

It should also be remembered that although peak pleasure, as indexed by chills, is associated with maximal activity in the ventral striatum, its activity is also elevated, albeit to a lesser degree, in the moments just prior to the peak pleasure (Figure 8.3B) (Salimpoor et al. 2011). Furthermore, subjective pleasure ratings also show elevation during periods leading up to the peak response (Salimpoor et al. 2009). Therefore, the anticipatory phase preceding an important musical event is itself pleasant and associated with reward responses in both dorsal and ventral striatum: it's the "sweet" in David Huron's book, *Sweet Anticipation*.

This account of the anatomo-functional dissociation within the striatum is related to other interpretations of the dorsal/ventral distinction coming from more basic neuroscience. For example, neuroimaging evidence shows that dorsal and ventral striatal areas can be characterized in terms of more executive versus more motivational processing loops, respectively (Seger et al. 2010). Executive functions would include temporal processes, whereas motivational processing would be more closely related to emotional reactivity in line with the distinction between anticipation versus experience of musical peak pleasure. A similar view, coming from animal neurophysiological studies of reinforcement learning (Lee et al. 2012), proposes that the dorsal striatum is more involved in computing predictions, whereas the ventral striatum is more involved in computing prediction error. Again, this dissociation can be mapped onto music, in terms of prediction producing the anticipatory response and (positive) prediction error generating the pleasurable chills.

Precision Weighting

As we saw in Chapter 6, and also introduced in Chapter 1, the precision with which an upcoming event can be predicted influences the response of dopamine neurons in the reward system. This concept is related to how much uncertainty is present and the degree to which the appearance of the potentially rewarding event will reduce that uncertainty, that is, how much information it will provide. Recall also from Chapter 6 that in a situation with learned cues leading to a rewarding stimulus, there are two responses in the reward system: the faster, phasic response is directly and linearly related to the probability of reward—the greatest response occurs when the reward is unexpected, while the smallest response comes when it is entirely predicted.

But there is also a second, sustained response to a predictive cue that follows an inverted-U shape and is of particular interest to us here because it is related to precision weighting of predictions (see Figure 6.6). To understand this effect, we need to consider the degree of entropy or disorder that may be present: high entropy means a lot of uncertainty about what will happen next, whereas low entropy means that there is sufficient structure to make predictions, leading to greater certainty. In conditions of high certainty (or low entropy), where predictions can be quite precise, there is less response to the outcome because it is mostly anticipated and hence is not surprising; it does not convey much information (recall the banana example in Chapter 6). Whereas in a medium uncertainty context, when a cue does provide some information, but the appearance of the reward is not guaranteed, there is a gradual ramping up of activity in dopamine neurons during the anticipatory period, with a maximal response occurring when the reward finally comes. The reward is therefore experienced as more powerful because it is harder to predict (recall the melon example) and hence carries more information. As such, it can be used to update or enhance the internal model and improve future prediction.

But if the context is so high in entropy that it is random, and thus completely devoid of any predictability at all, then the sustained response in the reward system shuts down again. That's because no prediction about the appearance of the reward is possible, and hence it cannot be incorporated into any model. This phenomenon is observed in striatal dopamine neurons in monkeys (Fiorillo et al. 2003) and also in human fMRI data in the striatum and midbrain nuclei (Preuschoff et al. 2006), as we saw in Chapter 6 (see Figure 6.7). We may therefore consider this neural response as reflecting the precision with which a prediction is made.

These ideas were developed further in the context of music in a theoretical paper by Stefan Koelsch, Peter Vuust, and Karl Friston, who suggested that precision weighting could be considered a second-order prediction, that is, a prediction about the prediction (Koelsch et al. 2019). According to this view, the nervous system not only makes predictions about the content of an upcoming event but also generates predictions about the precision that should be ascribed to those first-order predictions. The first-order predictions would depend upon computations derived from interactions between ascending (prediction error) and descending (prediction) neural activity, whereas the second-order predictions would depend upon more diffuse modulatory influences.

How would this concept of precision weighting apply empirically to music? Koelsch and colleagues suggest that musical context provides information about the precision with which to expect a given event: in conventional Western tonal music, for instance, certain harmonic relationships are very common. So if one hears a chord sequence of, let's say a tonic, a subdominant, and then a dominant seventh chord, it is very likely that the tonic chord will appear next, because such a familiar tonal context allows one to be pretty certain that this outcome is likely. In that situation, little or no prediction error would occur, and the outcome would be judged as conventional, rather than exciting or especially interesting. Thus, well-structured musical contexts allow the generation of high-precision predictions.

Such musical environments can be considered as having low entropy, that is, low uncertainty. Hence, deviations from the highly expected events are all the more surprising and do generate prediction errors. For example, in the chord progression just mentioned, hearing a submediant chord instead of the tonic at the end, generates the classic "deceptive cadence," which is schematically unexpected and typically experienced as pleasurable. However, in more entropic musical contexts, for instance, where the composer has not established a clear metrical or tonal key structure, the listener won't be able to make as precise predictions, leading to an overall flatter prediction function. In such cases, hearing a dominant seventh followed by a tonic might be considered more pleasant because it couldn't be predicted so well in that environment.

Experimental evidence showing that the precision of the prediction influences both behavior and neural responses, comes from a clever study by Dominique Vuvan and colleagues from Isabelle Peretz's lab, who asked listeners to judge whether or not a tonal melody contained wrong notes (Vuvan et al. 2018b). After each trial, they were informed whether their answer was right or wrong; but in the experimental group—unbeknownst to them, of course—this feedback was entirely random. As such, the feedback could be considered as providing noisy, low-precision information, and indeed it led both to worse task performance, as well as lower subjective confidence in this group, compared to a control group that received correct feedback. Interestingly, the early right frontal electrophysiological response that usually accompanies unexpected tones (Koelsch and Friederici 2003) was attenuated in amplitude in the experimental group. If we take this response to index the presence of a prediction error, then this finding shows that the ascending feedback input was no longer considered very reliable and was thus discounted, in accord with the precision-weighting idea (Koelsch et al. 2019). The behavior elicited in this study was also reminiscent of what happens in amusia, which is in keeping with the idea, reviewed in Chapter 3, that amusia represents a disorder of feedforward/feedback connections.

Altering feedback is one way to modify precision. But another, arguably more direct, way to do so is to modify the musical context—the statistical relationships within the musical environment itself. As we have seen, the claim is that low-entropy/high-certainty environments lead to higher precision weighting on predicted outcomes and vice versa for high-entropy/low-certainty environments, where predictions are less precise or noisier and are therefore down-weighted. This idea can now be tested directly, thanks to advances in computational modeling of musical structures based on information theory, as developed by Marcus Pearce and his colleagues in London. This model, known as Informational Dynamics of Music (IDyOM), has been used in numerous studies, although there are competing models as well (Verosky and Morgan 2021).

Modeling Predictability

This mathematical approach creates a model that can predict upcoming events as a musical piece unfolds, based upon the cumulative statistics of many past events; in other words, it generates schematic expectancies. To do so, rather than relying on explicit music-theoretic rules and principles as has been traditionally done, IDyOM mimics the implicit statistical learning that is presumed to happen in the listener's minds. The model provides quantitative estimates of two variables: (i) the complexity, or information content, of a given musical event within a sequence, based on how well it conforms to learned statistical priors (i.e., how predictable/surprising it is—the greater the surprise, the greater its information content); and (ii) an estimate of its entropy, which can be conceptualized as the precision with which the musical environment allows predictions to be made in the first place.

To generate these values, the model is first trained on a large corpus of musical samples to emulate long-term implicit knowledge of the statistical regularities of a given musical system. In addition, IDyOM is also trained on the structure of the specific musical items that are used, in any given experiment, to account for the fact that listeners update their knowledge of a given piece of music in real time as they listen to it (Oram and Cuddy 1995). The outcome variables, namely information content and entropy, reflect the degree of surprise experienced upon hearing a given sound event and the degree of uncertainty prior to experiencing that event, respectively. As such, surprise is more retrospective, while uncertainty is more prospective. These two variables nicely map onto prediction error and precision weighting, respectively.

In practice, these two variables tend to be correlated, although it is possible to find musical pieces that are relatively higher on one and lower on the other (see Figure 8.4). Note that the model generates these two values for any given piece of music but that they are based not only on the music's features but also on how those features would be interpreted by a listener with a given, prior implicit musical knowledge, as modeled by IDyOM. The same music, if presented to listeners with different exposure, would not necessarily be expected to have the same level of surprise/uncertainty. Thus, these estimates of complexity

FIGURE 8.4 Complexity and uncertainty in real musical excerpts. Four musical excerpts of varying levels of complexity (surprise) and uncertainty (entropy) are plotted as a function of those two variables. From bottom left to top right: Joachim Andersen, Studies for Flute, Op 41, no 18; Marin Marais, Folies d'Espagne, no 8; Robert Schumann, Fantasiestücke No 1; Kazuo Fukushima, Mei. Stimuli used in (Gold et al. 2019b).

and uncertainty are a joint product of musical features and listener knowledge, unlike classical music theory, which only considers musical structure, as notated in a score.

Armed with this more formal and quantifiable way to model the role of these factors, several labs have tested how they influence perceptual and neural responses. To examine specifically the role of entropy at a behavioral level, Niels Hansen and Marcus Pearce had listeners rate the expectedness of a tone that appeared at the end of musical sequences (drawn from hymn tunes and Schubert songs) that had been selected to have either high or low entropy (Hansen and Pearce 2014). Listeners experienced greater uncertainty in the high-entropy musical context. In other words, they were less sure of what the continuation tone ought to be, thus validating the idea that the precision of the prediction was weaker in those "noisier," less certain environments. The effect was present for both musically trained and untrained people, indicating that formal musical training is not required to be able to apply implicit musical knowledge. But the effect of entropy was more prominent for the musicians, suggesting that training enhances the learned representations from which predictions can be made.

To examine the neural correlates of this phenomenon, investigators at Aarhus University compared the MEG response to pitch deviants, when presented within very simple repetitive melodic contexts (low entropy), compared to the same deviants in somewhat less predictable contexts (higher entropy) (Quiroga-Martinez et al. 2019). They found a mismatch response coming from the right auditory cortex for pitch deviants (as expected, based on data presented in Chapters 3 and 5). But the more important finding was that this response was attenuated for the high entropy conditions, again showing that the uncertainty associated with the less predictable (lower-precision) context translates into a

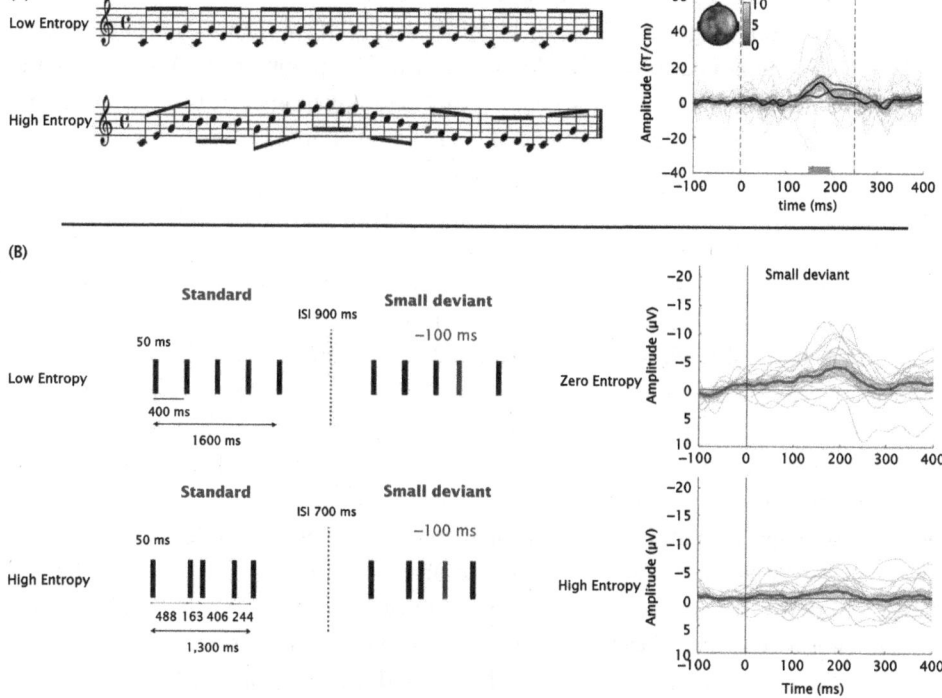

FIGURE 8.5 Neural responses to deviants as a function of low or high entropy. A: Left side: Pitch deviants (50 cents out of tune, red notes) were presented in either a low entropy context (top) or a higher entropy context (bottom). Right side: mismatch responses were greater in magnitude for deviants presented in low entropy (high certainty) econtexts (red trace) compared to the same deviant in a higher entropy (less certainty) context (blue trace. Blac trace represents difference between conditions. Reproduced from (Quiroga-Martinez et al. 2019) with permission from Elsevier.

B: Left side: Standard patterns for low entropy (top, events are presented isochronously) and for high entropy (bottom, events are presented nonisochronously); deviants in each case are the same 100 ms displacement from the standard pattern. Right side: mismatch responses were greater to the deviant presented in the low entropy context (top) compared to the high entropy context (bottom). Used with permission of John Wiley & Sons—Books, from (Lumaca et al. 2019); permission conveyed through Copyright Clearance Center, Inc.

less salient prediction error signal (see Figure 8.5A). A comparable effect was also documented in another study for rhythmic, as opposed to melodic, patterns such that deviants occurring within more complex (higher entropy) rhythms elicited lower-amplitude mismatch responses than the same deviants presented in a low-entropy isochronous sequence (Figure 8.5B). This finding shows that the down-weighting of prediction errors in uncertain contexts occurs in the temporal domain as well (Lumaca et al. 2019).

Predictability and Uncertainty
Relation to Pleasure

This work tells us about the role of precision weighting on sensory prediction error responses. But how is this related to pleasure? If the central claim of this book– that the mechanisms related to prediction are responsible for pleasure responses—is correct, then precision weighting ought to influence hedonic aspects of music and not only perceptual ones. Two studies have recently addressed exactly this point, one by Vincent Cheung working in Stefan Koelsch's group (Cheung et al. 2019) and another by Ben Gold from my lab (Gold et al. 2019b). Both studies used similar computational models based on IDyOM and examined the interaction of complexity (information content) and entropy on pleasure ratings. However, whereas the first team focused strictly on isochronous harmonic progressions derived from pop songs without using their melodies, our group selected brief melodic patterns without harmonization taken from real music of various styles (everything from Telemann to Chick Corea). These differences make the two studies quite complementary.

At a behavioral level, both studies found almost identical results. Subjective ratings of pleasure were strongly modulated by both information content (surprise) and entropy (uncertainty) but in an interesting way that reflected the interaction between these variables. First, looking at the information content variable alone, the highest pleasure was associated with intermediate values, such that people preferred musical structures that were neither too complicated (difficult to make predictions about, and hence, hard to follow) nor too simple (too predictable, lacking in surprise, and therefore boring). This resulted in the classic inverted-U curve discussed above but with the advantage that the complexity dimension was based on an objective and theoretically motivated measure.

But the novel finding that consistently emerged from the two studies was that entropy modulated this function, so that in a musical context that was very uncertain (high entropy), listeners preferred less surprise and more predictable patterns; whereas in an environment that offered greater certainty (lower entropy), there was a preference for at least some level of surprise. This interaction is demonstrated in Figure 8.6. It's as if people become more conservative when the situation becomes more chaotic and uncertain, preferring the comfort and safety of known, unsurprising outcomes. On the other hand, in a more structured, safer situation, they are willing to take a bit of a risk and actually enjoy experiencing some surprise. This outcome is reminiscent of studies discussed in Chapter 6, where rodents are more likely to explore to obtain unknown rewards if the environment is safe and orderly, rather than if it's uncertain and potentially dangerous. Our listening preferences are not so different, perhaps, from the preferences that drive very basic behaviors that we share with other creatures, and that's because they are based on fundamentally similar principles.

How do the convergent findings from these two studies fit with the effect of precision on prediction error? We know from all the studies just reviewed in this section, as well as in Chapter 6, that more certain predictions can lead to greater prediction errors because a deviation from expectation is more easily detected when the expectation is sharply defined. In contrast, in the case of an uncertain prediction, lower-magnitude prediction errors are generated since the expectation was not clear in the first place. But a critical factor here is that the opposite is the case when the predicted event actually occurs exactly as expected. Under those circumstances, there is a relatively low prediction error for a high-certainty prediction, but greater (positive) prediction error for low-certainty situations because, by definition, the prediction was not so well-specified in those latter cases. So, if uncertainty is high, low-surprise, conventional or standard patterns are preferred.

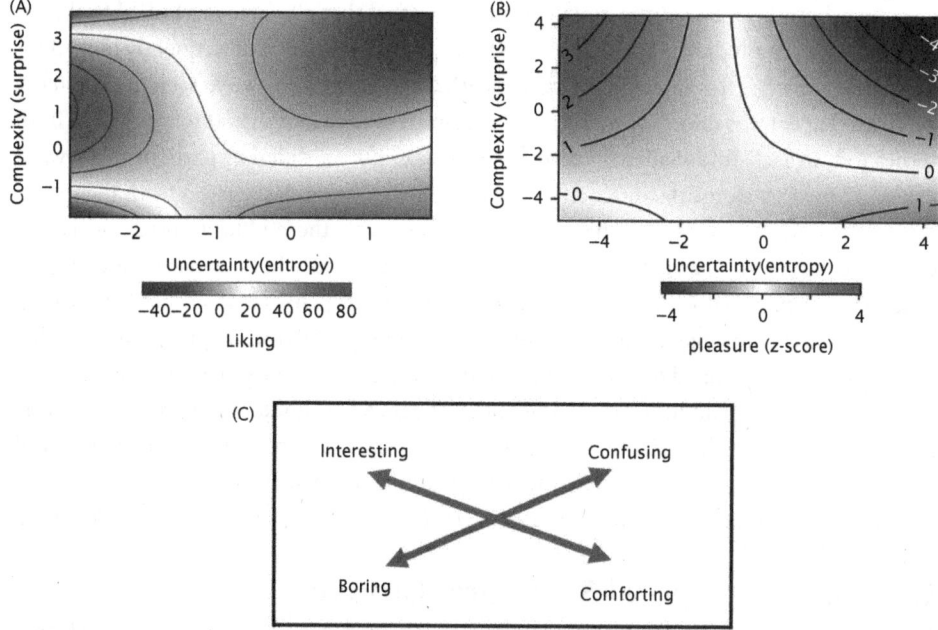

FIGURE 8.6 Preference as a function of complexity and uncertainty in melodic and harmonic contexts. A: Liking rate (red = more liked, blue = less liked) for melodies varying in uncertainty (entropy) and complexity (surprise) plotted as contours. Data redrawn from (Gold et al. 2019). Reproduced under the terms of the Creative Commons Attribution 4.0 International license (CC-BY).

B: Pleasure rating (red = more pleasure, blue = less pleasure) for harmonic progressions varying in uncertainty (entropy) and complexity (surprise) plotted as contours. Reproduced from (Cheung et al. 2019) with permission from Elsevier.

C: Theoretical explanation of the patterns seen in A and B: Top left and bottom right corners of the uncertainty-complexity space are preferred because they lead to interesting or comforting outcomes, respectively; bottom left and top right corners of the space are not preferred because they lead to either boring or confusing outcomes.

These ideas help to explain how preference and pleasure are expressed as a function of the two variables. If we look at the diagrams in Figure 8.6, we see that positive affect is roughly associated with the top-left and bottom-right corners, corresponding to low uncertainty–high surprise and high uncertainty–low surprise, respectively. I would interpret these positive hedonic responses as reflecting the desire for interesting versus comforting events in the two cases, both of which yield better-than-expected reward prediction error. In a predictable context (say, standard tonal music), a bit more surprise is needed to elicit pleasure; in a less predictable context (such as provided by a less clear tonal texture), we prefer more comforting outcomes. On the opposite axis, negative affect is associated with boredom or confusion when the stimuli are either too predictable or too unpredictable, respectively, yielding worse-than-expected reward prediction error by virtue of their lack of useful information (either too little or too much). Real music would most likely traverse this space at different time points within a given piece, providing enhancement of tension and resolution by contrasting moments that differ along the two dimensions. Comfort is more valuable when you've just been confused or disoriented, while interest is more valuable when you've recently been lulled or bored.

Since both Koelsch's group and my group observed that pleasure was linked to the interaction between predictability/surprise and uncertainty, both also subsequently looked for a neural signature of this interaction, using fMRI. Given everything we've seen in Chapter 7, one might expect this effect to show up in the reward system and in auditory regions. Both studies did, in fact, find evidence for this conclusion but in different parts of the reward system. In the first study, the interaction between the two variables derived from the model was expressed in the amygdala/hippocampus, along with the auditory cortex bilaterally (Cheung et al. 2019). The ventral striatum was involved but only in response to uncertainty, rather than surprise, which is, well, surprising, especially as many prior studies have shown that surprise reliably engages the striatum (Fouragnan et al. 2018). In a new sample of listeners, Ben Gold replicated the same behavioral interaction as in the other two studies (Gold et al. in press). But, in addition, there was some evidence for an interaction between uncertainty and surprise in the striatum. Thus, both studies show that neural responses in reward and auditory networks are sensitive to surprise and uncertainty, even if some of the details of which structures respond to harmonic versus melodic features remain to be worked out.

Relation to Temporal Structure: Groove

Why does music so often make us want to tap, swing, or dance (Large 2000)? And why are those movements usually so enjoyable? Almost all the research discussed up to this point in this chapter has focused on predictability associated with melodic or harmonic structure. But the same concepts apply to temporal structure and, in particular, are very relevant to the concept of groove. Although this term is used in specific ways by musicians within certain musical styles, a technical definition of groove has been developed in the cognitive science of music, where it refers to the pleasurable sensation of wanting to move to music (Janata et al. 2012).

Groove is consistently experienced by people from a variety of different cultures (Madison 2006). The phenomenon is intimately tied to sensory-motor coupling in the dorsal stream, as was discussed in Chapter 4. But it is also linked to hedonic responses, as shown by the finding that the more people express a desire to move to music, the more they report enjoying it (Janata et al. 2012). Therefore, according to the ideas promoted in this book, there should be a link between motor and reward systems that would explain that relationship. To explore this effect, we can examine the same kinds of predictive processes that have proven important to understanding the involvement of the reward circuitry in the other musical domains described in this chapter.

One important clue about groove is that it is related to the presence of syncopation. Syncopation can be defined in terms of violations of metrical expectations (Temperley 2010), that is, the presence of events at time points where they were not entirely expected (especially off the beat), and their concomitant absence when they would be expected (on strong beats). This musical definition of syncopation puts it squarely in the camp of predictive mechanisms. And since groove rhythms contain more surprise compared to metrically regular rhythms (because events occur at less-expected moments), we might expect that they would obey the same Wundt curve that we saw earlier for melodies and harmonies, if the same principle of preference for intermediate complexity holds.

That outcome is indeed what Maria Witek and colleagues at Aarhus University were able to document. They asked listeners from a variety of backgrounds to listen to drum breaks, mostly taken from funk music, that varied systematically in their degree of syncopation (Witek et al. 2014). As seen in Figure 8.7, ratings of pleasure, as well as ratings of wanting to move, both peaked at intermediate levels of syncopation, indicating that sensory-motor (covert) entrainment is most pleasurable when there is some degree of violation of the simple, regular metrical structure. When syncopation is entirely absent, the music is too regular, and metrical expectations are largely confirmed, meaning that there is little prediction error and hence little pleasure. On the other hand, when syncopation is too high, expectations are too weak to be violated because the complexity disrupts the ability to form an idea of the underlying meter, also leading to a reduction in pleasure. The parallels with the pitch-based finding described in the previous section are obvious. In this experiment, although musical training did not influence the results, those with high levels of dance experience showed significantly higher ratings of both the desire to move and of experienced pleasure. But, in a further study, musical training was specifically found to sharpen the shape of the Wundt curve to groove rhythms (Matthews et al. 2019). Therefore, experience of one sort or another results in learning, which in turn generates more precise

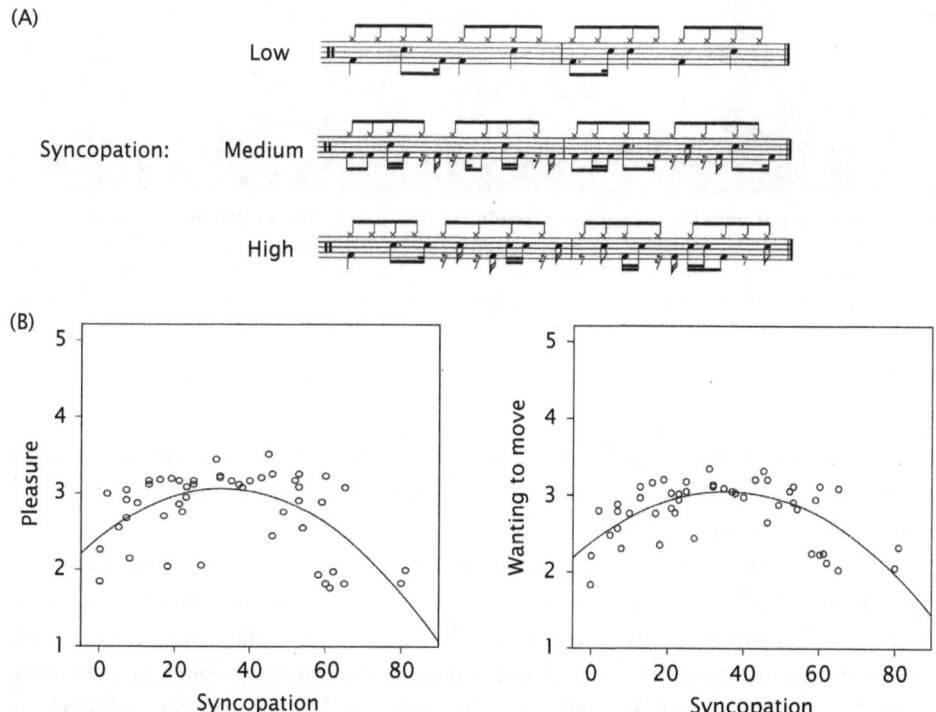

FIGURE 8.7 Preference curves for rhythmic stimuli. A: Rhythms having different levels of syncopation. Medium syncopation is typically referred to as groove. B: Ratings of pleasure (left) and wanting to move (right) follow an inverted U- shape as a function of syncopation. Images courtesy of M Witek (see: (Witek et al. 2014)). Reproduced under the terms of a Creative Commons Attribution License.

internal models and can also lead to greater enjoyment. The more you dance, the groovier you get.

The next step in this research domain, of course, was to test the hypothesized involvement of the reward system in relation to groove. Tomas Matthews, working with Virginia Penhune and the Aarhus group, set out to address this question by using fMRI while volunteers listened to stimuli that varied in their syncopation/groove (Matthews et al. 2020). Replicating prior behavioral studies, rhythms with intermediate syncopation were rated as both more pleasant and as inducing a greater desire to move compared to those with high levels of syncopation, thus validating the stimuli. Analysis of the brain activity elicited by the intermediate versus the high complexity rhythms showed activity in essentially the entire auditory-motor dorsal stream: dorsal premotor cortex, supplementary motor area, intraparietal sulcus, and dorsolateral frontal regions, as shown in Figure 8.8. Furthermore, a strong response was found in portions of the basal ganglia, particularly the putamen. These findings are of interest because there was no overt movement during the presentation of the rhythms; yet, the motor network was engaged, just as we saw in many studies in Chapter 4.

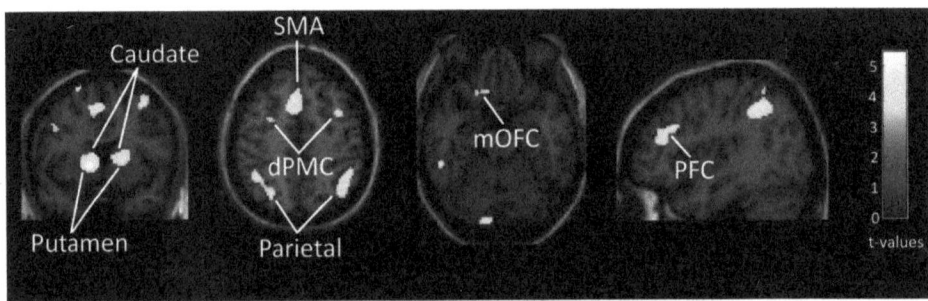

FIGURE 8.8 Neural activity for groove stimuli. Functional MRI data comparing medium complexity (higher groove) rhythmic stimuli to higher complexity (lower groove) stimuli shows activation in caudate (dorsal striatum) and putamen, supplementary motor area (SMA), dorsal premotor cortex (dPMC) and parietal cortex, medial orbitofrontal cortex (mOFC), and prefrontal cortex (PFC). Reproduced from (Matthews et al. 2020) with permission from Elsevier.

But unlike most of those studies, the activity was not simply related to more complexity in the metrical structure, but rather it was related to the intermediate complexity characteristic of groove stimuli.

Further analysis of structures within the reward system revealed greater activity for the groovier stimuli in the ventral and dorsal striatum and in the orbitofrontal cortex, confirming the hypothesis that the stimuli with intermediate syncopation should engage the reward system. But remember that groove is defined as comprising both a pleasure component and a wanting-to-move component. To understand the contributions of these two distinct factors, Tomas computed the relationship between ratings of those variables, for each stimulus within the different regions of interest, to see how much of the variance in the response could be accounted for by factors above and beyond the beat strength of the stimuli themselves. Although there was a great deal of shared variance for both pleasure and movement ratings, the neural responses within the ventral striatum tended to be more related to reported pleasure, whereas the responses in the dorsal striatum tended to be more related to the movement ratings.

Putting it all together, these findings are consistent with the global concept that the groove phenomenon is related to the interaction between motor and reward systems. More specifically, in line with what we saw in Chapter 4, the study supports the idea that the dorsal-stream cortical components, together with the basal ganglia, are responsible for generating predictions about beat structure (*when* events are expected to happen) and that when these predictions are violated to a moderate extent, they generate prediction errors. Those error signals would propagate from dorsal motor cortical regions to the dorsal striatum, as we know from the anatomy described in Chapter 6 (Figure 6.3), and thence to the ventral striatum where a valenced (positive) reward signal would be generated if the deviation is neither too large nor too small. The desire to move would, therefore, emerge more from the motor system interactions with dorsal striatum and putamen, while the more hedonic component would emerge from the ventral striatum. This interpretation parallels the explanation for valenced reward signals in the case of tonal patterns, which, as we saw in the earlier part of this chapter, entail interactions between the auditory ventral stream and the reward system.

Relation to Musical Elements and Styles

How might the ideas proposed so far apply to real music, with all its variety of styles and structures? The thesis I defend is that sensory prediction errors, generated in the auditory pathways, also spread to the reward system, where they result in valenced (positive or negative) reward prediction error signals, which in turn generate pleasure (or displeasure). Under this scenario, then, the preference for medium surprise in low uncertainty environments can be explained by the relative enhancement of prediction error signals, compared to their down-weighting in high uncertainty environments (Lumaca et al. 2019, Quiroga-Martinez et al. 2019). As we saw from previous neurophysiological and imaging studies, when a better-than-expected outcome occurs, there is a maximal reward system response.

By the same token, in an uncertain, high-entropy context which generates less precise predictions, there is a preference for confirmation (little or no surprise) because even when events occur just as predicted, the prediction was uncertain to begin with thus generating some degree of sensory prediction error. In turn, if the events, despite not being highly predicted, nevertheless provide valid information that can update an internal model, they would be expected to generate a positively valenced reward prediction error signal in the reward system.

A lot of music seems to seek the optimal middle ground, where there is a balance between sufficient structure (to be able to make good predictions) and sufficient surprise (to maintain interest and generate pleasure). This conclusion is in accord with the concepts of optimal arousal potential/challenge and hedonic value that have been developed in many contexts (Wundt 1904, Abuhamdeh and Csikszentmihalyi 2012), including the context of musical aesthetics (Chmiel and Schubert 2017). However, there is little doubt that it's difficult to find this ideal sweet spot, which is one reason why not every piece of music ever written is a hit—not by any stretch of the imagination.

But musicians have also figured out how to exploit the entire range of uncertainty, to expand ways to generate aesthetic appeal. Some styles of music may take advantage of the certainty provided by low entropy: an illustration of this idea is provided by minimalist music, which is very low in uncertainty, due to its highly repetitive nature and its use of a limited number of elements. Low entropy leads to very precise predictions, with sharp boundaries about what's expected. Once you've heard, let's say, many repetitions of a single chord for a long time, the expectancy is set up that it will be repeated some more with exactly the same rhythm or timbre (I'm thinking of Steve Reich's now-classic piece of minimalism from 1970, *Four Organs*). Under these circumstances, composers can employ very small deviations in the pattern, (e.g., adding one more note to the same chord) or the timing (e.g., presenting a new sound just off the beat), which, because of the high precision of the predictions, can become quite salient and generate pleasure, at least in those listeners who possess a high tolerance for repetition.

The opposite situation might be obtained in some styles of jazz, in which a theme is stated and then the musical texture can progress to a very high level of entropy, leaving the listener unsure of what will happen next. In many cases, there is an eventual return to the original theme near the end of the piece, which is often met by sighs of recognition and appreciative clapping from the audience. In such cases, the familiar theme serves as a confirmation, generating positive reward prediction error. Something similar happens in a classical sonata form, where the theme returns in the recapitulation after a development section which often contains a lot of unpredictable material higher in entropy than the surrounding sections. In these cases, the more entropic section serves as a contrast, to generate a higher reward value to the subsequent confirmation.

The interplay between uncertainty and surprise can play out, simultaneously and in real time, across different elements of music. For instance, an expected melodic event occurring in a metrical context where the beat is unclear; or conversely, an unexpected timbral change might occur but within a predictable harmonic progression. As we saw in earlier chapters, the brain can track multiple simultaneous streams of ongoing sound (Fujioka et al. 2005). So, the way that these different levels of prediction play out across features can provide an infinite variety of possibilities in real music. Another possibility, which has not been studied much to my knowledge, is that the intermediate levels of complexity and uncertainty that listeners prefer could be manifest at different timescales within a piece of music. In other words, there may sometimes be a kind of fractal organization, so that a local element (say, a chord progression) may be high or low in complexity in terms of the relationship between adjacent sounds, yet it might play a different role across longer time frames (when that same chord progression forms part of a higher-order musical structure, which itself has a different level of complexity). As we saw earlier, longer-scale music structure is tracked by more distal portions of the cortical processing streams. Therefore, it would be interesting to see how prediction tracking at these different levels could generate multiple, embedded predictions, at different timescales.

Meanwhile, in still other styles of music, there is not necessarily any intent to resolve tension or reduce uncertainty at all. Instead, the musician may actively desire to convey a sense of anxiety, desperation, or angst. For this purpose, chaotic, high entropy passages are excellent: some types of heavy metal music may fit into this category, as they are well-suited for conveying or expressing anger and other darker emotions (Thompson et al. 2019). Composers of horror movie soundtracks are particularly adept at deploying this high-entropy approach, to enhance the terror felt by the viewer. Some of these concepts may apply to atonal music as well (Mencke et al. 2019). Even in these situations, though, there may still be some kind of higher-order prediction at play—someone who goes to hear a heavy metal concert or to watch a horror film *expects* to hear music that expresses anger, anxiety, or fear, and fans of such genres often report pleasure or other positive emotions from the experience (Thompson et al. 2019). This is an area that seems promising for further in-depth study.

Relation to Learning

The role that information can play as a reward is intimately tied to learning, as we have seen in many experimental situations reviewed in this book. If that is the case, then the amount of information one has previously acquired—either implicitly via statistical learning and/or explicitly via formal or informal training—should modulate the rewarding value of information. This effect of knowledge accumulation, or expertise, could be implemented via changes to predictions by making them sharper. More precise predictions would lead to enhanced neural responses when sounds deviate from those predictions, in turn leading to more effective updating of internal models. Accordingly, synergies would emerge between pleasure (mediated via the reward system), memory (mediated by the medial-temporal lobe system), and cognitive/perceptual processes (mediated via the dorsal and ventral corticocortical loops).

In Chapter 2, we noted that even at relatively early stages of processing, people with musical training show more precise encoding of acoustical information. The consequences of training are demonstrated, for example, in the frequency-following response, which is of higher amplitude and lower variability in musicians (Musacchia et al. 2007) and which correlates with musicians' better pitch discrimination ability (Bidelman et al. 2011, Coffey et al. 2016b). In a prediction framework, this phenomenon can be restated as indicating that prediction errors in musicians would be generated and transmitted to later levels of processing even for small changes in frequency, as compared to the average listener. In turn, this would lead to higher precision in the prediction, that would be fed back to earlier levels. As a corollary of this idea, musicians ought to be better than average at detecting small deviations in pitch, which, of course, is generally known to be true and has also been shown experimentally (Micheyl et al. 2006).

At later levels of hierarchical processing, mismatch responses are also sensitive to learning. As reviewed in Chapter 3, these responses encompass a role for context, since they display sensitivity to how well a target sound fits into the pattern established by earlier sounds. Hence, they represent integration of information over a longer time window. Many studies have reported that mismatch responses can emerge over the course of learning within an experiment, demonstrating that the precision of predictions can be improved with appropriate feedback over the short term (Tervaniemi et al. 2001, Reetzke et al. 2018). It is also well-established that mismatch responses originating in auditory cortical areas are larger in amplitude and shorter in latency among musically trained people for both harmonic changes (Koelsch et al. 1999, Brattico et al. 2009) and metrical changes (Vuust et al. 2009); and that such differences emerge in children as they acquire musical expertise (Putkinen et al. 2014). Furthermore, the electrical potentials originating in frontal cortical regions that can be measured to deviations from expected musical syntax, and that therefore depend on long-term representations of musical regularities of the sort modeled by IDyOM, are also enhanced in musically trained individuals (Koelsch et al. 2002b).

All this empirical evidence supports the idea that the better the knowledge base from which sensory predictions are derived, the more precise the predictions can be. That knowledge base would become more extensive as one's musical performance training grows because of the interactions between motor and auditory systems that would be more developed in those who learn to play an instrument, as we discussed in Chapter 4 (Alluri et al. 2017). There is also intriguing evidence for greater functional connectivity between dorsal auditory-motor regions and dorsal striatum in musicians (van Vugt et al. 2021), which could play an important role in mediating enhanced hedonic responses.

We should recall, however, that musical sophistication can also be acquired via listening and acculturation, even in the absence of formal training (Bigand and Poulin-Charronnat 2006). Therefore, some of the effects attributed to formal training may be related to musical exposure and/or interest, independent of the ability to play an instrument. It is also likely that personality traits, such as novelty-seeking or openness to experience, would play a role in learning about music. This would mean that those individuals with greater curiosity not only might listen to music more attentively but may also expose themselves to a wider range of music (and perhaps, would be more likely to seek out training), all of which would sharpen their internal musical representations even further.

The advantage of musical training for prediction was shown behaviorally in a study in which musicians were shown to be overall more certain about how a given melodic sequence would continue, compared to untrained listeners, thus indicating that their more precise representations generated better predictions (Hansen and Pearce 2014). That effect also interacted with entropy values of the stimuli, such that for high entropy stimuli, the musicians did not differ much from the nonmusicians, since the stimuli did not afford very high possibilities for strong expectations. But in the low-entropy stimuli, the musicians were better able to use the clearer structure to generate stronger predictions. A comparable effect was observed in Ben Gold's experiment but in relation to pleasure (Gold et al. 2019b). Listeners whose responses followed the Wundt curve in that experiment were more likely to be musically sophisticated and also showed sharper (more peaked) functions. We interpreted this finding as evidence that their pleasure was derived from a more precise internal model.

In a broader context, these ideas may also be applied to implicit knowledge of different musical systems that would vary across populations, depending on their musical culture. An advantage of approaches such as IDyOM, that learn contingencies based on the statistics of the sounds it is exposed to, is that they can adapt, at least in principle, to the sounds of any musical culture, unlike traditional music-theory concepts, which are rooted in a single particular musical system. One attempt to analyze this question empirically compared the predictive judgments of expert jazz musicians versus classically trained musicians and nonmusicians when listening to jazz music (Hansen et al. 2016). The authors found that the jazz experts' explicit judgments of expectedness differed from those of the others, indicating that they had better conscious access to internalized knowledge that was pertinent to that particular musical system. However, even the nonexperts still demonstrated some implicit knowledge of the jazz style. Whether this is because of their occasional exposure to jazz, or because of shared features of the two systems, remains to be seen. But it does raise the interesting (and highly plausible) hypothesis that people are able to hold more than one probabilistic framework in long-term memory, perhaps just as multilingual individuals can apply the relevant rules of the languages they know to whichever one they happen to be listening to at any given moment.

With regard to enculturation of non-Western musical styles, although there has been scant work to date, IDyOM can be trained on musical systems outside the conventional Western system. When this is done, for example, with Chinese musical examples, IDyOM returns very different sets of predictions (Pearce 2018), as one might expect. In future, it will be interesting to see how well these models predict the behavioral and neural responses of listeners who are enculturated in one system to sounds of a completely different and unknown system, and how responses to the same system differ as a function of knowledge and exposure.

An important aspect of learning is that it depends on the motivation to obtain information, that is, curiosity. In Chapter 6, we reviewed extensive evidence that individual differences in personality traits related to curiosity, including openness to experience and novelty-seeking, are directly linked to the dopaminergic reward system. Consequently, people with higher novelty-seeking traits show greater reward-system reactivity to various rewarding stimuli, including even euphoria-producing drugs (Depue and Collins 1999, Smillie 2008). The concept of reward sensitivity is also related to curiosity and openness to experience. In fact, in psychological questionnaire studies, aesthetic sensitivity, in particular, shows the strongest relation with openness to experience (Smolewska et al. 2006); also, people with high scores on openness to experience have musical chills more frequently (Nusbaum and Silvia 2011). These individual personality differences are reflected in the context of music and learning.

Across many cognitive domains, items that are better liked and that activate dopaminergic reward circuitry, are also usually better remembered—an effect mediated via interactions between hippocampus and reward structures (Shohamy and Adcock 2010,

Gruber et al. 2014). This effect has also been found for musical stimuli by Laura Ferreri and colleagues (Ferreri and Rodriguez-Fornells 2017). They also found that memory enhancement is modulated by each individual's sensitivity to musical reward, as measured with the Barcelona Music Reward Questionnaire, such that higher musical hedonia scores predict better memory. This result makes sense because we know that higher sensitivity to music is associated with greater engagement of the reward system (Martinez-Molina et al. 2016, Sachs et al. 2016). Therefore, this effect constitutes yet another example of the importance of individual personality factors in reward processing.

To test whether dopamine plays a causal role in this relationship, Laura and the rest of the Barcelona team carried out another pharmacological manipulation study, in which healthy volunteers were given drugs that modulated dopamine levels up or down, together with placebo, in a double-blind protocol (Ferreri et al. 2021a). They reported that after 24 hours, memory for musical excerpts was better for the preferred items, as expected, but when dopamine transmission was pharmacologically disrupted, this effect disappeared. However, the drug primarily affected people with lower musical reward sensitivity, suggesting that those who had greater reward-related responses to music in the first place were less susceptible to the decrease in dopaminergic transmission, perhaps because their intrinsic motivation and/or curiosity about music can compensate for the drug effect.

It seems likely that, over longer time periods (including during the course of a lifetime), these kinds of interactions between hedonic and mnemonic systems would lead to better musical memory among individuals who particularly enjoy music, even if they have little or no formal training. Enhanced long-term memory representations for music, in turn, would generate better predictive capacity, and more precise models as we saw above, leading to more enjoyment in a kind of virtuous cycle. It's hardly surprising, therefore, that music lovers tend to be people with a lot of exposure to music!

When music is presented repeatedly, it tends to become better liked (Peretz et al. 1998, Schellenberg et al. 2008)—the so-called "mere exposure effect"—which is also observed for many other classes of stimuli (Bornstein 1989). This phenomenon can be understood based on the same logic: repetition leads to better memory representations, which, in turn, leads to better predictive models, thus enhancing pleasure. However, in keeping with the data reviewed above about the inverted-U function for pleasure, too much exposure eventually can lead to an overly precise model, such that the listener becomes bored and no longer experiences any surprise or much pleasure to the repeated music. This effect is often seen in repetition experiments, where the peak in pleasure is reached after a moderate number of repeats and decreases thereafter (Schellenberg et al. 2008). It has also been known for years that much the same happens with some forms of popular music: too much repetition of the same top 10 hits often leads to a decline in their popularity (Jakobovits 1966).

It's also highly relevant to mention that this effect of repetition on liking is modulated by the personality factor of openness to experience, such that people with greater openness report liking unfamiliar pieces of music more and after fewer exposures, but also tire of them more quickly, compared to those who do not score as high on that personality scale (Hunter and Schellenberg 2011). The interaction with this variable nicely fits into prior findings that openness and novelty-seeking are associated with greater responsivity in the reward system, leading to both greater initial responses and quicker satiation.

The memory enhancement associated with pleasant and rewarding stimuli also extends to the motor domain. Roberta Bianco, working in Virginia Penhune's lab at Concordia University, examined this question using a clever experimental design in which people without prior piano training learned to play a sequence of tones to complete a melody they were listening to (Bianco et al. 2019). The sequences to be learned were all of similar complexity and difficulty; but the context they were embedded in differed according to melodic complexity/predictability, based on the IDyOM model. The learning of the movements on the keyboard was facilitated when the motor sequence formed part of a more predictable melody, compared to a less predictable one. However, melodies that were better liked also led to better learning, even if they were unpredictable. These findings suggest that both structural features (predictability) and hedonic value (liking) influence learning. But both factors can be thought of in relation to enhanced reward-system activity, which would lead to better memory formation, in line with the evidence for dopamine-mediated learning already reviewed (Gruber et al. 2014).

Reprise

Musical pleasure depends critically on interactions between cortico-cortical auditory processing loops and the reward system. Cortical systems generate neural signals whenever an input does not match expectations. Those signals are, in turn, transmitted to the reward system, where they can be evaluated for their hedonic value. Thus, a sensory prediction error becomes a reward prediction error. More specifically, hierarchically organized cortical systems enable perception via the interaction of ascending pathways, that transmit information from the environment, and descending control signals that modulate those inputs based on internal models. The descending influences can be thought of as predictions, which can happen at local levels or at more global levels. The ascending signals can be thought of as generating prediction errors at each hierarchical level when there is a mismatch between expectation and reality, depending on what aspect of the input is unexpected.

Based on the known anatomical connections of the two principal cortico-cortical loops with the reward system, reward prediction errors are computed in distinct networks. For inputs related to temporal expectancies, arriving from dorsal cortical regions, errors are computed in the dorsal striatum; while for inputs related to pattern-based expectancies, arriving from the auditory ventral stream, errors are computed in the ventral striatum. The ventral striatum seems to be most important for the hedonic pleasure response, based on its ability to compute valenced reward prediction errors in terms of whether an event is better or worse than expected.

The concept of "better-than-expected" in the context of music is linked to the notion that information is itself rewarding, even when it is abstract. Thus, better-than-expected outcomes can be conceived of as those that provide useful information that reduces uncertainty and serves to update internal models. Valenced reward prediction errors would happen whenever a musical event occurs that was not entirely predicted. These conclusions are supported by many studies, showing that subjective pleasure, together with physiological indications of arousal, are generated by novelty or surprise in musical passages.

Precision weighting of predictions is an important concept that can be quantified via computational models based on statistical regularities in music. Precision is related to the degree of uncertainty or entropy, which is to say, how well the context of a given piece of music allows predictions to be made. A separate measure is the complexity or amount of surprise present in a given passage of music. Many studies have shown that pleasure from music (as well as many other human activities) follows an inverted-U shaped function that peaks at intermediate values of surprise, where complexity is neither too high nor too low. This relationship is modulated by the precision of the prediction, so that in a musical environment where precision is high and predictions are certain, listeners prefer intermediate-to-higher complexity, with some level of surprise. Whereas when precision is low, listeners prefer less surprise because even the most likely outcome is still somewhat uncertain, and thus confirmation can still generate a positive prediction error.

The connection between reward and learning is fundamental. Biologically, the hedonic feelings generated within the reward system can be viewed as signals that allow an organism to learn about the environment, and how to interact with it, to maximize fitness and survival. Since music frequently generates these positively valenced responses, it also enables learning. Consequently, implicit and explicit knowledge of musical relationships grows with more exposure to it. This greater knowledge, in turn, leads to more precise models of the musical environment. By the same token, repeated presentation of a given musical item generally leads to higher pleasure because repetition enables more precise mental models. But, if repeated too often, enjoyment declines as the opportunities for surprise diminish when the knowledge becomes too precise.

Better knowledge of musical structures acquired via learning or exposure leads to better encoding of musically relevant sound features, already at early levels of auditory processing. Better predictions also help to explain why musical knowledge facilitates detection of subtle deviations in complex musical patterns at later levels of processing, which are accompanied by physiological mismatch responses. Learning is also the vehicle for enculturation, since listeners can only learn the statistical regularities of the sounds that they are used to hearing.

The motivation to explore and learn—curiosity—varies significantly from person to person and is linked to broader personality factors, such as openness to experience and, in the context of music, to music reward sensitivity. Differences in this trait modulate the influence of reward on memory. Thus, preferred music is generally better remembered, but this effect is stronger in those with greater reward responses to music. Similarly, pharmacologically blocking dopamine disrupts this effect of pleasure on memory but not in those with high music reward sensitivity. These interactions arise because greater curiosity is associated with higher reactivity in the reward system. Therefore, all these factors—exposure, training, curiosity, openness, and musical sensitivity—modify our musical knowledge and determine how any given individual engages with music, thus helping to explain both the commonalities and variability in human hedonic responses to music.

9

Pleasure and Beyond

If I had to live my life again, I would have made a rule to read some poetry and listen to some music at least once every week; for perhaps the parts of my brain now atrophied would thus have been kept active through use. The loss of these tastes is a loss of happiness, and may possibly be injurious to the intellect, and . . . to the moral character, by *enfeebling the emotional part* of our nature.
—Charles Darwin, *Autobiography* (1881)

This remarkable quote (emphasis added) from Darwin's autobiography has always struck me as a poignant, plaintive regret from one of the greatest intellects of the 19th century. We know that at the end of his life Darwin was quite depressed and suffered from anhedonia (he writes "I have also almost lost my taste for pictures or music"), probably due to many factors, including his continuing illness and especially because of the loss of his beloved daughter, Annie, aged 10. But what's notable in the passage is Darwin's belief that had he been able to maintain greater interest in artistic pursuits, including music, during his life, his brain would have atrophied less (thus presaging the idea that cognitive/music training could prevent age-related decline). Even more striking is that he explicitly recognizes the value of these "tastes" because their absence weakens "the emotional part of our nature." And with this insight, Darwin, in effect, answers the question he had found so puzzling ten years earlier, about the apparent lack of evolutionary value of music, quoted at the start of this book. Whereas earlier he had argued that music was of no particular use, he now acknowledges that it is indeed, in some sense, essential for our emotional well-being.

This concept is a good way to introduce this final chapter, whose goal is to expand upon the phenomena and proposed mechanisms discussed throughout the previous chapters and to think about how the role of the reward system in music may be considered in the broader context of emotion, well-being, and human happiness, a concept which is gaining ground in the literature (Stark et al. 2018, Saarikallio et al. 2019). It is sometimes presumed by those with a superficial acquaintance with the research on music and reward that what we propose is some simple reflex-like effect, so that music just inherently generates mindless

From Perception to Pleasure. Robert Zatorre, Oxford University Press. © Oxford University Press 2024.
DOI: 10.1093/oso/9780197558287.003.0009

pleasure, kind of like "soma" in Huxley's *Brave New World*. If our explanations about music's power were no more than that, the critique would indeed be accurate. But, as I hope has been made clear throughout this book, musical pleasure emerges from the relations of perceptual, cognitive, and mnemonic mechanisms with prediction and reward processes. It is precisely because of these complex interactions between many mental functions that musical experiences can be considered beyond the sole context of the pleasure elicited to also encompass the rich emotional responses that we experience to music.

What Is Music for?

This question may seem either obvious or impossible. But I mean it only in the empirical sense of a psychologist who must rely on people's thoughts and actions since that is what we ultimately seek to explain. So, if we simply ask that question, or some variant of it such as "Why do you listen to music?" the answers people provide turn out to be instructive about the broader role that music can play in their (and all our) lives. Of course, it's neither easy nor perhaps even desirable to obtain a single set of answers to such an open question, especially as the responses will vary enormously based on the characteristics of the individual and of their social and cultural context. Responses will also depend on what exact questions are asked and how they are interpreted by the respondent. Despite these limits, there is surprisingly broad agreement on several ways in which people engage with music that can provide insight into the questions that interest us. Perhaps the most relevant point is that pleasure, along with regulation or expression of emotion, frequently appears as among the most salient reasons for why people listen to music in almost all empirical studies.

As already noted in Chapter 1, music consistently emerges among the top activities that are most often described as a source of pleasure (Dubé and Le Bel 2003). A study of North American, European, Asian, and Latin American respondents reported that emotional self-regulation was the most important use of music on a personal level; other common responses included social bonding and the expression of cultural identity (Boer and Fischer 2011). Another report from the UK concluded that people listen to music "primarily to manage/regulate their moods" (Lonsdale and North 2011). Yet another study asked Swedish volunteers across a wide age range to indicate why they listen to music (Juslin and Laukka 2004). The answers are informative: "to express, release, and influence emotions" (47%); "to relax and settle down" (33%); "for enjoyment, fun, and pleasure" (22%); "as company and background sound" (16%); "because it makes me feel good" (13%); "because it's a basic need, I can't live without it" (12%); "because I like/love music" (11%); "to get energized" (9%); "to evoke memories" (4%). Patrik Juslin (2019), from Uppsala University, summarizes this large and diverse literature with the pithy statement "wherever there are human beings there is music; and wherever there is music there is emotion" (p. 3).

I would add one more point to this idea: that many, if not all, of these responses are ultimately related in some way to the reward system and its associated circuitry. Hence, they may also be linked to the predictive mechanisms that underlie the engagement of the reward system, if perhaps indirectly. The conclusion that the reward system is engaged is evident from the language people use in their descriptions of why they listen to, or otherwise engage with, music. Note the prominent idea that music makes people feel good, or that they love it, or that they get pleasure from it. These are essentially the same sorts of responses that appear in studies of reward-system modulation by a variety of stimuli, as we saw in Chapter 6. It is not a coincidence that many of the responses just enumerated align nicely with the five separable categories that emerged from factor analysis of the Barcelona Music Reward Questionnaire (BMRQ), as detailed in Chapter 7, which included mood regulation and emotion evocation as well as music seeking and sensory-motor aspects. These factors can all be directly or indirectly linked to reward-system engagement.

Another point stressed by many authors, and which also emerged as a separate factor in the BMRQ, is the importance of social factors in music (Nummenmaa et al. 2021). Everyone knows that musical experiences can feel quite different depending on the social context (hearing an unfamiliar song for the first time by yourself, for instance, is usually not the same as hearing the same song together with a friend who knows it and likes it); and vice versa, music can influence the social interactions between people (think of the bonding effect on sports fans singing the team's song together during a match, and perhaps, even the effect on the team itself). This socially mediated effect is in addition to synchrony-induced social effects, discussed below. Without going into the large psychological literature on the topic of music and social interactions, I would merely make the point that positive affect related to social phenomena is mediated, at least in part, via the reward system (Feldman 2017), as we also saw in Chapter 6. Therefore, any social modulation of music-induced responses most likely also involves interactions with reward circuits.

For my thesis to hold, I neither wish to nor need to claim that the engagement of the reward system necessarily underlies *all* possible musical affective responses. There may well be aspects of musical behaviors that are related to independent neural systems, just as there probably are many different ways in which music can generate emotions (Juslin and Västfjäll 2008). Some very basic arousal or alerting responses (to loud, abrupt sounds, for instance) may bypass the entire cortical-reward loop, generating physiological responses on their own via brainstem mechanisms (Koelsch 2014). And perhaps there are musical systems already in existence, or yet to be invented, that could generate hedonic responses in some totally different way from that described in this book. It would be interesting to find that out. But the evidence that we have available does strongly point to the critical involvement of the reward system in the generation of musical emotions. Therefore, as a next step, let us turn to a consideration of music-induced emotions themselves to better understand how they may be linked to reward.

Music-Related Emotion

The topic of how and why music may generate emotions and feelings is a complex one, and it's certainly not a question I will definitively resolve here. Some scholars have even denied that music can generate genuine emotions, such as sadness or happiness, at all (Kivy 1990, Konečni 2003). Obviously, much depends on the definitions not only of what exactly constitutes emotion (about which there is barely a modest consensus) but also of how we can even tell whether someone is feeling an emotion.

The latter problem, about subjective experience, is a common one in the context of other covert phenomena, such as we encountered in the domain of imagery, in Chapter 3 for example. But the empirical answer is always the same: we infer the presence of these phenomena based on external evidence of a presumed internal process. The inference is our scientific model, subject to revision and refutation, based on the testable predictions it generates, at least if it's any good. And, of course, in neuroscience we have the distinct advantage of being able to read out some of the neural correlates of those internal processes and link them to overt behaviors. This approach does not solve the legitimate philosophical debate about the ontological status of musical emotions or how to classify them. But it does give us an empirical way forward, allowing us to collect data, which should eventually inform such theoretical questions.

What kinds of musical emotions exist? A systematic assessment of the variety of musical emotions has been proposed by Marcel Zentner and colleagues (Zentner et al. 2008), who identified nine descriptors based on factor analysis. These descriptors—wonder, transcendence, tenderness, nostalgia, peacefulness, power, joyful activation, tension, and sadness—capture the variance in listener responses across several musical genres (classical, jazz, pop/rock, techno, and Latin-American). This classification scheme, specifically designed for Western music, seems to better account for responses than a more general emotion model consisting of the two major axes of valence and arousal, which has been often used in the psychology of emotion (Watson et al. 1999, Russell 2003).

However, the contribution of arousal to musical emotion cannot be ignored, and several studies have suggested that emotional arousal itself—irrespective of which emotional descriptor may apply—is a primary component of musically induced emotion. Behavioral evidence in favor of this conclusion comes from a study reporting that emotion ratings to music from one group of listeners were predictable on the basis of a separate group's ratings of the same music, in terms of 'liking' and 'arousal potential' (North and Hargreaves 1997). Physiologically, as we already saw in earlier chapters, markers of autonomic nervous system arousal, such as skin conductance, heart rate, and respiration, are associated with strong subjective expressions of felt pleasure to music (Grewe et al. 2007, Salimpoor et al. 2009). Also, ratings of subjective emotionality, specifically in the context of surprising musical moments, are accompanied by psychophysiological modulations that are indicative of arousal (Steinbeis et al. 2006). These findings therefore suggest a direct link between emotional arousal and musical emotions. Furthermore, we also saw evidence in preceding chapters that the opioid system seems to be systematically linked to physiological arousal to music, as opposed to valence. Therefore, a distinction between arousal and valence may be useful to understand the differential contributions of the opioid versus the dopamine system to the emotions generated by aesthetic rewards such as music, even though such an idea remains only a hypothesis for now.

Music-Related Emotion: Perceiving Versus Feeling

One issue which repeatedly comes up in this literature is the distinction between perceived and felt emotions. Most of the research studies on pleasure derived from music that were described in the previous two chapters involve a situation in which the listener is actively feeling some kind of emotion, usually pleasurable, as indicated by verbal reports, rating scales, psychophysiology, or monetary valuation (not to mention, brain activity). But some psychology and neuroscience studies have focused more on judgments of the emotional content that the music intends to portray as opposed to attempting to induce an emotion in the listener. So in those cases, it's more about perceiving or recognizing the emotional content of the music rather than the individual's emotional experience to that content (for further discussion of these nuances, see [Juslin 2019]).

Emotion-recognition judgments are informative because they tell us about how certain musical features may relate to the expression of emotion in a piece of music. But they do not necessarily tell us directly about the state that the listener may be in upon perceiving the music. In practice, it is often difficult to tell the difference between the two processes—recognition and experience—because hearing music that conveys an emotion may also lead to at least some feeling of the emotion too. Nonetheless, it is also possible to distinguish between the two (Gabrielsson 2001, Kallinen and Ravaja 2006). This dissociation is seen most notably in the case of musical anhedonia. People who experience no emotional response to music are nonetheless usually able to identify the intended emotion (Mas-Herrero et al. 2014, Mas-Herrero et al. 2018b), at least for musical excerpts that convey straightforward emotions such as happiness, sadness, fear, or peacefulness (Vieillard et al. 2008). This result specifically implies that emotion recognition can occur without access to the reward system. However, the converse need not hold: the reward system might still be implicated in emotion recognition, even if no emotion is felt—at least under some circumstances, as we shall see below. Another way to think of the problem would be in terms of a separate category of aesthetic emotions, which are associated with both real, subjectively felt pleasure and with perceived and evaluated aesthetic appeal (Koelsch 2018). This approach purposely blurs the lines between the perception and experience of emotion; time will tell if this idea proves useful in the case of music and other artistically relevant emotions.

In a more general way, perhaps, we can consider the issue of perceiving versus feeling as reflecting the two parties involved in any communicative act, the sender and the receiver. Music may be thought of as the medium for transmission of emotions from one to another. As such, music embodies or portrays in some way the emotive state of the composer/performer, or at least, it depicts the state that the musician wishes to convey to the listener. For communication to occur, the receiver must be able to decode the content—that's the recognition part—following which, the listener may or may not enter into a similar emotional state as that of the sender—that's the experience or feeling of the emotion part. Thinking of it in this way, we can perhaps better integrate the process of emotion recognition as one component of the broader aspect of emotional communication.

Among the more notable conclusions from experimental judgments about musical emotion recognition is that there is significant agreement about what emotions are expressed in music across individuals and even across cultures. This consistency is true, at least for basic emotion descriptors such as happy, sad, and scary and maybe even for more complex ones, such as heroic, annoying, or amusing (Cowen et al. 2020). One might have expected that such judgments would be entirely and exclusively determined by arbitrary conventions pertaining to culturally determined musical gestures. However, even if that is certainly true to a large extent, commonalities can nonetheless be found which are highly instructive for our purposes.

For instance, listeners from an African society with little or no exposure to Western music were nonetheless able to classify the intended emotion in Western-style musical excerpts well above chance levels (Fritz et al. 2009). Another study carried out with a similar group of listeners did not find any evidence of similarity in valence ratings between African and Western groups; however, that study did find a positive correlation between subjective arousal ratings across the two groups that was accompanied by increases in psychophysiological indices (heart rate, skin conductance, and respiration rate) (Egermann et al. 2015). Therefore, it's possible that acoustical features related to arousal may be more likely to be shared across cultures than those related to valence.

Conversely, as mentioned earlier in this book, Western listeners were able to categorize songs from many different parts of the world according to their intended use in the culture from which they were taken (lullabies, healing songs, dance, and love songs), despite having no knowledge of the cultural precepts of the groups in question (Mehr et al. 2019). In this study, specific musical features (including tempo, distribution of musical interval sizes, and metrical organization,) were identified that could be used to distinguish the song types. However, it's also possible that such judgments could be based on inferred emotions from those songs, since acoustical or musical features are typically correlated with intended emotional content (Juslin and Laukka 2004, Gomez and Danuser 2007) and since the emotion conveyed in a lullaby, for instance, is no doubt rather different than the one conveyed in a dance.

None of this implies that cultural influences aren't strong; quite the contrary. As I've repeatedly emphasized, learning the regularities of the sound system one is exposed to underlies the ability to create mental models within that system, which in turn is critical for predictions, surprise, and enjoyment. But it seems that there may nonetheless be some aspects of emotion that are conveyed in preferred ways by certain sounds over others, even across musical systems that are historically or geographically unrelated to one another. Therefore, it behooves us, as scientists, to understand that phenomenon better.

Cues and Mechanisms for Music-Related Emotion

Movement

Why might certain acoustical cues be associated with certain emotions in very widespread ways? A compelling idea is that this relationship may be related to the tight interplay between musical sounds and the motor system phenomena that we discussed in Chapter 4 (Molnar-Szakacs and Overy 2006). For example, tempo is strongly associated with happy versus sad musical emotion (Dalla Bella et al. 2001, Gagnon and Peretz 2003) and/or with arousal (Gomez and Danuser 2007) in Western listeners. Both of these can be easily linked to the speed and nature of typical movements associated with those emotional states: if you're sad or depressed your movements will tend to be slow and of limited excursion; but when you're excited and/or happy your movements will likely be fast and jumpy (Michalak et al. 2009). If you sing or play an instrument slowly and softly, with small pitch changes, it will tend to sound sadder compared to the sound generated by singing or playing quickly and loudly, with large jumps in pitch. So, music may convey the relevant emotion via mimicry of the motion associated with a particular state.

This relationship between perception of movement and emotion and its relation to music in a cross-cultural context was experimentally demonstrated in a clever way by Beau Sievers and colleagues from Dartmouth College. They devised software that could manipulate the specific movement parameters (rate, jitter, smoothness, and so forth) of an animated bouncing ball, and also of a musical pattern, such that the dynamic contours of both the moving object and the music were directly linked (Sievers et al. 2013). They then asked two groups of American college students to set those parameters so that either the bouncing ball (one group) or the music (the other group) would express one of five emotions: happy, sad, peaceful, angry, and scared. The results showed a clear correspondence between the settings for each expressed emotion across the moving object and the music. The authors then repeated the entire experiment with people living in an isolated village in Cambodia. Remarkably, the correspondence of movement parameters and emotions for music or visual input was similar across the two populations, Americans and Cambodians, despite the huge differences in their cultural traditions (Figure 9.1). Thus, movement seems to be directly linked to these basic emotions, for both visual and musical stimuli, in a way that is not exclusively determined by cultural factors.

If music can portray emotion, at least in part via its relationship to movement, as shown by the results just presented, then we have another possible link to the engagement of the reward system in processing emotion from music. As previously reviewed, the anatomical connectivity between the auditory dorsal stream, which serves as the interface to transform sounds into movements and movements into sounds, sends information to the dorsal portions of the reward system. Moreover, we have seen various examples in prior chapters of how temporal predictions are processed via this subsystem, leading to reward-related activity. Therefore, engagement of movement representations within the dorsal auditory-motor stream by music whose temporal properties parallel those of emotionally meaningful movements could also generate pleasurable emotions via this interaction.

We have already seen evidence in the previous chapter that "groove" music, which is defined as music that elicits the pleasant sensation to move, not only engages the dorsal motor cortical regions but also the dorsal striatum, hence linking motor and reward circuits. Additional relevant evidence on this point was provided in an imaging experiment showing an interaction between musical pleasantness and its metrical organization, such that judgments about events on strong beats were more accurate for pleasant, consonant

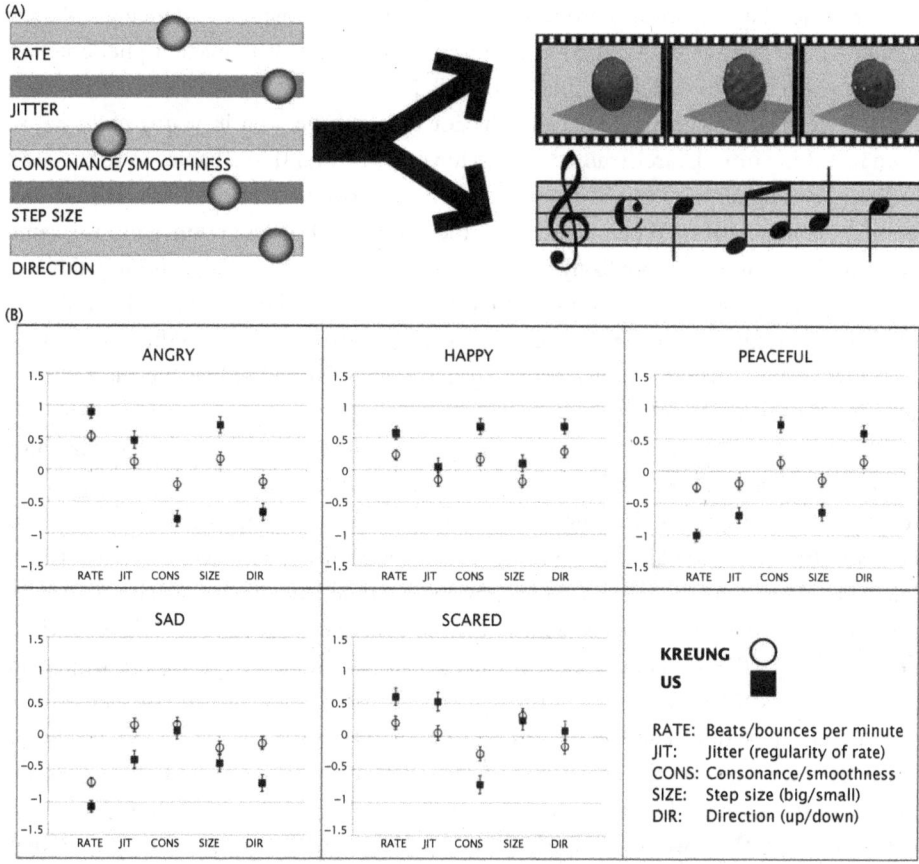

FIGURE 9.1 Shared expression of emotion in auditory and visual features. A: Task requiring adjustment of movement parameters (rate, jitter, etc.) for either a moving ball or a musical phrase. B: Settings selected to express various emotions (angry, happy, peaceful, etc.) by people from the United States (US) and from a village in Cambodia (Kreung), averaged across musical and visual movement conditions. Both groups chose similar settings to express each emotion in both modalities. Adapted with permission from (Sievers et al. 2013).

music compared to unpleasant dissonant versions (Trost et al. 2014). This behavioral interaction corresponded to the greatest brain responses in the dorsal striatum, consistent with its role in temporal predictions, thus suggesting that it expresses the joint computation of beat strength and pleasantness.

Movement may also pertain to emotion in another way: via synchronization of groups of people. Social scientists have proposed that synchronized or mimicked movements across pairs or groups of individuals, even outside the context of music, promote pro-social behavior or feelings of affiliation (Hove and Risen 2009) and generate positive emotions and a sense of belonging to the group (Wiltermuth and Heath 2009). Such coordination is very often elicited in the context of music (couples dancing to club music, armies marching to martial music, religious groups swaying to gospel music, and so forth). And so, even though music is not strictly necessary for these kinds of social synchrony effects, it is very likely to generate them in many circumstances. Several experimental studies have shown that group music activities that require synchronization, such as singing together or dancing, enhance social closeness (Pearce et al. 2015) and can even raise pain thresholds, a phenomenon thought to be related to social bonding (Tarr et al. 2016).

The effect of synchrony on social behavior can even be seen in young children, as shown in a study from Laurel Trainor's group in which 14-month old babies were bounced along to music, either in or out of synchrony with the movement of another person facing them (Cirelli et al. 2014). When tested later, the children who were bounced in the same way as the assistant were more likely to help that person solve a problem (presumably because they liked them or felt a bond with them) than if they had been bounced out of sync with them. Therefore, music may indirectly generate positive emotions via this socially mediated movement entrainment effect, adding to the list of mood-enhancing mechanisms that music is capable of exploiting.

Synchronization of movement to music, whether alone or together with others, also has effects on physiology, which in turn can affect mood and emotion. For example, in some circumstances, heart rate and respiration can entrain with the tempo of music, leading to changes in those physiological variables that also contribute to affective responses (Etzel et al. 2006). Music can also directly alter physiology, independently of synchrony per se. For example, Thomas Fritz and colleagues from the Max Planck Institute in Leipzig devised an exercise device that generates music interactively based on the movements of the person exercising. For a control condition, music was played passively during exercise without any link between the actions and the sounds (Fritz et al. 2013b). When music was produced directly by the actions on the exercise machine, oxygen consumption was more efficient, and volunteers reported less exertion during the workout. Thus, in this auditory-motor feedback context, there were both objective and subjective effects of music on physiology. Further work with the same approach has shown changes in pain tolerance (Fritz et al. 2018) and mood (Fritz et al. 2013a) associated with the generation of music from the exercise as opposed to mere passive listening, thus supporting a link between movement and emotion, independent of temporal entrainment.

Vocal Sounds

Another potential source of acoustical information in music that conveys emotion may come from its relationship to the sounds made by the human vocal tract. Several authors have suggested that acoustical features associated with certain emotions expressed in vocalizations bear similarities to how emotion is portrayed in music (Scherer 1995, Juslin and Laukka 2003). Intensity, rate, contour, and roughness are among the dimensions that most clearly distinguish emotional states in both vocal speech and music (Ilie and Thompson 2006, Coutinho and Dibben 2013). Dale Purves and colleagues from Duke University have gone further, proposing that major and minor musical scales are associated with different affective profiles (notably along the happiness-sadness continuum) because their acoustical frequency spectra bear some similarity to the spectra of speech vocalizations recorded when people speak in happy versus sad emotional states (Bowling et al. 2010). They have also shown that this relationship holds, to some extent, for both the Western musical system and for music from South India (Bowling et al. 2012). Therefore, it does not seem to be a phenomenon exclusive to only a single system/culture, although it may not be found in every culture either (Athanasopoulos et al. 2021).

The communication of musical emotion via cues that are also used in vocalizations implies that there exists some commonality in the processing of these signals within the brain, a likely candidate being the voice-sensitive regions of the temporal cortex that we talked about in Chapter 3 (Frühholz and Belin 2018). But the phenomenon could also be construed as being related to motor systems and movement since vocal expressions are produced via contractions of the laryngeal and vocal musculature.

A test of these ideas was provided by two independent studies with very similar approaches. Matthew Sachs and colleagues from the University of California (Sachs et al. 2018), and Sébastien Paquette and colleagues working with Pascal Belin's group in Marseille (Paquette et al. 2018) both compared functional MRI (fMRI) brain activity patterns elicited by emotional nonverbal vocalizations that portrayed happiness, fear, or sadness with sounds generated by a clarinet or violin that had matched acoustical profiles associated with each affective state (Paquette et al. 2013). The brain activity was fed into a machine-learning algorithm that was able to correctly classify the emotion in either class of stimulus.

Most interestingly, in both studies, the result held when the algorithm was trained on brain activity from voices and then was tested on the emotion depicted in the brain activity, elicited by the musical instruments or vice versa. These results support the existence of a shared neural code across vocal and musical emotion portrayal, as shown in Figure 9.2A. The regions whose activity encoded those different emotions were found bilaterally in the auditory cortex, as expected, indicating that the acoustical profiles of voices and musical instruments are processed in a parallel manner. This finding supports the idea that musical passages that mimic vocal emotion could convey that emotion without any special learning, if they rely on the same perceptual processing that is already in place for vocalizations.

But an important finding in Paquette's study was that significant classification of emotions from voices and instruments was also possible from activity in the dorsal premotor cortex (see Figure 9.2B), thus implicating both perceptual and motor systems in the processing of emotions. Other studies of affective nonverbal communication also find recruitment of motoric regions (Warren et al. 2006), underlining the potential importance of auditory-motor interactions for emotion processing. On the other hand, in the Sachs study, significant classification was also observed within the posterior insula. This result is of interest given the insula's prominent role in mapping changes in physiological inputs onto feeling states (Damasio et al. 2013). There is also evidence that the medial temporal structures of the amygdala and hippocampus are sensitive to both musical and vocal emotions under various circumstances (Frühholz et al. 2014), which, together with the insula findings, suggests that several distinct neural pathways may underlie the response to such emotion-bearing stimuli.

Roughness and Dissonance

An interesting specific example of how vocal cues convey emotion is provided by the sounds of vocal screams. Luc Arnal and colleagues in Geneva showed that when decomposed into spectral and temporal modulations, the acoustical signature of screaming encompasses

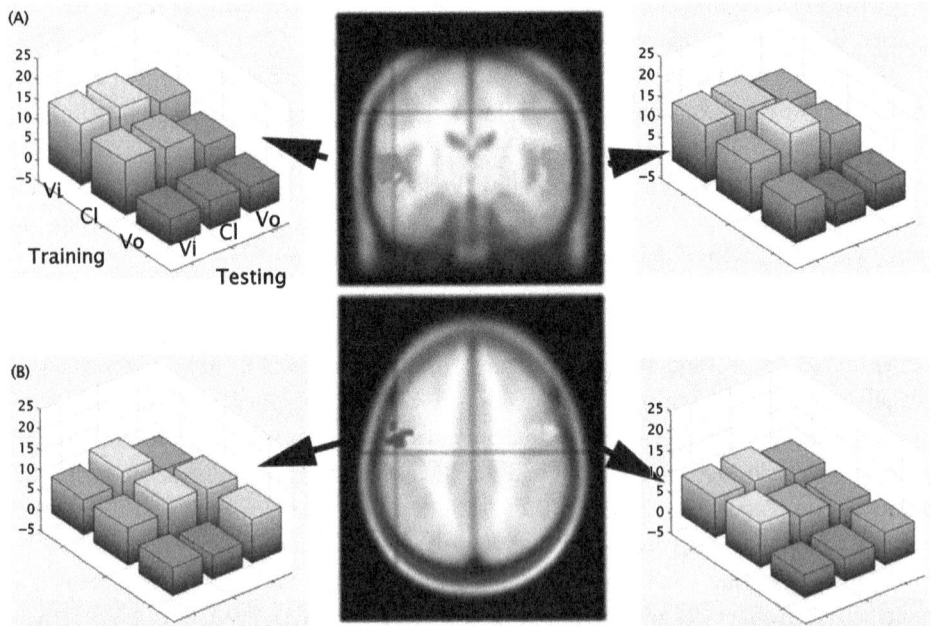

FIGURE 9.2 Emotion processing for voice and instruments. A: Emotion classification accuracy (left and right panels) using machine learning based on fMRI data from left and right auditory cortex (green and blue regions, respectively, in center panel) for various combinations of training and testing with different stimuli (Vi = violin timbre, Cl = clarinet timbre, Vo = vocal timbre). Classification accuracy (percent above chance) is similar whether trained and tested with the same timbre or across voice/instruments. B: Same as A but for two premotor cortex regions (red and yellow). Used with permission of John Wiley & Sons—Books, from (Paquette et al. 2018); permission conveyed through Copyright Clearance Center, Inc.

very fast temporal modulations that do not overlap acoustically with most other sounds in the environment (including normal speaking voices or music). This feature makes vocal screams salient from the background (Arnal et al. 2015). These highly arousing and fear-inducing sounds preferentially activated the amygdala, as shown by functional imaging. To make the link to music explicit, these authors recently showed that acoustical analysis of instrumental music from horror movies often contains similar acoustical features as do actual screams (Trevor et al. 2020), as shown in Figure 9.3. This conclusion will not surprise anyone who has heard Bernard Hermann's famous soundtrack for the Alfred Hitchcock film *Psycho*.

The acoustics of screams are also related to the phenomenon of roughness, which may be (roughly) defined in physical terms as the effect produced when two or more periodic sounds that are very close in frequency are presented together. Such a situation results in high rates of temporal modulation in the amplitude envelope, referred to as beating, and consequently causes interference patterns in the cochlea. Going back at least to Helmholtz (Helmholtz 1863/2009), dissonance has been thought to be related to roughness. Dissonance, however, turns out to be more complex than that single factor, as it is also strongly related to the harmonic relationships of the complex tones that make up a chord (McDermott et al. 2010). There are also very important contextual and cognitive

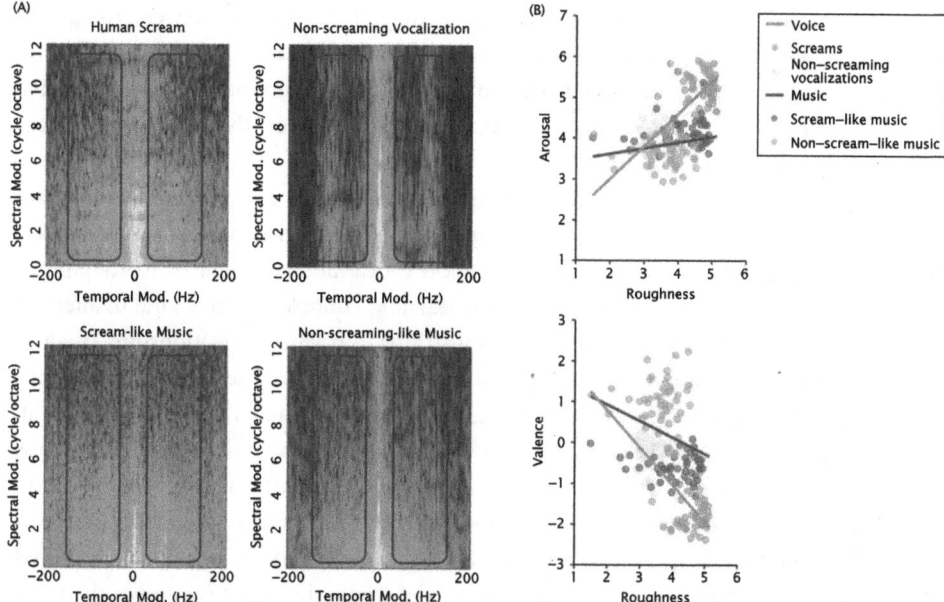

FIGURE 9.3 Screams and instrumental music from horror films. A: Spectrotemporal modulation plots of human screams and non-scream vocalizations (top left and right) and of scream-like or non-scream-like music (bottom left and right). Both screams and scream-like music contain energy at high temporal modulation rates (red boxes). B: Ratings of arousal (top) and valence (bottom) for vocalizations and music as a function of acoustical roughness (high temporal modulation). More roughness is associated with higher arousal and lower valence for both types of stimuli. Reprinted with permission from (Trevor et al. 2020). Copyright 2020. Acoustical Society of America.

factors that affect the response to dissonance (Harrison and Pearce 2020), especially those related to knowledge and experience with different musical styles (Popescu et al. 2019). Screams can be part of artistic expression even (Belin and Zatorre 2015). But all else being equal, tones that are close together in frequency are reliably perceived as unpleasant when presented simultaneously, most likely due to their roughness, which may also relate to the negative emotions evoked by screams or horror-film music.

Roughness does not seem to depend much on learning: musical training has little influence on it (McDermott et al. 2010), and even people from cultures whose music is entirely homophonic (i.e., lacking harmony), and who therefore are largely indifferent to Western concepts of harmony or dissonance, still show a reliable dislike of roughness (McDermott et al. 2016). It's also interesting that people with congenital amusia dislike roughness to the same extent as control listeners, yet they show no preference for harmonic over inharmonic chords, unlike controls (Cousineau et al. 2012). Neurophysiologically, responses of auditory cortex neurons in monkeys show phase-locked oscillatory activity to simultaneous tones close in frequency, such as minor and major seconds, which generate roughness; and these responses correlate with human perceptual ratings of dissonance (Fishman et al. 2001b). Since it would be hard to argue that monkeys have much of a musical culture, those results could not be explained by acculturation, although they might be related to exposure to the animal's own vocalizations, which like many natural sounds, contain harmonic relationships. All these findings suggest that the link between roughness and negative affect is most likely an innate phenomenon, which could explain why it plays a prominent role in aversive sounds that are universally used to indicate distress or fear, like screams. Furthermore, it also explains why such sounds are frequently used to induce negative affect in certain styles of music.

Harmonicity, on the other hand, might be more related to exposure and learning and hence may be more relevant for an understanding of higher-order musical aesthetic preferences. It may also be a determinant of emotion evocation, as compared to roughness, which instead may relate to a more primal, less learning/context-sensitive kind of affective response. That said, preferences for tone combinations in harmonic relationships (such as perfect fifths or octaves) over those that are more inharmonic (like tritones and minor ninths), are already present in infants between two and four months of age (Trainor et al. 2002) even though neither of those particular tone combinations produce much roughness since they are not close together in frequency. An even more striking demonstration of very early preference for consonance over dissonance was provided by a study showing that two-day-old infants looked longer at a visual stimulus associated with hearing a normally played Mozart sonata than they did to a stimulus associated with a dissonant version of the same sonata (Masataka 2006). This effect was even true for hearing infants born to deaf parents, who presumably had more limited exposure to voiced speech sounds in utero. Since the dissonance manipulation was applied between two musical lines separated by more than an octave, there would not be much roughness, but plenty of dissonance. So, a possible innate preference for harmonicity cannot be entirely ruled out, even though we also know that statistical learning can occur rather quickly, including in infants, when appropriate environmental examples are provided.

Consonance and dissonance need not map in a simple way to pleasantness and unpleasantness, regardless of perceptual predispositions that may exist due to biological constraints. Rather, it is probably more fruitful to consider them as a dimension of musical sounds that can be dynamically and expressively exploited to modulate emotional responses. Importantly for the prediction error hypothesis (at least in tonal music) dissonance is frequently used to generate tension in a musical texture, thus leading to a prediction that it will resolve to a consonant sound. The musician can then manipulate the degree and timing of any resolution, to engender positive reward prediction errors and hence induce pleasure. This idea is related to the concepts of entropy and complexity, as discussed in the previous chapter, so that dissonance might be considered in that context in a similar way to how melodic expectancies have been modeled. Thus, the constant interplay of consonance and dissonance in music is very important to the generation of various emotional responses. Next, we turn to the neural pathways by which dissonance can engender these affective responses.

Interactions with Amygdala

Whether dissonance arises from acoustical roughness or from inharmonicity, we have already seen in earlier chapters that medial temporal-lobe structures, particularly including the amygdala and the parahippocampal area, play an important role in their processing.

This conclusion is supported by functional imaging data (Blood et al. 1999, Koelsch et al. 2006), lesion studies (Gosselin et al. 2006), and intracranial recordings (Dellacherie et al. 2009). So the finding about amygdala responses to the acoustics of screams (Arnal et al. 2015) fits well with this literature. In fact, it was already known from earlier studies that aversive sounds, such as nails on a blackboard (which most likely have similar acoustical features) activate the amygdala, along with other limbic structures (Zald and Pardo 2002).

However, the situation is more complex than a simple link between negatively valenced sounds and the amygdala because positive emotional vocalizations, such as laughter or sighs of pleasure, can also activate the amygdala in certain cases (Fecteau et al. 2007). Consonant and dissonant music both also generate responses within the amygdala (Ball et al. 2007), but different nuclei within the structure have different response patterns, which may explain why results from neuroimaging studies that do not have the resolution to distinguish anatomical subregions of the amygdala are not always consistent. More generally, these outcomes indicating amygdala responses to more than just negatively valenced items would be in keeping with the amygdala's proposed role as a detector of emotionally salient stimuli across modalities (Sander et al. 2003) as well as its broader role in emotional learning and emotion regulation (Phelps and LeDoux 2005).

How do the relevant acoustical features of a sound that determine its affective valence, whether positive or negative, reach the amygdala? Does the information go via the auditory cortical system or is there some direct input? This question was addressed in an important study by Sukhbinder Kumar and colleagues (Kumar et al. 2012). These authors used fMRI with a large set of (nonmusical) stimuli that varied in their acoustical structure and their affective valence, from mildly positive or neutral (bubbling water, phone ringing) to unpleasant (crying babies, dentist's drill, and such). With this set of sounds, they were able to distinguish between brain responses to the acoustics (spectrotemporal modulations) and the brain responses to the emotion elicited (as indicated by subjective ratings). They found that activity in certain subregions of the amygdala corresponded to the acoustical features, while others corresponded to the emotional ratings, and the same was true for the auditory cortex.

But the most important finding in that study, for our purposes, was how those two structures communicated with each other. Using a causal modeling approach, the authors observed a reciprocal modulation in the direction of functional connectivity, such that responses to the acoustical information of the sounds was relayed to the amygdala from the auditory cortex, whereas the emotional valence information flowed in the opposite direction, from amygdala to auditory cortex. This result implies that the affective impact of a sound emerges from the interaction between these structures so that initial perceptual analysis is required before the valence can be assessed by the amygdala, which then sends feedback signals to auditory areas. Importantly, modeling this phenomenon found no evidence of a direct functional input to the amygdala from the auditory thalamus, unlike what occurs in the case of simple fear conditioning, for example (LeDoux et al. 1984).

Although this study did not use music, we can presume that something similar may happen: as passages of music vary over time in their dissonance or other more abstract features that may signal various types of negative emotions (fear, anxiety, heartbreak—all things that music is good at conveying), the auditory pathways would send and receive information to/from the amygdala, which would generate changes in the evoked emotional state. This concept is highly reminiscent of the ideas developed in earlier chapters about the interactions between auditory pathways and the reward system where reciprocal modulation, based on expectancies and prediction error, results in pleasurable responses. In both cases, exchange of information between auditory cortical loops and subcortical structures (amygdala or striatum) enables affective evaluation of complex auditory patterns. Hence, music can derive some of its remarkable power to influence and arouse our emotions via distinct neural subsystems that encode both positive and negative emotions.

Memory

Most studies on musically induced pleasure and emotion that have been discussed in Part II of this book explicitly avoided situations in which the affective response was related to a memory elicited by the music. This design choice was predicated on the reasonable grounds that such a response would not necessarily be directly related to properties of the music itself, but rather to the memory associated with it. Since the aim of most of these studies was to understand the more fundamental link between musical structure and pleasure or emotion, effects mediated by memory would constitute an unwanted confound in that context.

However, that does not mean that memory is not an important component of how music generates emotional responses; it's just potentially a different, or parallel, process that needs to be understood separately from the dynamics of musical structures themselves and from the predictive mechanisms discussed in Chapter 8. But even if there is a distinct mechanism involved, interactions between music and memories also seem to involve links to the reward system.

Music frequently evokes episodic memories—recollections of past personal events, especially from earlier years in one's life. Several studies have found that when people are presented with popular music from their past, it frequently reminds them of autobiographical episodes (Janata et al. 2007, Krumhansl 2017). Music may not necessarily be unique in this regard, as memories are often stronger when paired with various kinds of emotionally salient or pleasurable stimuli (Buchanan 2007). But what's interesting is that music appears to serve as a trigger for memories due to its own rewarding properties, which in turn can lead to strong associations with an unrelated event. This effect is sometimes referred to as the "Darling, they're playing our tune" phenomenon. And as we saw in Chapter 7, music's ability to engage the reward system modulates memory systems, leading to enhanced memory formation.

The ability to recall events from one's past represents a complex cognitive skill that engages numerous brain systems, in particular the default mode network (which includes ventromedial frontal cortex), as well as dorsolateral frontal areas and hippocampal circuitry (Spreng and Grady 2010, Rissman et al. 2016). Although only a few studies have examined the functional neuroanatomy of autobiographical memories elicited by music, they consistently report engagement of the ventromedial frontal cortex in relation to emotionally salient autobiographical memories, together with various other regions, including more dorsomedial frontal areas (Janata 2009) and hippocampus (Ford et al. 2011).

The ventromedial frontal region is of particular relevance in this context because not only does it form part of the default mode network, which is thought to involve introspective functions, including retrieval of autobiographical memories (Spreng and Grady 2010), but also because it is a core structure in the reward system, as discussed in Chapter 6. Specifically, ventromedial and/or orbitofrontal areas encode reward value of different stimuli. This function has also been described in the context of retrieval of autobiographical memories, such that brain activity in that region was correlated with the personal significance and emotional intensity of the memory (Lin et al. 2016).

Putting these threads together, it seems that music, by virtue of its ability to engage dopaminergic reward circuitry and generate emotional arousal, strengthens memories of concurrently occurring events (Ripollés et al. 2018). If those events are emotionally salient in and of themselves, there would likely be a synergistic effect, solidifying the link between the music and the event. This link can then manifest itself, even years later, via the network of regions involved in autobiographical memory retrieval. The ventromedial frontal cortex likely plays a key role in this phenomenon since it encodes the reward value and/or emotional salience, both at the time of the co-occurrence of the music and the emotional event as well as at the time of retrieval.

By virtue of these connections between music and memory, music can evoke many emotions that are intertwined with the memory, ranging from longing and nostalgia about earlier times, to feelings of happiness and love for others with whom one might have shared good memories, or even fear and loathing if the music is associated with some horrible event from the past. The precise mechanisms behind these complex interactions remain to be understood in greater detail; but they point to the myriad ways that music can elicit emotional responses, via the interactions between cortical networks (dorsal and ventral auditory-frontal loops and default mode network) and largely subcortical networks (reward system, amygdala, medial-temporal structures, and other limbic areas).

Music Preference: Exposure, Adolescence and Personality

We have talked a great deal throughout Part II of this book about enjoyment and pleasure derived from music and the mechanisms that underlie them. However, those ideas don't directly explain individual musical preferences. One could analyze an operatic aria and a country and western ballad and conclude that they both manipulate melodic, harmonic, or rhythmic expectancies so as to generate positive reward prediction errors. But that would not explain why I might like one but not the other. Therefore, to understand a bit more about what drives such preferences we need to look at other factors. There is a well-developed literature in the psychology of music about this topic (Schäfer and Sedlmeier 2009), but here we will only consider exposure, along with age and personality, among the various factors that could influence preference, because of their possible links to the reward system.

The most obvious factor that drives musical preference is simply exposure. We already saw evidence that liking of specific music pieces increases with repeated playing

(Schellenberg et al. 2008), which can be explained based on the idea that exposure leads to better internal models, thus generating an enhanced predictive capacity. This idea can be expanded to entire genres, in so far as a given style of music will express certain regularities which the listener will internalize via statistical learning over long time periods; these are what generate the schematic expectancies that were discussed in Chapter 1. Hence, it's hardly surprising that people generally like styles of music with which they are familiar.

But this exposure-related liking interacts with age in an interesting way. When we are very young, our musical exposure depends entirely on what the adults who care for us decide to listen to. This is how we initially become familiarized with the music of the culture that surrounds us. But as we get older, we usually develop more control over our musical listening choices. And upon reaching adolescence, there is very often a dramatic break with parental preferences in favor of peer preferences—a phenomenon familiar to every teenager and their parents (my parents, for instance, were aghast when I came home with the Woodstock album; they were merely perplexed when I started to play Bartók in the basement). These preferences formed during adolescence tend to last for the rest of our lives.

Experimentally, several studies show that people tend to prefer music from their adolescence more than music heard earlier or later in life (Janssen et al. 2007). Importantly, and linking back to music's role in autobiographical memory (see Figure 9.4), these studies often emphasize that the preference for music from one's youth is tied to emotionally salient events during that period of life (Schulkind et al. 1999, Krumhansl 2017). The link between adolescence and development of preferences is not confined to music, but is also seen for other items, such as books or movies, although it seems to be most prominent for music (Janssen et al. 2007). Carol Krumhansl showed that not only is there a relative preference for music from adolescence but also that once they are adults, people demonstrate a secondary preference for the music that they heard as children, which itself was the favorite music of their parents during *their* adolescence (Krumhansl and Zupnick 2013). So, there is a kind

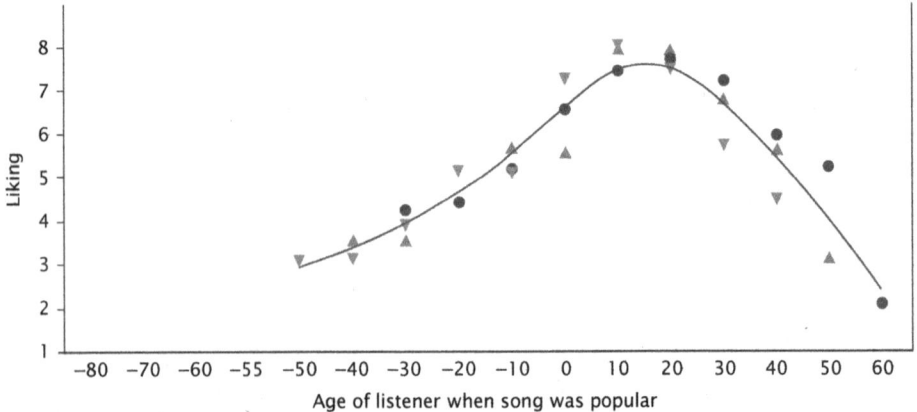

FIGURE 9.4 Music preference and age. Liking ratings for cohorts of people born in the 1940s, 50s, or 60s as a function of the age of the listener when the song was popular. Preference peaks in adolescence (between ages 10 and 20) and is lower for music popular before or after that time. Adapted from (Krumhansl 2017). Reproduced under the terms of the Creative Commons Attribution (CC BY) license.

of cascade effect, where the preferences formed in adolescence are visited upon subsequent generations.

What accounts for this bump in preference associated with music heard during those formative years in the transition from childhood to adulthood? No doubt there are many influences, especially social factors associated with peer bonding and mate selection. It is probably no coincidence that this is the time of life when we first fall in love—with other people and also with music. It should come as no surprise that considerable evidence implicates maturational changes in the dopaminergic reward system in the way that adolescents respond to various stimuli, including music, all of which leads to heightened emotional arousal and laying down of emotional memories.

These changes are complex and should not be collapsed into a unidimensional view that only focuses on this one system. However, for our purposes, it is striking that experimental evidence shows that the ventral striatum responds more during gambling/risk-taking tasks in adolescence, than at earlier or later times in life (Braams et al. 2015), as shown in Figure 9.5. Several developmental models propose that certain aspects of adolescent behavior are driven, at least in part, by this peak in dopamine activation in the striatum and other reward-system structures, which subsequently declines in later adulthood (Luciana et al. 2012, Luna et al. 2015). There is also a link between the increased behavioral reward sensitivity observed in adolescence and anatomical measures of the ventral striatum, which also show a peak in volume during late adolescence followed by a decline at later ages (Urošević et al. 2012).

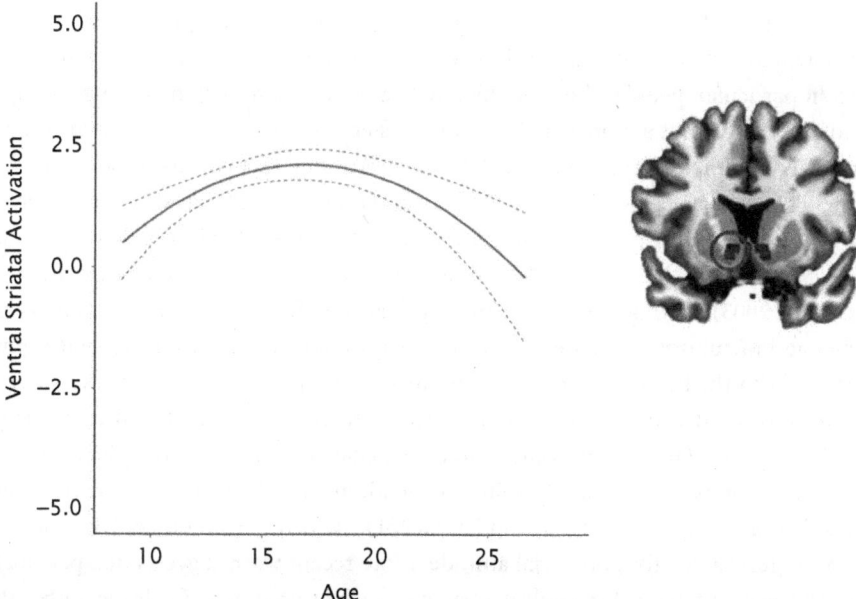

FIGURE 9.5 Striatal reactivity in adolescence. Left: fitted curve showing peak in striatal activation during a gambling task between ages 15 and 20. Right: location of ventral striatum region of interest for data shown in A. Adapted with permission of the Society for Neuroscience, from (Braams et al. 2015); permission conveyed through Copyright Clearance Center, Inc.

This naturally occurring modulation in reward-system function and structure has been linked to features of adolescent behavior, including impulsivity, and is related to various disorders that adolescents are prone to, including addiction and other risk-taking behaviors (Schneider et al. 2012). But it's not all negative, as this dopamine surge is also very likely important for adaptive behaviors, like exploring new environments (physical or cognitive—epistemic curiosity), establishing peer bonds, and achieving sexual maturity, all of which are necessary for future life as an adult. Therefore, and given the importance of the reward system in generating musical pleasure, it makes sense to propose that these age-sensitive processes that enhance reward-related responses (to good and/or to dangerous things) can explain the great attachment we develop for music experienced during this critical time period. Conversely, at later points in life, preferences tend to be less changeable.

Evidence supporting this latter idea comes from comparison of music-related responses in older (60–80) versus younger (18–35) people, which revealed higher reactivity to music in the younger group, along with less extreme emotional ratings in the older people (Pearce and Halpern 2015). Similarly, responses for all factors from the BMRQ were highest for the adolescent/young adult group and declined with increasing age in the other cohorts (Mas-Herrero et al. 2013). This finding was confirmed in a later study, which also reported that the greatest age-related decline was in the music seeking category; but at the same time, there was no age effect on a separate measure of aesthetic experience (Belfi et al. 2022). The conclusion from these studies then is that the heightened responses that we experience during adolescence are strongly linked to emotional arousal and influence our preferences for musical genres for the rest of our lives.

Turning to personality as another factor of interest, we saw in previous chapters that individual differences in personality traits can influence the engagement of the reward system in a variety of situations. Hence, these differences can also change responses to music. In particular, people who score high on traits such as sensation-seeking and openness to experience have a more reactive reward system, which is reflected in their response to music. For example, music-induced chills are more often reported by persons with high openness to experience (Nusbaum and Silvia 2011). Evidence about whether personality traits influence musical preference for certain specific styles of music is, however, more limited. Some studies have found that genre preference is linked to personality (Rentfrow and Gosling 2003). So traits such as openness (Vella and Mills 2016) and sensation-seeking (Schäfer and Mehlhorn 2017) are linked to preference for more arousing, complex music, in keeping with the heightened need for stimulation in this group. Their greater susceptibility to chills could also lead to a preference for music with prominent chill-inducing features (Grewe et al. 2007). On the other hand, the personality trait of agreeableness leads to preferences for music expressing happiness or tenderness and a relative aversion to music expressing fear or anger (Vuoskoski and Eerola 2011), which is in accord with the view that this trait reflects a positive, prosocial attitude. More recent work suggests that personality traits influence preference for musical attributes, such as arousal and valence, rather than musical genres per se (Greenberg et al. 2016).

It should be noted, however, that these associations are not always statistically very powerful. Therefore, such personality factors may play a comparatively small role in determining musical preference (Schäfer and Mehlhorn 2017). Nevertheless, such traits can interact with reward-system activation to music, possibly influencing preferences for certain musical styles. For example, people scoring higher on empathy trait show more fMRI activity in reward-related areas when listening to familiar music compared to lower-empathy people (Wallmark et al. 2018). Other neuroimaging studies have reported that various personality traits influence neural responses to different musical emotions, especially in the insula and reward regions (Montag et al. 2011, Park et al. 2013), although the findings are not terribly consistent.

So, although there remains some uncertainty about the role of individual personality traits in influencing preferences for musical pieces or genres, it seems likely that such effects are mediated, at least in part, via differential responsivity of the reward system to the emotional cues provided by music. A question for future research is how these personality differences may interact with maturational changes during adolescence, and with exposure-related factors, to generate the unique individual preferences and responses to music that each person expresses.

Music, Mood Modulation, and Well-Being

Emotion Regulation

As indicated by the surveys mentioned at the start of this chapter, emotion or mood regulation represents a major reason that people listen to music. The uses of music that have been documented across dozens of cultures, including common types of music such as lullabies, love songs, and healing songs (Mehr et al. 2019), can be construed, among other things, in terms of influencing other's moods (calm the baby, attract the lover, soothe the sick). Many of the topics mentioned in this chapter that touch upon music's capacity for emotional arousal (movement, vocal features, consonance/dissonance, memory), can also be thought of in the context of emotion regulation. By manipulating these different factors, we can create music that has the power to change, influence, or enhance our own emotions and also those of others.

The use of music to modify emotions and induce certain mood states has been around since music has been around. In fact, it is perhaps even responsible for music's evolution and continued existence. In antiquity, the mythical Orpheus was supposed to possess the remarkable power to charm living and even nonliving things by his musical gifts, thereby ascribing to music a kind of magical power that can influence others' emotions. A few millennia later (in 1697) English playwright William Congreve wrote the often-misquoted line "Music hath charms to soothe the savage breast" (not "beast"), which refers to music's ability to calm disturbed feelings. Modern psychological research confirms these ancient observations, albeit less poetically.

Survey studies have found that healthy adults often indicate that music serves as one of the main ways to regulate their moods. In one report from California, for example, listening to music was rated as the second-most effective means of successfully changing a bad mood (after exercise—which may not apply outside California), as well as raising energy and reducing tension (Thayer et al. 1994). A poll of Israeli adults reported that 85% listened to music when they felt in a bad mood (Shifriss et al. 2015). In this sample, some preferred sad and some preferred happy music to deal with the emotional state, but the latter choice was much more common in older individuals. In a large Canadian sample of people over age 60, 70% reported listening to music daily and rated music as very important to their lives (Cohen et al. 2002). Not surprisingly, the emotion-regulation effects of music are also related to individual factors, such as musical experience or sophistication (Pelletier 2004). This makes sense, based on the evidence that musical knowledge helps people to establish better internal models of music from their culture, leading to better predictive capacity and more reward-system engagement.

Using experimental approaches rather than surveys, music preferred by the volunteers was found to reduce anxiety after a stress-inducing task (Walworth 2003). It also increased feelings of relaxation, which were also accompanied by changes in skin conductance measures (Labbé et al. 2007). But it's important to note that mood changes induced by music do not necessarily always result in a positive psychological outcome—it depends on the music and the situation. Thus, calming music can improve driving performance, but fast music has the opposite effect, leading to more aggressive and dangerous driving (Brodsky 2001). Totalitarian states are well-known to use music to manipulate the masses to their ends. And we should not forget the disturbing fact that music can be used as a form of torture or degradation in various contexts (Grant 2013). Music's power is such that it not only can be used to enhance emotions, but, unfortunately, it can also be misused to generate emotions that can cause psychological harm.

On a more positive side, therapeutic emotion regulation via music can be effective in different populations. There is a very large literature in the clinical domain evaluating the uses of music to improve mood, reduce anxiety, and enhance well-being in many different clinical groups, including psychiatric disorders (Gebhardt et al. 2014), depression (Maratos et al. 2011), stroke (Särkämö et al. 2008), heart disease (Bradt et al. 2013), and dementia (Guetin et al. 2011). For systematic reviews and meta-analysis see (Sihvonen et al. 2017) and (de Witte et al. 2020). Furthermore, despite the huge role that music plays in the lives of adolescents, as discussed above, music use in aging populations is also often associated with reports of enhanced emotional well-being and lower levels of loneliness (Saarikallio 2011).

These sorts of clinical findings are part of a much larger and very promising initiative to develop evidence-based therapeutic interventions using music in many different clinical and nonclinical settings throughout the world (see, e.g., the efforts of the National Institutes of Health in the USA, led by its then-director Francis Collins and famed soprano Renée Fleming [Cheever et al. 2018]). These developments are well beyond our scope here. But, in so far as some of these health-related applications target emotion regulation as part of the intervention, they do raise the very pertinent question of what the mechanisms may be by which music exerts beneficial effects on emotion, in particular, and well-being, in general (Chen et al. 2022).

Mechanisms

At a psychological level, current models of emotion regulation often distinguish between two general mechanisms to cope with negative emotions: one focused on finding ways to reevaluate the situation and generate more positive affect; the other focused on avoiding, ignoring, or suppressing negative emotions (John and Gross 2007). Emotion regulation can be either explicit or implicit in nature (Gyurak et al. 2011). But, in general, coping with negative emotions via reappraisal—an explicit strategy—is considered to lead to better outcomes than suppression (Gross and John 2003) and tends to be the more common strategy according to surveys (Saarikallio 2008). This conclusion was partly supported by a large-sample Australian study that reported that music engagement for purposes of emotion regulation led to enhanced measures of well-being among those who used reappraisal strategies. However, the use of music coupled with suppression strategies led to less positive outcomes (Chin and Rickard 2014).

At a neural level, several experiments have studied the neural networks involved in emotion regulation, but almost all of them have used visual stimuli to elicit emotion. A consistent finding is that most types of active emotion regulation engage various frontal cortical structures (Ochsner et al. 2012), which is in keeping with the well-established regulatory role of frontal cortex on other cortical and subcortical structures (Petrides 2005). Different emotion regulation approaches make use of cognitive functions such as working memory, selective attention, and response inhibition to varying extents. These functions depend on some general and some specific frontal cortical circuitry, helping to account for the diversity and partial overlap of brain regions associated with different strategies (Morawetz et al. 2017).

Of particular relevance to the theme of this chapter, some neuroimaging studies of emotion regulation (outside of the realm of music) have documented interactions between frontal cortical regions and subcortical structures, including amygdala and striatum. With regards to amygdala modulation, several studies have shown that when people view unpleasant pictures and are then asked to consciously diminish the negative emotion they elicit, it results in decreased amygdala responses, along with increased brain activity in lateral and dorsal prefrontal areas (Ochsner et al. 2004, Ohira et al. 2006). The influence of top-down frontal activity on the amygdala during active reduction of negative emotion is most likely mediated via the medial frontal cortex, as indicated by functional connectivity analysis, which revealed downregulation of the amygdala as a function of increased medial frontal activity (Urry et al. 2006).

In addition to influences on the amygdala, other studies of emotion regulation have uncovered interactions between frontal cortex and the striatum, such that when regulating negative emotions, correlation of activity in inferior frontal cortex and ventral striatum was linked to reappraisal success. On the other hand, correlation between inferior frontal cortex and amygdala was linked to reduced reappraisal success (Wager et al. 2008). Thus, in this context, the higher the engagement of the reward system, the better the outcome of emotion regulation. A similar result, implicating top-down influences of frontal cortex on amygdala and ventral striatum, was reported in a study of smokers who were asked to inhibit their craving upon viewing images of cigarettes. Successful reduction of craving, which probably also involved regulation of emotions, was associated with correlated increases in frontal cortex and decreases in the ventral striatum activity (Kober et al. 2010).

Another relevant set of studies has approached the self-regulation of the striatum via neurofeedback. In these studies, brain activity from reward regions is measured via fMRI, and the activity level is then displayed in real time as a visual signal (such as a colored bar that moves up or down with the level of activity). Volunteers are trained to use this feedback to willfully modulate their reward system, with the goal that such abilities might eventually be applicable for treatment of reward-related disorders. These neurofeedback approaches have proven successful in allowing people to learn how to modulate the activity in different reward-related structures. One study found that neurofeedback allowed volitional modulation of the ventral tegmental area, corresponding to the midbrain dopamine region, and that this capacity persisted after the end of training (MacInnes et al. 2016). Another study found that the ventral striatum could be similarly regulated via neurofeedback (Greer et al. 2014). This study also reported that functional connectivity between medial frontal cortex and the ventral striatum was enhanced during feedback training, suggesting that top-down modulation engaged this pathway, like what has been observed with modulation of the amygdala. Also, in this study, successful upregulation of the striatum was associated with positive behavioral arousal responses, as one might expect. These findings converge on the idea that self-regulation of the reward system is possible via purely internal, cognitively driven inputs; but they also suggest that music could serve such a function too because of its influence on the reward system.

Hardly any neuroimaging studies have directly examined emotion regulation using music, and the results would likely be complex given the many individual listener characteristics and the diverse ways in which music can be used for emotion regulation (Carlson et al. 2015). Nevertheless, the types of cortical-subcortical interactions described in the nonmusical emotion regulation studies just reviewed are reminiscent of several kinds of results encountered in the literature on musically evoked emotions, even outside of a true emotion regulation context. For instance, although the pattern of recruitment of the amygdala by music is complicated, we have already seen that some studies observed reduced amygdala activity and increased frontal activity for more pleasant compared to less pleasant music (Blood and Zatorre 2001, Koelsch et al. 2006).

These results allow us to expect that the sorts of changes in emotions elicited by music that have been discussed in this chapter most likely occur via the same frontal-lobe-mediated modulation of subcortical structures, including amygdala and striatum. Looking first at the amygdala, several fMRI studies indicate that there might be a reduction in amygdala reactivity induced by positively valenced music. One study used the intersubject correlation technique and reported that activity in the amygdala was synchronized across listeners perceiving the same music, indicating a common modulation of this structure by musical features (Trost et al. 2015). Moreover, positive valence ratings were associated with decreases in the amygdala, suggesting a top-down suppression as music is more positively evaluated.

A second paper exploring amygdala connectivity showed that listening to music from one's childhood decreased behavioral and physiological stress, induced by a speeded math task, and that this effect was associated with the degree of functional connectivity between medial frontal cortex and the amygdala, supporting the idea that stress reduction via music involves downregulation of the amygdala from frontal sources (Gabard-Durnam et al. 2018). Another study examined responses to scary music, finding that both the amygdala and inferior frontal areas were co-activated to the negatively valenced music (Lerner et al. 2009). But with eyes closed, which is associated with more focused processing of the musical emotion, there was greater directed influence from the amygdala to the frontal cortex than vice versa, suggesting that interactions between these structures can go in one or another direction, depending on how much attention one places on the emotional content.

Turning to the striatum and other reward-related regions, an obvious question would be how all the knowledge that forms the core of this book could be harnessed in the context of active regulation of the reward system. One recent report on music-based interventions that addresses this question tested healthy older adults who were enrolled in a music listening protocol, and who were scanned with fMRI before and after that experience (Quinci et al. 2022). Results showed that after the listening intervention, connectivity between medial frontal cortex and auditory cortex increased, and that this change was strongest for the most preferred items. These findings thus support the idea that functional interactions between auditory and reward-related areas (in this case, medial frontal areas) are related to enhanced ability to engage with music.

In order to address the question of causal modulation of the reward system with music directly, my former postdoc Neomi Singer developed a new protocol that takes advantage of neurofeedback methods already well-established to modulate the amygdala by Talma Hendler's group at Tel-Aviv University (Keynan et al. 2019); Our key innovation was to adapt these methods to make use of music as the carrier of the feedback information, providing a kind of double whammy, since music serves both as information-bearing input and as a direct trigger to the reward system (Singer et al. 2023). In a recent study, we manipulated pieces of real music selected by the volunteers in such a way that it could be made to vary along a continuum, from more to less pleasant, via filtering out certain frequency bands. Next, we measured activity in the ventral striatum using an electroencephalography (EEG) "fingerprint" that was validated via fMRI. Then, closing the loop, the EEG signal which indexes reward-system reactivity was fed into the algorithm that generates the music, such that the higher the signal the better the music sounded and vice versa, creating a positive feedback loop. We randomly assigned two groups of people to receive either the neurofeedback or a sham condition, in a double-blind procedure. We trained them over the course of several weeks to attempt to enhance their reward activation by doing whatever they could to make the music sound as pleasant as possible. Results showed that people in the active but not the sham group were able to increase activity in the ventral striatum at will, based on pre- and post-training fMRI measures. Furthermore, measures of positive affect increased, while measures of anhedonia decreased in the feedback group, but not in the sham group. These outcomes show that music can be combined with neurofeedback techniques to help people learn to modulate their own reward activity volitionally, with positive affective outcomes. Although applications of these methods to real clinical disorders of the reward system are still a ways off, such results hold considerable promise in the realm of emotion regulation, above and beyond the way people have used music to "self-medicate" for millennia.

Music in the Time of Covid

Shortly after I began to write this book, the world was shaken by the spread of the coronavirus. The pandemic brought with it not only a great deal of physical pain and suffering but also very significant psychological distress. Some of that was due to the illness itself and its social and financial consequences; and some of it was due to the confinement that was required in order to contain the virus, which resulted in stress, loneliness, and isolation for many. Surveys indicated significant increases in anxiety and depression among people in many countries, including China (Wang et al. 2020), Spain (Rodríguez-Rey et al. 2020), and the USA (Barzilay et al. 2020).

Even as early as the first weeks of the first wave, numerous media reports emerged of people using music to cope with the psychological problems induced by the pandemic. Who hasn't seen the many YouTube videos of people, especially in Mediterranean countries, singing, dancing, or playing music from their balconies? In light of the research reviewed in this chapter, it should not come as a surprise that music could serve as a way to relieve stress and regulate emotions. But these anecdotal reports about people playing music do not constitute good scientific evidence. We don't know to what extent music, as opposed to other activities, was really used by people or whether it was even useful in coping with covid-induced psychological distress.

The situation called for some serious investigation, and scientists around the world quickly stepped up to the plate. I was fortunate to be able to work with an international team of friends and colleagues to address some of these issues (Mas-Herrero et al. 2023). We put together an online questionnaire targeted at people in three of the worst-hit countries during that first wave—Italy, Spain, and the USA—to ask about their use of music as a coping mechanism, along with many other possible activities that people might engage in (including cooking/food, exercise, talking to friends, social media, entertainment/movies/reading, prayer/meditation, sex, and alcohol/drugs). We sampled more than 1000 people, making sure that the sampling was not in any way biased toward music by not mentioning music at all in the recruitment ads or in the title of the research.

To our delight, even though participants did not know our research focus a priori, music came out ranked as the number one activity that helped people best cope with the psychological effects of the pandemic and confinement (closely followed by entertainment, exercise, talking to friends, and cooking/food). However, not all these various options were equally useful in improving psychological well-being. In fact, we found that greater music use during the pandemic was linked to lower levels of depression, whereas none of the other items, except cooking/food, showed such an association.

Based on a lot of the material reviewed in this and prior chapters, we had, of course, posited that music's beneficial effects would be related to engagement of the reward system. Although we would have loved to put our volunteers in a scanner to answer this question, that was not feasible; but we did administer personality questionnaires. A mediation analysis on these questionnaire results revealed that the positive influence of music on depression was directly linked to sensitivity to reward: those individuals who were more likely to experience pleasure from various stimuli in general were the ones who most benefitted from music during the pandemic. In contrast, a similar analysis for food-related activities indicated that its benefit was mediated by individual differences in emotional suppression: those individuals who could avoid inhibiting their emotions were the ones for whom food and cooking were most helpful. In a follow-up analysis of the same data (Ferreri et al. 2021b), we also found that people preferred to listen to more happy and new music during the confinement and that doing so was specifically associated with high scores on the emotion-regulation scale of the BMRQ and with the cognitive reappraisal facet of a general emotion regulation questionnaire (Gross and John 2003). Therefore, positively valenced music seemed to be used during the pandemic to regulate emotions, in keeping with a lot of the literature reviewed in this chapter.

Several related studies, with even larger and more diverse samples than ours, came to similar conclusions. For example, one survey of over 5000 people from different countries across three continents documented extensive use of music for emotion regulation and for social cohesion (Fink et al. 2021). Importantly for the interpretation that music aids emotion regulation, this study also found that those who experienced more negative emotions during the pandemic were the ones more likely to use music to reduce negative emotions, such as stress and loneliness. Consistent with the latter conclusion, in a study conducted in Brazil, music was found to be especially beneficial for mood management, particularly among those who were suffering severe depression during the covid emergency (Ribeiro et al. 2021). Similarly, another study with large samples taken from several European and South American countries, along with the USA and China, reported that music was the most effective activity for three well-being goals, enjoyment, venting negative emotions, and self-connection and that cultural differences were relatively minor (Granot et al. 2021). Lest we think that the use of music to cope with a pandemic is strictly a contemporary phenomenon, there is evidence of similar activities during the 1576 bubonic plague outbreak in Milan, including singing from windows and balconies and even illicit dance parties (Chiu 2020).

Collectively, these studies converge on the conclusion that music can be one of the best means to inhibit negative affect and enhance positive affect. While this is something which happens on an everyday basis, it was brought out dramatically during a time of worldwide crisis, when many people experienced exceptionally trying psychological difficulties. Different psychological mechanisms seem to be involved for the various coping activities that people described using, which depend on many factors; but there seems to be consensus that, for music, reappraisal is one of the most prominent psychological mechanisms involved. Based on much of the research described in this chapter, I would suggest that the neural mechanisms behind these phenomena that emerged during the pandemic involve top-down regulation from frontal-lobe regions onto structures such as the amygdala (which would be downregulated) and onto the reward system (which would be upregulated). These findings, therefore, not only provide some vindication of our global model about music and the reward system but also point to how this knowledge is relevant for health and well-being.

Reprise

The principal reasons given by most people for listening to music revolve around emotional arousal: pleasure and enjoyment as experienced through music; but also sadness, nostalgia, wonder, and much else. These felt emotions frequently serve to regulate one's internal affective states. Music also serves as a medium to transmit emotion from one person to another and, hence, acts as a means to influence or modulate other's emotions as well. Many of these observations have been made throughout history, and modern psychological methods largely confirm the importance of emotion, and especially emotion regulation, in various musical practices around the world.

Neuroscience allows us to go beyond this descriptive approach, to start to understand the mechanistic reasons why music can have such profound emotional impacts. A principal reason for music's power is because it induces emotional arousal based on interactions between the many complex cortical circuits involved in perception, memory, and other cognitive functions, with various subcortical structures, including, in particular, the striatum and associated reward circuitry as well as the amygdala and other limbic regions. These neural processes consistently manifest themselves in relation to common factors, including movement, vocal cues, roughness/dissonance, and memory. Indeed, despite the vast differences in musical systems across cultures, certain specific patterns of sound are recurrently associated with particular emotions, and these factors are likely to form the sources of this commonality.

The role of movement in evoking emotion is well-established behaviorally. It is likely related to parallels between emotion-related actions and musical features, such that acoustical parameters of music can mimic movements associated with certain affective states. Synchrony of movements across individuals induced by music can also enhance social emotions, such as empathy or group cohesion. The neural bases for these effects are varied but likely involve the auditory dorsal pathway, with its ability to generate temporal predictions, which in turn feeds into the dorsal portions of the striatum and generate spleasure.

Musical emotions can also be conveyed via mimicry of emotional vocal features, not only in singing but even with instruments, whose sounds can imitate vocal patterns associated with affective vocalizations, such as happiness, sadness, fear, and so forth. Neuroimaging studies show that vocalizations and instrumental sounds share an underlying neural coding of brain activity within auditory cortex, especially in voice-sensitive areas. But some motor regions are also engaged, arguing for the idea that vocal emotional features exert their effects via both auditory and motor systems. Vocalizations can also engage the amygdala, especially when they contain rapid temporal modulations, or roughness, that are typically found in human screams, and result in highly negative affective reactions. Modeling of directed functional connectivity patterns associated with unpleasant sounds indicates that acoustical cues are sent from auditory cortex to amygdala, whereas affective valuation is computed in the amygdala and transmitted back to auditory cortex.

Roughness and harmonic relationships are distinct components of dissonance. Even outside of the context of vocal sounds, roughness generally evokes negative emotional responses, including in infants who are a few days old, suggesting that it represents a rather basic form of emotion transmission. The role of inharmonic sounds in creating negative affect from dissonance is more nuanced and is most likely modifiable based on exposure, even though it can also manifest itself early in life. Dissonance can play a key role in generating tension, without which resolution could not occur. Thus, it fits well with the idea of predictive coding, providing opportunities for expressive modulation of negative and positive emotions and setting up possibilities for manipulation of reward prediction.

Music is frequently linked to episodic memory, especially autobiographical events. As a result, it can generate a vast array of complex emotions via this associative mechanism. Learning and memory are intimately linked to the action of the dopaminergic reward system; and since music engages this system when it co-occurs with some emotionally arousing event, the two reinforce one another, leading to emotional arousal that can be triggered by the music at later times. Retrieval of such memories likely involves interactions with the default-mode network and, in particular, the ventromedial frontal cortex, which encodes value.

Music from one's adolescence tends to have a stronger influence on formation of genre preferences than music from other time periods, in part, because of the strong emotional arousal experienced during that stage of life. In turn, this sensitivity is linked to a dopaminergic surge in the striatum that happens during adolescence and which is associated with increased risk-taking behaviors, curiosity, and sexual activity. These factors, together with peer social interactions, likely influence the formation of strong and long-lasting musical preferences. In addition, individual differences in personality traits also influence musical preferences. Thus, higher scores on openness to experience are associated with greater reactivity in the reward system and lead to experiencing greater emotional arousal (such as chills) to music.

Music is frequently used to regulate one's emotions. Behavioral studies indicate that this regulation is often accomplished via the adaptive psychological mechanism of reappraisal. Neuroscience studies implicate top-down modulation from frontal-lobe areas onto amygdala and the reward system as the mechanism by which music can modify emotions. Thus, negative affect can be reduced by downregulating the amygdala with positive music, while neurofeedback studies show that the reward system can be upregulated with preferred music.

A prominent example of emotion regulation via music is provided by the global Covid-19 pandemic. Many behavioral studies confirmed the importance of music for dealing with the negative psychological consequences of the disease and/or the associated confinement in groups of people from many diverse countries. Notably, those who benefitted most from music to reduce anxiety or depression, were more likely to have high sensitivity to reward, once again linking emotion regulation to modulation of the reward system.

Coda
The Miracle of Music

Gracias a la vida que me ha dado tanto	*Thanks to life that has given me so much*
Me ha dado la risa y me ha dado el llanto	*It's given me laughter and given me tears*
Así yo distingo dicha de quebranto	*It's how I distinguish happiness from sorrow*
Los dos materiales que forman mi canto	*The two elements that make up my song*
Y el canto de ustedes que es el mismo canto	*And your song which is the same song*
Y el canto de todos que es mi propio canto.	*And everybody's song which is my own song.*

—Violeta Parra, *Gracias a la vida* 1966

We've been through a journey—from perception to pleasure—starting with some very fundamental aspects of hearing, then moving through more complex processes, and then all the way to the intricate mechanisms of the reward system. But I don't want to lose track of the bigger picture: the miracle of music. There's always a potential danger of reductionism in science; by focusing on the constituent parts of a phenomenon we may fail to see the way those parts work together. Yet, understanding the function of each part is also essential. So, what I've tried to emphasize throughout this book is the interactive nature of the neural processes that allow us humans to create, perceive, and enjoy music. The whole in this case really is more than the sum of its parts.

I believe that by achieving a certain level of scientific understanding we deepen not only our knowledge but also our appreciation, even awe, for the object of our study. When I stare at the sky on a starry night, the wonder of my experience is enhanced by what I know about astronomy; indeed, I wish I knew much more than I do, so that I could achieve greater insights. And I feel the same way about music. Anyone can experience the powerful forces that music stirs within us—no scientific knowledge required. But the more we know about how and why music exerts its influence, the more we should acknowledge how astonishing it really is.

We've seen throughout this book how music exploits some of our most fundamental neural and cognitive mechanisms. It depends on our remarkable auditory capacities, specialized for the optimal processing of the characteristic sounds of greatest importance to our species: speech and music. It exploits our brain's mechanisms to distinguish an important

From Perception to Pleasure. Robert Zatorre, Oxford University Press. © Oxford University Press 2024.
DOI: 10.1093/oso/9780197558287.003.0010

sound from an irrelevant background and to encode its pitch. It makes use of our ability to hold sounds in memory as they unfold over time, which in turn allows us to understand the relationships between those sounds and infer regularities in their patterns. It takes advantage of the highly developed motor control we have over the larynx, hands, and fingers, which allows us to sing and play instruments. It integrates sound signals with actions in many ways promoting not only movement and dance but also enabling aspects of rhythm and meter.

Most importantly, music engages the human brain's vast predictive powers, essential for all perception, but achieving in music the particular status that enables us not only to perceive patterns but also to derive pleasure from them despite their abstract nature. Because it is via the reward system that subtle variations between what is expected and what is heard allow us to experience pleasure and emotion. Musicians have intuitively known of this push and pull between the known and unknown since the beginning of music, of course; else it would not exist. But now science can uncover some of the reasons how and why this works. And even though science's answers are always provisional, I still find those discoveries exhilarating.

It's fascinating too that whereas the reward system is phylogenetically ancient—we largely share its characteristics with other creatures, from monkeys to rodents—the circuitry that allows abstract cognitive representations and predictions to occur is phylogenetically recent, encompassing corticocortical loops in the frontal lobes, which are most highly developed in humans. So, music, by virtue of how it enables interactions between these two systems, links in a sense our most basic biology with our most advanced mental capacities. This idea suggests a reason why music can feel somehow primal yet depends fundamentally on learning, experience, knowledge, and culture.

We humans have learned how to harness the power of music not only to generate pleasure but also to modulate emotions. Music is no panacea. If your lover has left you for another, music alone is unlikely to make them return; but there's a song for that. If you are a victim of oppression, music by itself will not stop your persecution; but there's a song for that. If you are sad, lonely, depressed, angry, or anxious, there are songs for each mood, just like there are songs for love, joy, play, meditation, or relaxation. The fact that people have made music for those situations from prehistory testifies to music's power to soothe and heal but also to provide strength and share feelings with others. Music will not fix what's wrong with the world and can even be used for ill; but we cannot deny its uncanny ability to move us.

Every time we hear a piece of music, we are experiencing it through the filter of our own past knowledge, acquired through a lifetime of listening, and bequeathed to us by our culture. It is that knowledge that allows us to anticipate what is likely to happen, based on what has happened before. And it is that same knowledge that permits us the pleasure of surprise when the musician does something interesting and unexpected. Every act of music is at once a recapitulation of the past and a prediction of the future. It is a microcosm of the world—a simulation if you will—but one with immense power to express and influence our deepest thoughts and feelings. Music captures our illusions and disillusions, loves and losses, encapsulating both the pain and the pleasure of life itself, just as Violeta Parra said in her song.

Acknowledgments

The research upon which this book is based reflects the hard scientific work of many people in many places. In particular, I must emphasize my appreciation for the many students and trainees that I have been fortunate to work with in my lab over several decades. It's literally true that this book could not have been written if it weren't for their scientific contributions. I cannot name all of them individually, but there is no doubt that I've learned more from them over the years than vice versa.

Special thanks go to those who really pushed the envelope on studying music and reward: Anne Blood, whose insistence that we really needed to explore musical pleasure and emotion, and not just perception, first propelled us onto that fruitful path; Valorie Salimpoor, whose enthusiasm for new and innovative experiments allowed us to make major advances in understanding the mechanisms of music-induced reward (and thank you for the beautiful *setar* that still sits on my shelf); Ben Gold, who explored the phenomena from a novel, more cognitive and computational perspective (thanks for the hat, too); Neomi Singer, whose very original contributions in the neurofeedback domain are taking us in new and exciting directions (and who introduced me to the beautiful Negev desert); and Ernest Mas-Herrero, who—apart from his remarkable discovery of specific music anhedonia—also made critical advances to understanding the causal role that the reward system plays in music (and whose outdoor paella-making skills are much prized). Thank you all so much for sharing the joy of discovery with me.

In addition to my former students working on this topic, I've had the pleasure of working with numerous other colleagues and students whose contributions to music and reward were essential. The Barcelona group, in particular, has provided tremendous impetus, both to the experiments themselves and to their theoretical implications. Josep Marco-Pallarés and Antoni Rodríguez-Fornells have been instrumental in this regard for many years, and I am thankful for their insights and friendship, as well as for the contributions made by their great students, Noelia Martínez-Molina, Pablo Ripollés, Alberto Ara Romero, and Laura Ferreri. I look forward to more sharing of ideas, tapas, and cava in the coming years. ¡Muchas gracias, amigos! I'm also indebted to my MNI colleague Alain Dagher, who has taught me a great deal about the anatomy and functionality of the reward system.

Beyond the work on music and reward, this book is also based on much other work on auditory neuroscience and music cognition, carried out by many scientists throughout the world; they are too numerous to name here, but I have learned enormously from all of them (even when I occasionally disagree with them). I wish to point out in particular some of those who worked in my lab on these topics, specifically because they each brought their own unique perspectives that opened up new avenues of research. Without them, my research domain would be much narrower (and more boring!), and I would not have been able to draw the links between the cortical networks and the reward system that are at the heart of this book.

My very first doctoral student, Séverine Samson, really helped to get things started with foundational work on the effect of temporal-lobe lesions on musical processing that set the stage for a lot of the later work using brain imaging. Merci bien, Séverine. Marc Schönwiesner first introduced me to the topic of spectrotemporal modulations, which turned out to be critical to our understanding of how the auditory cortex works. Vielen dank, Marc. I was also fortunate to work with Pascal Belin, who discovered the voice-sensitive regions in the auditory ventral stream and helped me develop theoretical models of auditory function (I'm looking forward to the next banjo concert in Marseille, Pascal).

Our first foray into understanding the dorsal stream was motivated by Joyce Chen, who brought our lab into the world of sensory-motor processing and rhythm (and of Celtic fiddling). Thank you for breaking new ground and for your continued friendship, Joyce. I never would have thought of studying singing were it not for Jean Zarate, whose insights as both performing artist and scientist opened up that area. Along with Boris Kleber, who did very original further work in that domain, we have gained important insights into this basic human ability. Songs of gratitude are in order! Jamila Andoh first introduced our lab to brain stimulation, which has proven to be an important tool that we have continued to exploit in many contexts; I'm thankful for that (and for the squirrel, too). Thanks to all of you for continuing to be part of the extended research family.

I feel particularly lucky to count Emily Coffey among my best colleagues and among my best friends. Her brilliant work on periodicity encoding in the auditory cortex and subcortex and other topics, including learning and hearing in noise, have helped build a better understanding of these critical processes. And, as if that were not enough, her remarkable artistic talent is on display in this book in the many beautiful anatomical illustrations that she carefully and patiently drew for me. Thank you, Emily, for everything. I anticipate, with great pleasure, continuing to work together (and to playing more Handel duets and occasionally sliding on frozen lakes). Many ideas developed in Part I of this book can be attributed to Philippe Albouy, who generated amazing insights into the functionality of the ventral stream via his work on amusia, opened up novel understandings of the dorsal stream via his work on oscillatory mechanisms for working memory, and has also provided the most solid data for hemispheric differences that we have been able to obtain. He also kindly made some excellent figures for me to use in the book. Merci pour tout ça, Philippe. I'm excited about more good science and fine dining in Québec City with you.

I also want to acknowledge the important role played by some institutions and the people behind them. One of the most important is the BRAMS lab, the brainchild of my colleague and friend, Isabelle Peretz. It's hard to overstate the transformative impact of setting up the world's first international consortium devoted to music and neuroscience, and it would not have happened without Isabelle's leadership. BRAMS is now in good hands under the co-directorship of Simone Dalla Bella, and I expect much great research to continue to emanate from it. I also would like to acknowledge the Centre for Research in Brain, Language, and Music, now headed by my good friend and colleague Denise Klein, for continued research support. Another important institution that has contributed enormously to the growth of music neuroscience is the Mariani Foundation for Pediatric Neurology in Milan and its vice president, Maria Majno. Their sponsorship of the Neuromusic meeting over many years has been instrumental in creating a worldwide community of scientists, musicians, and clinicians devoted to music and neuroscience. Grazie mille, Maria!

And speaking of institutions, I cannot fail to thank my own home base, the Montreal Neurological Institute. The Neuro is more than a place to work; it's a way of thinking, a culture, without which I could never have done the research I talk about in this book, I am sure. Its successive directors have been very supportive of my work over many years, as have been my many colleagues and friends in the Cognitive Neuroscience Unit from whom I have learned, and continue to learn, so much. Many thanks also go to Annie LeBire, who tirelessly, cheerfully, and efficiently has helped me deal with mountains of bureaucracy over the years; my students and I could not get a thing done without her help. Merci infiniment, Annie. I also want to express my appreciation for my colleagues at the McGill psychology department and in the faculty of music, with whom I have enjoyed interacting for many years.

As is obvious from reading this book, my work has depended strongly on neuroimaging techniques, and none of that could have happened without the strong and ongoing support of my colleagues in the Brain Imaging Centre of the Neuro. I would especially like to thank Alan Evans for his critical contributions to our very first studies using these techniques. More recently, I have benefitted from Julien Doyon's leadership of the Centre, and from Sylvain Baillet, who has provided essential support and ideas for our research using the MEG platform. Outside McGill, a very special appreciation goes to my collaborator and friend of longest standing, Andrea Halpern, who has steadfastly worked with me in the domain of musical imagery for over four decades. Thank you, Andrea, for sharing the fun of science (and let's try those Neapolitan songs again soon).

Going back in time, I must also acknowledge the enormous role played by my mentors in my development as a scientist. The late Peter Eimas, my PhD supervisor at Brown, played a critical role in teaching me about psychoacoustics, speech, experimental psychology, and, more generally, how to think. In the days before personal computers, he would selflessly drive me two hours to the Haskins labs, work all night until dawn, and drive back, just to create the stimuli for my experiments. Most of all, he gave me the freedom to explore my own interests, even though he didn't necessarily share them, and taught me that the best work is done when people pursue their own passions. Sheila Blumstein provided some of my first glimpses of brain function in the context of speech, for which I am still grateful. Brenda Milner generously accepted me as a postdoc in her lab way back in the early 1980s. She and her team introduced me to the field of human neuroscience, with her careful work dissecting the behavioral effects of focal lesions, and with her stringent methodology for establishing brain–behavior relationships. Thank you, Brenda, for giving me that opportunity which launched my independent career (and for teaching me the wonders of French cheeses).

Writing this book has been helped along by several people. A big thank-you goes to Farhan Ahmed for his valuable assistance with editing the text and especially for his hard work preparing many of the illustrations. I also must thank Evan Zatorre and Adam Karman for their free legal advice about publishing contracts. Joan Bossert, Martin Baum, and Mary Funchion from Oxford University Press gladly answered my many questions during the course of writing, for which I am grateful. David Huron, Marc Schönwiesner, Niels Disbergen, Ben Gold, and Maria Witek kindly provided data and/or figures to include in the book, and Michael Petrides provided helpful information on neuroanatomy.

I have been fortunate to receive ongoing support for many decades from various funding agencies, in particular the Canadian Institutes of Health Research and the Natural Sciences and Engineering Research Council of Canada, which have allowed me to do much of the research work presented in this book. I've also benefitted from funding provided by the *Fonds de Recherche du Québec* and from the Canada Fund for Innovation, which allowed us to create and equip the BRAMS lab. I also acknowledge support from the *Institut d'Études Avancées*, Université Aix-Marseille, where I was a visiting scientist when I first began writing this book. More recently, I have been fortunate to receive additional financial support from the Canadian Institute for Advanced Research, the Canada Research Chairs Program, the *Grand Prix Scientifique* from the *Fondation pour l'Audition* (Paris), and the deCarvalho-Heineken Prize in Arts and Sciences (Amsterdam).

The last and most important person to thank is my constant and closest companion, my wife, Virginia Penhune. Without her love, steadfast support, continuous encouragement, cheerful attitude, and wise advice, this book would never have even been started, let alone finished. In addition to sharing many exciting scientific adventures during the years, including with our fantastic, shared students, she also read through many versions of the text, listened patiently to my grumblings and rantings, and provided many good ideas and suggestions for every aspect of the book. Thank you, my dear spouse, for all that and so much more.

References

Aalto, S., Ingman, K., Alakurtti, K., Kaasinen, V., Virkkala, J., Någren, K., Rinne, J. O. and Scheinin, H. (2015). Intravenous ethanol increases dopamine release in the ventral striatum in humans: PET study using bolus-plus-infusion administration of [11c]raclopride. *Journal of Cerebral Blood Flow & Metabolism* **35**(3): 424–431.

Abivardi, A. and Bach, D. R. (2017). Deconstructing white matter connectivity of human amygdala nuclei with thalamus and cortex subdivisions in vivo. *Human Brain Mapping* **38**(8): 3927–3940.

Abrams, D. A., Ryali, S., Chen, T., Chordia, P., Khouzam, A., Levitin, D. J. and Menon, V. (2013). Inter-subject synchronization of brain responses during natural music listening. *European Journal of Neuroscience* **37**(9): 1458–1469.

Abuhamdeh, S. and Csikszentmihalyi, M. (2012). The importance of challenge for the enjoyment of intrinsically motivated, goal-directed activities. *Personality and Social Psychology Bulletin* **38**(3): 317–330.

Adcock, R. A., Thangavel, A., Whitfield-Gabrieli, S., Knutson, B. and Gabrieli, J. D. (2006). Reward-motivated learning: Mesolimbic activation precedes memory formation. *Neuron* **50**(3): 507–517.

Adrián-Ventura, J., Costumero, V., Parcet, M. A. and Ávila, C. (2019). Reward network connectivity "at rest" is associated with reward sensitivity in healthy adults: A resting-state fMRI study. *Cognitive, Affective, & Behavioral Neuroscience* **19**(3): 726–736.

Agus, T. R., Paquette, S., Suied, C., Pressnitzer, D. and Belin, P. (2017). Voice selectivity in the temporal voice area despite matched low-level acoustic cues. *Scientific Reports* **7**(1): 11526.

Aharon, I., Etcoff, N., Ariely, D., Chabris, C. F., O'connor, E. and Breiter, H. C. (2001). Beautiful faces have variable reward value: fMRI and behavioral evidence. *Neuron* **32**(3): 537–551.

Alain, C., Arnott, S. R., Hevenor, S., Graham, S. and Grady, C. L. (2001). "What" and "where" in the human auditory system. *Proceedings of the National Academy of Sciences* **98**(21): 12301–12306.

Alain, C., Woods, D. L. and Knight, R. T. (1998). A distributed cortical network for auditory sensory memory in humans. *Brain Research* **812**(1): 23–37.

Albouy, P., Benjamin, L., Morillon, B. and Zatorre, R. J. (2020). Distinct sensitivity to spectrotemporal modulation supports brain asymmetry for speech and melody. *Science* **367**(6481): 1043–1047.

Albouy, P., Caclin, A., Norman-Haignere, S. V., Leveque, Y., Peretz, I., Tillmann, B. and Zatorre, R. J. (2019a). Decoding task-related functional brain imaging data to identify developmental disorders: The case of congenital amusia. *Frontiers in Neuroscience* **13**: 1165.

Albouy, P., Mattout, J., Bouet, R., Maby, E., Sanchez, G., Aguera, P.-E., Daligault, S., Delpuech, C., Bertrand, O. and Caclin, A. (2013). Impaired pitch perception and memory in congenital amusia: The deficit starts in the auditory cortex. *Brain* **136**(5): 1639–1661.

Albouy, P., Mattout, J., Sanchez, G., Tillmann, B. and Caclin, A. (2015). Altered retrieval of melodic information in congenital amusia: Insights from dynamic causal modeling of MEG data. *Frontiers in Human Neuroscience* **9**: 20.

Albouy, P., Mehr, S. A., Hoyer, R. S., Ginzburg, J. and Zatorre, R. J. (2023). Spectro-temporal acoustical markers differentiate speech from song across cultures. *bioRxiv*: 2023.2001. 2029.526133.

Albouy, P., Peretz, I., Bermudez, P., Zatorre, R. J., Tillmann, B. and Caclin, A. (2019b). Specialized neural dynamics for verbal and tonal memory: FMRI evidence in congenital amusia. *Human Brain Mapping* **40**(3): 855–867.

Albouy, P., Weiss, A., Baillet, S. and Zatorre, R. J. (2017). Selective entrainment of theta oscillations in the dorsal stream causally enhances auditory working memory performance. *Neuron* **94**(1): 193–206.e195.

Aleman, A., Nieuwenstein, M. R., Böcker, K. B. E. and De Haan, E. H. F. (2000). Music training and mental imagery ability. *Neuropsychologia* **38**(12): 1664–1668.

Alho, K., Woods, D. L., Algazi, A., Knight, R. and Näätänen, R. (1994). Lesions of frontal cortex diminish the auditory mismatch negativity. *Electroencephalography and Clinical Neurophysiology* **91**(5): 353–362.

Alicart, H., Cucurell, D. and Marco-Pallarés, J. (2020). Gossip information increases reward-related oscillatory activity. *NeuroImage* **210**: 116520.

Alluri, V., Toiviainen, P., Burunat, I., Kliuchko, M., Vuust, P. and Brattico, E. (2017). Connectivity patterns during music listening: Evidence for action-based processing in musicians. *Human Brain Mapping* **38**(6): 2955–2970.

Alluri, V., Toiviainen, P., Jääskeläinen, I. P., Glerean, E., Sams, M. and Brattico, E. (2012). Large-scale brain networks emerge from dynamic processing of musical timbre, key and rhythm. *NeuroImage* **59**(4): 3677–3689.

Amalric, M. and Dehaene, S. (2018). Cortical circuits for mathematical knowledge: Evidence for a major subdivision within the brain's semantic networks. *Philosophical Transactions of the Royal Society B: Biological Sciences* **373**(1740): 20160515.

Amiez, C., Hadj-Bouziane, F. and Petrides, M. (2012). Response selection versus feedback analysis in conditional visuo-motor learning. *NeuroImage* **59**(4): 3723–3735.

Ammirante, P., Patel, A. D. and Russo, F. A. (2016). Synchronizing to auditory and tactile metronomes: A test of the auditory-motor enhancement hypothesis. *Psychonomic Bulletin & Review* **23**(6): 1882–1890.

Andersen, R. A. (1995). Encoding of intention and spatial location in the posterior parietal cortex. *Cerebral Cortex* **5**(5): 457–469.

Anderson, B., Southern, B. D. and Powers, R. E. (1999). Anatomic asymmetries of the posterior superior temporal lobes: A postmortem study. *Neuropsychiatry, Neuropsychology, & Behavioral Neurology* **12**: 247–254.

Anderson, L. A., Christianson, G. B. and Linden, J. F. (2009). Stimulus-specific adaptation occurs in the auditory thalamus. *Journal of Neuroscience* **29**(22): 7359–7363.

Andoh, J., Matsushita, R. and Zatorre, R. J. (2015). Asymmetric interhemispheric transfer in the auditory network: Evidence from TMS, resting-state fMRI, and diffusion imaging. *Journal of Neuroscience* **35**(43): 14602–14611.

Angulo-Perkins, A., Aubé, W., Peretz, I., Barrios, F. A., Armony, J. L. and Concha, L. (2014). Music listening engages specific cortical regions within the temporal lobes: Differences between musicians and non-musicians. *Cortex* **59**: 126–137.

Ara, A. and Marco-Pallarés, J. (2020). Fronto-temporal theta phase-synchronization underlies music-evoked pleasantness. *NeuroImage* **212**: 116665.

Ara, A. and Marco-Pallarés, J. (2021). Different theta connectivity patterns underlie pleasantness evoked by familiar and unfamiliar music. *Scientific Reports* **11**: 18523.

Archakov, D., DeWitt, I., Kuśmierek, P., Ortiz-Rios, M., Cameron, D., Cui, D., . . . Rauschecker, J. P. (2020). Auditory representation of learned sound sequences in motor regions of the macaque brain. *Proceedings of the National Academy of Sciences* **117**: 15242–15252.

Arnal, L. H., Doelling, K. B. and Poeppel, D. (2014). Delta–beta coupled oscillations underlie temporal prediction accuracy. *Cerebral Cortex* **25**(9): 3077–3085.

Arnal, L. H., Flinker, A., Kleinschmidt, A., Giraud, A.-L. and Poeppel, D. (2015). Human screams occupy a privileged niche in the communication soundscape. *Current Biology* **25**(15): 2051–2056.

Arnal, L. H. and Giraud, A.-L. (2012). Cortical oscillations and sensory predictions. *Trends in Cognitive Sciences* **16**(7): 390–398.

Assmann, P. F. and Summerfield, Q. (1994). The contribution of waveform interactions to the perception of concurrent vowels. *Journal of the Acoustical Society of America* **95**(1): 471–484.

Athanasopoulos, G., Eerola, T., Lahdelma, I. and Kaliakatsos-Papakostas, M. (2021). Harmonic organisation conveys both universal and culture-specific cues for emotional expression in music. *PLOS ONE* **16**(1): e0244964.

Atiani, S., Elhilali, M., David, S. V., Fritz, J. B. and Shamma, S. A. (2009). Task difficulty and performance induce diverse adaptive patterns in gain and shape of primary auditory cortical receptive fields. *Neuron* **61**(3): 467–480.

Attneave, F. and Olson, R. K. (1971). Pitch as a medium: A new approach to psychophysical scaling. *American Journal of Psychology* **84**: 147–166.

Ayotte, J., Peretz, I. and Hyde, K. (2002). Congenital amusia: A group study of adults afflicted with a music-specific disorder. *Brain* **125**(2): 238–251.

Bachem, A. (1950). Tone height and tone chroma as two different pitch qualities. *Acta Psychologica* **7**: 80–88.

Badre, D. and D'Esposito, M. (2009). Is the rostro-caudal axis of the frontal lobe hierarchical? *Nature Reviews Neuroscience* **10**(9): 659–669.

Baer, L. H., Park, M. T. M., Bailey, J. A., Chakravarty, M. M., Li, K. Z. H. and Penhune, V. B. (2015). Regional cerebellar volumes are related to early musical training and finger tapping performance. *NeuroImage* **109**: 130–139.

Balezeau, F., Wilson, B., Gallardo, G., Dick, F., Hopkins, W., Anwander, A., Friederici, A. D., Griffiths, T. D. and Petkov, C. I. (2020). Primate auditory prototype in the evolution of the arcuate fasciculus. *Nature Neuroscience* **23**(5): 611–614.

Ball, T., Rahm, B., Eickhoff, S. B., Schulze-Bonhage, A., Speck, O. and Mutschler, I. (2007). Response properties of human amygdala subregions: Evidence based on functional MRI combined with probabilistic anatomical maps. *PLOS ONE* **2**(3): e307.

Bangert, M., Peschel, T., Schlaug, G., Rotte, M., Drescher, D., Hinrichs, H., Heinze, H.-J. and Altenmüller, E. (2006). Shared networks for auditory and motor processing in professional pianists: Evidence from fMRI conjunction. *NeuroImage* **30**(3): 917–926.

Barbas, H. (1993). Organization of cortical afferent input to orbitofrontal areas in the rhesus monkey. *Neuroscience* **56**(4): 841–864.

Bartra, O., Mcguire, J. T. and Kable, J. W. (2013). The valuation system: A coordinate-based meta-analysis of BOLD fMRI experiments examining neural correlates of subjective value. *NeuroImage* **76**: 412–427.

Barzilay, R., Moore, T. M., Greenberg, D. M., Didomenico, G. E., Brown, L. A., White, L. K., Gur, R. C. and Gur, R. E. (2020). Resilience, COVID-19-related stress, anxiety and depression during the pandemic in a large population enriched for healthcare providers. *Translational Psychiatry* **10**(1): 1–8.

Bastian, A. J. (2006). Learning to predict the future: The cerebellum adapts feedforward movement control. *Current Opinion in Neurobiology* **16**(6): 645–649.

Baumann, S., Petkov, C. and Griffiths, T. (2013). A unified framework for the organization of the primate auditory cortex. *Frontiers in Systems Neuroscience* **7**: 11.

Baumeister, R. F. (2005). *The cultural animal: Human nature, meaning, and social life*. Oxford University Press.

Beaman, C. P. and Williams, T. I. (2010). Earworms (stuck song syndrome): Towards a natural history of intrusive thoughts. *British Journal of Psychology* **101**(4): 637–653.

Bechara, A., Damasio, H. and Damasio, A. R. (2000a). Emotion, decision making and the orbitofrontal cortex. *Cerebral Cortex* **10**(3): 295–307.

Bechara, A., Tranel, D. and Damasio, H. (2000b). Characterization of the decision-making deficit of patients with ventromedial prefrontal cortex lesions. *Brain* **123**(11): 2189–2202.

Becker, G. M., Degroot, M. H. and Marschak, J. (1964). Measuring utility by a single-response sequential method. *Behavioral Science* **9**(3): 226–232.

Behroozmand, R., Oya, H., Nourski, K. V., Kawasaki, H., Larson, C. R., Brugge, J. F., Howard, M. A. and Greenlee, J. D. (2016). Neural correlates of vocal production and motor control in human Heschl's gyrus. *Journal of Neuroscience* **36**(7): 2302–2315.

Belfi, A. M., Evans, E., Heskje, J., Bruss, J. and Tranel, D. (2017). Musical anhedonia after focal brain damage. *Neuropsychologia* **97**: 29–37.

Belfi, A. M., Moreno, G. L., Gugliano, M. and Neill, C. (2022). Musical reward across the lifespan. *Aging & Mental Health* **26**: 932–939.

Belin, P. and Zatorre, R. J. (2015). Neurobiology: Sounding the alarm. *Current Biology* **25**(18): R805–R806.

Belin, P., Zatorre, R. J., Lafaille, P., Ahad, P. and Pike, B. (2000). Voice-selective areas in human auditory cortex. *Nature* **403**(6767): 309–312.

Belyk, M. and Brown, S. (2017). The origins of the vocal brain in humans. *Neuroscience & Biobehavioral Reviews* **77**: 177–193.

Belyk, M., Pfordresher, P. Q., Liotti, M. and Brown, S. (2015). The neural basis of vocal pitch imitation in humans. *Journal of Cognitive Neuroscience* **28**(4): 621–635.

Bendor, D., Osmanski, M. S. and Wang, X. (2012). Dual-pitch processing mechanisms in primate auditory cortex. *Journal of Neuroscience* **32**(46): 16149.

Bendor, D. and Wang, X. (2005). The neuronal representation of pitch in primate auditory cortex. *Nature* **436**(7054): 1161–1165.

Bendor, D. and Wang, X. (2006). Cortical representations of pitch in monkeys and humans. *Current Opinion in Neurobiology* **16**(4): 391–399.

Bengtsson, S. L., Csíkszentmihályi, M. and Ullén, F. (2007). Cortical regions involved in the generation of musical structures during improvisation in pianists. *Journal of Cognitive Neuroscience* **19**(5): 830–842.

Bengtsson, S. L., Ehrsson, H. H., Forssberg, H. and Ullén, F. (2004). Dissociating brain regions controlling the temporal and ordinal structure of learned movement sequences. *European Journal of Neuroscience* **19**(9): 2591–2602.

Bengtsson, S. L., Nagy, Z., Skare, S., Forsman, L., Forssberg, H. and Ullén, F. (2005). Extensive piano practicing has regionally specific effects on white matter development. *Nature Neuroscience* **8**(9): 1148–1150.

Bengtsson, S. L., Ullén, F., Henrik Ehrsson, H., Hashimoto, T., Kito, T., Naito, E., Forssberg, H. and Sadato, N. (2009). Listening to rhythms activates motor and premotor cortices. *Cortex* **45**(1): 62–71.

Bennett, D., Bode, S., Brydevall, M., Warren, H. and Murawski, C. (2016). Intrinsic valuation of information in decision making under uncertainty. *PLoS Computational Biology* **12**(7): e1005020.

Bergmann, T. O. and Hartwigsen, G. (2021). Inferring causality from noninvasive brain stimulation in cognitive neuroscience. *Journal of Cognitive Neuroscience* **33**(2): 195–225.

Berkowitz, A. L. and Ansari, D. (2008). Generation of novel motor sequences: The neural correlates of musical improvisation. *NeuroImage* **41**(2): 535–543.

Berlyne, D. E. (1971). *Aesthetics and psychobiology*. Appleton Century Crofts.

Bermudez, P., Lerch, J. P., Evans, A. C. and Zatorre, R. J. (2009). Neuroanatomical correlates of musicianship as revealed by cortical thickness and voxel-based morphometry. *Cerebral Cortex* **19**(7): 1583–1596.

Bernardini, F., Attademo, L., Blackmon, K. and Devinsky, O. (2017). Musical hallucinations: A brief review of functional neuroimaging findings. *CNS Spectrums* **22**(5): 397–403.

Berns, G. S. and Moore, S. E. (2012). A neural predictor of cultural popularity. *Journal of Consumer Psychology* **22**(1): 154–160.

Berridge, K. C. (2018). Evolving concepts of emotion and motivation. *Frontiers in Psychology* **9**: 1647.

Berridge, K. C. (2019). Affective valence in the brain: Modules or modes? *Nature Reviews Neuroscience* **20**(4): 225–234.

Berridge, K. C. and Kringelbach, M. L. (2015). Pleasure systems in the brain. *Neuron* **86**(3): 646–664.

Berridge, K. C. and Valenstein, E. S. (1991). What psychological process mediates feeding evoked by electrical stimulation of the lateral hypothalamus? *Behavioral Neuroscience* **105**(1): 3.

Berridge, K. C., Venier, I. L. and Robinson, T. E. (1989). Taste reactivity analysis of 6-hydroxydopamine-induced aphagia: Implications for arousal and anhedonia hypotheses of dopamine function. *Behavioral Neuroscience* **103**(1): 36.

Bewernick, B. H., Hurlemann, R., Matusch, A., Kayser, S., Grubert, C., Hadrysiewicz, B., Axmacher, N., Lemke, M., Cooper-Mahkorn, D. and Cohen, M. X. (2010). Nucleus accumbens deep brain stimulation decreases ratings of depression and anxiety in treatment-resistant depression. *Biological Psychiatry* **67**(2): 110–116.

Bey, C. and McAdams, S. (2002). Schema-based processing in auditory scene analysis. *Perception & Psychophysics* **64**(5): 844–854.

Bharucha, J. J. (1994). Tonality and expectation. In R. Aiello and J. A. Sloboda (Eds.), *Musical perceptions* (pp. 213–239). Oxford University Press.

Bianchi, F., Hjortkjaer, J., Santurette, S., Zatorre, R. J., Siebner, H. R. and Dau, T. (2017). Subcortical and cortical correlates of pitch discrimination: Evidence for two levels of neuroplasticity in musicians. *NeuroImage* **163**: 398–412.

Bianco, R., Gold, B., Johnson, A. and Penhune, V. (2019). Music predictability and liking enhance pupil dilation and promote motor learning in non-musicians. *Scientific Reports* **9**(1): 1–12.

Bidelman, G. M. (2018). Subcortical sources dominate the neuroelectric auditory frequency-following response to speech. *NeuroImage* **175**: 56–69.

Bidelman, G. M. and Alain, C. (2015). Hierarchical neurocomputations underlying concurrent sound segregation: Connecting periphery to percept. *Neuropsychologia* **68**: 38–50.

Bidelman, G. M., Krishnan, A. and Gandour, J. T. (2011). Enhanced brainstem encoding predicts musicians' perceptual advantages with pitch. *European Journal of Neuroscience* **33**(3): 530–538.

Bigand, E., Madurell, F., Tillmann, B. and Pineau, M. (1999). Effect of global structure and temporal organization on chord processing. *Journal of Experimental Psychology: Human Perception and Performance* **25**(1): 184.

Bigand, E. and Poulin-Charronnat, B. (2006). Are we "experienced listeners"? A review of the musical capacities that do not depend on formal musical training. *Cognition* **100**(1): 100–130.

Bilsen, F. (2006). Huygens on pitch perception; staircase reflections reconsidered. *Nederlands akoestisch genootschap* **178**: 1–8.

Bilsen, F. and Ritsma, R. (1969). Repetition pitch and its implication for hearing theory. *Acta Acustica united with Acustica* **22**(2): 63–73.

Bitterman, Y., Mukamel, R., Malach, R., Fried, I. and Nelken, I. (2008). Ultra-fine frequency tuning revealed in single neurons of human auditory cortex. *Nature* **451**(7175): 197–201.

Blank, H., Wieland, N. and Von Kriegstein, K. (2014). Person recognition and the brain: Merging evidence from patients and healthy individuals. *Neuroscience & Biobehavioral Reviews* **47**: 717–734.

Blood, A. J. and Zatorre, R. J. (2001). Intensely pleasurable responses to music correlate with activity in brain regions implicated in reward and emotion. *Proceedings of the National Academy of Sciences U S A* **98**(20): 11818–11823.

Blood, A. J., Zatorre, R. J., Bermudez, P. and Evans, A. C. (1999). Emotional responses to pleasant and unpleasant music correlate with activity in paralimbic brain regions. *Nature Neuroscience* **2**(4): 382–387.

Boemio, A., Fromm, S., Braun, A. and Poeppel, D. (2005). Hierarchical and asymmetric temporal sensitivity in human auditory cortices. *Nature Neuroscience* **8**(3): 389–395.

Boer, D. and Fischer, R. (2011). Towards a holistic model of functions of music listening across cultures: A culturally decentred qualitative approach. *Psychology of Music* **40**(2): 179–200.

Boileau, I., Assaad, J. M., Pihl, R. O., Benkelfat, C., Leyton, M., Diksic, M., Tremblay, R. E. and Dagher, A. (2003). Alcohol promotes dopamine release in the human nucleus accumbens. *Synapse* **49**(4): 226–231.

Bornstein, R. F. (1989). Exposure and affect: Overview and meta-analysis of research, 1968–1987. *Psychological Bulletin* **106**(2): 265.

Botinis, A., Granström, B. and Möbius, B. (2001). Developments and paradigms in intonation research. *Speech Communication* **33**(4): 263–296.

Bouton, S., Chambon, V., Tyrand, R., Guggisberg, A. G., Seeck, M., Karkar, S., Van De Ville, D. and Giraud, A.-L. (2018). Focal versus distributed temporal cortex activity for speech sound category assignment. *Proceedings of the National Academy of Sciences U S A* **115**(6): E1299–E1308.

Bowling, D. L., Gill, K., Choi, J. D., Prinz, J. and Purves, D. (2010). Major and minor music compared to excited and subdued speech. *Journal of the Acoustical Society of America* **127**(1): 491–503.

Bowling, D. L., Sundararajan, J., Han, S. E. and Purves, D. (2012). Expression of emotion in eastern and western music mirrors vocalization. *PLOS ONE* **7**(3): e31942.

Braams, B. R., Van Duijvenvoorde, A. C. K., Peper, J. S. and Crone, E. A. (2015). Longitudinal changes in adolescent risk-taking: A comprehensive study of neural responses to rewards, pubertal development, and risk-taking behavior. *Journal of Neuroscience* **35**(18): 7226.

Bradley, M. M., Greenwald, M. K., Petry, M. C. and Lang, P. J. (1992). Remembering pictures: Pleasure and arousal in memory. *Journal of Experimental Psychology: Learning, Memory, and Cognition* **18**(2): 379–390.

Bradshaw, J. L. and Nettleton, N. C. (1981). The nature of hemispheric specialization in man. *Behavioral and Brain Sciences* **4**(1): 51–63.

Bradt, J., Dileo, C. and Potvin, N. (2013). Music for stress and anxiety reduction in coronary heart disease patients. *Cochrane Database of Systematic Reviews* **12**: 1–87.

Brattico, E., Bogert, B., Alluri, V., Tervaniemi, M., Eerola, T. and Jacobsen, T. (2016). It's sad, but I like it: The neural dissociation between musical emotions and liking in experts and laypersons. *Frontiers in Human Neuroscience* **9**: 676.

Brattico, E., Pallesen, K. J., Varyagina, O., Bailey, C., Anourova, I., Järvenpää, M., Eerola, T. and Tervaniemi, M. (2009). Neural discrimination of nonprototypical chords in music experts and laymen: An MEG study. *Journal of Cognitive Neuroscience* **21**(11): 2230–2244.

Brauer, L. H. and De Wit, H. (1997). High dose pimozide does not block amphetamine-induced euphoria in normal volunteers. *Pharmacology Biochemistry and Behavior* **56**(2): 265–272.

Brechmann, A. and Scheich, H. (2004). Hemispheric shifts of sound representation in auditory cortex with conceptual listening. *Cerebral Cortex* **15**(5): 578–587.

Bregman, A. S. (1994). *Auditory scene analysis: The perceptual organization of sound.* MIT Press.

Bregman, M. R., Patel, A. D. and Gentner, T. Q. (2016). Songbirds use spectral shape, not pitch, for sound pattern recognition. *Proceedings of the National Academy of Sciences* **113**(6): 1666.

Brodsky, W. (2001). The effects of music tempo on simulated driving performance and vehicular control. *Transportation Research Part F: Traffic Psychology and Behaviour* **4**(4): 219–241.

Brody, A. L., Olmstead, R. E., London, E. D., Farahi, J., Meyer, J. H., Grossman, P., Lee, G. S., Huang, J., Hahn, E. L. and Mandelkern, M. A. (2004). Smoking-induced ventral striatum dopamine release. *American Journal of Psychiatry* **161**(7): 1211–1218.

Bromberg-Martin, E. S. and Hikosaka, O. (2009). Midbrain dopamine neurons signal preference for advance information about upcoming rewards. *Neuron* **63**(1): 119–126.

Brown, R. and Palmer, C. (2013). Auditory and motor imagery modulate learning in music performance. *Frontiers in Human Neuroscience* **7**: 320.

Brown, R. M., Chen, J. L., Hollinger, A., Penhune, V. B., Palmer, C. and Zatorre, R. J. (2013). Repetition suppression in auditory-motor regions to pitch and temporal structure in music. *Journal of Cognitive Neuroscience* **25**(2): 313–328.

Buchanan, T. W. (2007). Retrieval of emotional memories. *Psychological Bulletin* **133**(5): 761.

Büchel, C., Miedl, S. and Sprenger, C. (2018). Hedonic processing in humans is mediated by an opioidergic mechanism in a mesocorticolimbic system. *eLife* **7**: e39648.

Burger, B., London, J., Thompson, M. R. and Toiviainen, P. (2018). Synchronization to metrical levels in music depends on low-frequency spectral components and tempo. *Psychological Research* **82**(6): 1195–1211.

Burunat, I., Alluri, V., Toiviainen, P., Numminen, J. and Brattico, E. (2014). Dynamics of brain activity underlying working memory for music in a naturalistic condition. *Cortex* **57**: 254–269.

Calvert, G. A., Bullmore, E. T., Brammer, M. J., Campbell, R., Williams, S. C., Mcguire, P. K., Woodruff, P. W., Iversen, S. D. and David, A. S. (1997). Activation of auditory cortex during silent lipreading. *Science* **276**(5312): 593–596.

Camalier, C. R., D'angelo, W. R., Sterbing-D'angelo, S. J., De La Mothe, L. A. and Hackett, T. A. (2012). Neural latencies across auditory cortex of macaque support a dorsal stream supramodal timing advantage in primates. *Proceedings of the National Academy of Sciences* **109**(44): 18168–18173.

Cammoun, L., Thiran, J. P., Griffa, A., Meuli, R., Hagmann, P. and Clarke, S. (2015). Intrahemispheric cortico-cortical connections of the human auditory cortex. *Brain Structure and Function* **220**(6): 3537–3553.

Carbajal, G. V. and Malmierca, M. S. (2018). The neuronal basis of predictive coding along the auditory pathway: From the subcortical roots to cortical deviance detection. *Trends in Hearing* **22**: 1–23.

Carlson, E., Saarikallio, S., Toiviainen, P., Bogert, B., Kliuchko, M. and Brattico, E. (2015). Maladaptive and adaptive emotion regulation through music: A behavioral and neuroimaging study of males and females. *Frontiers in Human Neuroscience* **9**: 466.

Castro, D. C. and Berridge, K. C. (2017). Opioid and orexin hedonic hotspots in rat orbitofrontal cortex and insula. *Proceedings of the National Academy of Sciences* **114**(43): E9125–E9134.

Catani, M., Allin, M. P., Husain, M., Pugliese, L., Mesulam, M. M., Murray, R. M. and Jones, D. K. (2007). Symmetries in human brain language pathways correlate with verbal recall. *Proceedings of the National Academy of Sciences* **104**(43): 17163–17168.

Cavada, C., Compañy, T., Tejedor, J., Cruz-Rizzolo, R. J. and Reinoso-Suárez, F. (2000). The anatomical connections of the macaque monkey orbitofrontal cortex. A review. *Cerebral Cortex* **10**(3): 220–242.

Cha, K., Zatorre, R. J. and Schönwiesner, M. (2016). Frequency selectivity of voxel-by-voxel functional connectivity in human auditory cortex. *Cerebral Cortex* **26**(1): 211–224.

Chance, S. A. (2014). The cortical microstructural basis of lateralized cognition: A review. *Frontiers in Psychology* **5**: 820.

Chandrasekaran, B. and Kraus, N. (2010). The scalp-recorded brainstem response to speech: Neural origins and plasticity. *Psychophysiology* **47**(2): 236–246.

Chang, A., Kragness, H. E., Livingstone, S. R., Bosnyak, D. J. and Trainor, L. J. (2019). Body sway reflects joint emotional expression in music ensemble performance. *Scientific Reports* **9**(1): 205.

Chapman, L. J., Chapman, J. P. and Raulin, M. L. (1976). Scales for physical and social anhedonia. *Journal of Abnormal Psychology* **85**(4): 374.

Champod, A. S., and Petrides, M. (2007). Dissociable roles of the posterior parietal and the prefrontal cortex in manipulation and monitoring processes. *Proceedings of the National Academy of Sciences* **104**: 14837–14842.

Champod, A. S., and Petrides, M. (2010). Dissociation within the frontoparietal network in verbal working memory: a parametric functional magnetic resonance imaging study. *Journal of Neuroscience* **30**: 3849–3856.

Cheever, T., Taylor, A., Finkelstein, R., Edwards, E., Thomas, L., Bradt, J., Holochwost, S. J., Johnson, J. K., Limb, C., Patel, A. D., Tottenham, N., Iyengar, S., Rutter, D., Fleming, R., and Collins, F. S. (2018). NIH/Kennedy Center workshop on music and the brain: Finding harmony. *Neuron* **97**(6): 1214–1218.

Chelnokova, O., Laeng, B., Eikemo, M., Riegels, J., Løseth, G., Maurud, H., Willoch, F. and Leknes, S. (2014). Rewards of beauty: The opioid system mediates social motivation in humans. *Molecular Psychiatry* **19**(7): 746–747.

Chen, J. L., Penhune, V. B. and Zatorre, R. J. (2008a). Listening to musical rhythms recruits motor regions of the brain. *Cerebral Cortex* **18**(12): 2844–2854.

Chen, J. L., Penhune, V. B. and Zatorre, R. J. (2008b). Moving on time: Brain network for auditory-motor synchronization is modulated by rhythm complexity and musical training. *Journal of Cognitive Neuroscience* **20**(2): 226–239.

Chen, J. L., Rae, C. and Watkins, K. E. (2012). Learning to play a melody: An fMRI study examining the formation of auditory-motor associations. *NeuroImage* **59**(2): 1200–1208.

Chen, J. L., Zatorre, R. J. and Penhune, V. B. (2006). Interactions between auditory and dorsal premotor cortex during synchronization to musical rhythms. *NeuroImage* **32**(4): 1771–1781.

Chen, W. G., Iversen, J. R., Kao, M. H., Loui, P., Patel, A. D., Zatorre, R. J., and Edwards, E. (2022). Music and brain circuitry: Strategies for strengthening evidence-based research for music-based interventions. *Journal of Neuroscience* **42**: 8498–8507.

Cheung, V. K. M., Harrison, P. M. C., Meyer, L., Pearce, M. T., Haynes, J.-D. and Koelsch, S. (2019). Uncertainty and surprise jointly predict musical pleasure and amygdala, hippocampus, and auditory cortex activity. *Current Biology* **29**(23): 4084–4092.

Chevillet, M., Riesenhuber, M. and Rauschecker, J. P. (2011). Functional correlates of the anterolateral processing hierarchy in human auditory cortex. *Journal of Neuroscience* **31**(25): 9345–9352.

Chi, T., Ru, P. and Shamma, S. A. (2005). Multiresolution spectrotemporal analysis of complex sounds. *Journal of the Acoustical Society of America* **118**(2): 887–906.

Chib, V. S., Rangel, A., Shimojo, S. and Doherty, J. P. (2009). Evidence for a common representation of decision values for dissimilar goods in human ventromedial prefrontal cortex. *Journal of Neuroscience* **29**(39): 12315–12320.

Childress, A. R., Ehrman, R. N., Wang, Z., Li, Y., Sciortino, N., Hakun, J., Jens, W., Suh, J., Listerud, J. and Marquez, K. (2008). Prelude to passion: Limbic activation by "unseen" drug and sexual cues. *PLoS One* **3**(1): e1506.

Chin, T. and Rickard, N. S. (2014). Emotion regulation strategy mediates both positive and negative relationships between music uses and well-being. *Psychology of Music* **42**(5): 692–713.

Chiu, R. (2020). Functions of music making under lockdown: A trans-historical perspective across two pandemics. *Frontiers in Psychology* 11: 3628.

Chmiel, A. and Schubert, E. (2017). Back to the inverted U for music preference: A review of the literature. *Psychology of Music* 45(6): 886–909.

Cho, S. S. and Strafella, A. P. (2009). rTMS of the left dorsolateral prefrontal cortex modulates dopamine release in the ipsilateral anterior cingulate cortex and orbitofrontal cortex. *PloS One* 4(8): e6725.

Chobert, J., François, C., Velay, J.-L. and Besson, M. (2014). Twelve months of active musical training in 8-to 10-year-old children enhances the preattentive processing of syllabic duration and voice onset time. *Cerebral Cortex* 24(4): 956–967.

Cirelli, L. K., Einarson, K. M. and Trainor, L. J. (2014). Interpersonal synchrony increases prosocial behavior in infants. *Developmental Science* 17(6): 1003–1011.

Cisek, P. and Kalaska, J. F. (2004). Neural correlates of mental rehearsal in dorsal premotor cortex. *Nature* 431(7011): 993–996.

Cloutman, L. L. (2013). Interaction between dorsal and ventral processing streams: Where, when and how? *Brain and Language* 127(2): 251–263.

Coffey, E. B., Colagrosso, E. M., Lehmann, A., Schonwiesner, M. and Zatorre, R. J. (2016a). Individual differences in the frequency-following response: Relation to pitch perception. *PLoS One* 11(3): e0152374.

Coffey, E. B., Herholz, S. C., Chepesiuk, A. M., Baillet, S. and Zatorre, R. J. (2016b). Cortical contributions to the auditory frequency-following response revealed by MEG. *Nature Communications* 7: 11070.

Coffey, E. B. J., Arseneau-Bruneau, I., Zhang, X., Baillet, S. and Zatorre, R. J. (2020). Oscillatory entrainment of the frequency following response in auditory cortical and subcortical structures. *Journal of Neuroscience* 41: 4073–4087.

Coffey, E. B. J., Arseneau-Bruneau, I., Zhang, X. and Zatorre, R. J. (2019a). The music-in-noise task (MINT): A tool for dissecting complex auditory perception. *Frontiers in Neuroscience* 13: 199.

Coffey, E. B. J., Chepesiuk, A. M. P., Herholz, S. C., Baillet, S. and Zatorre, R. J. (2017a). Neural correlates of early sound encoding and their relationship to speech-in-noise perception. *Frontiers in Neuroscience* 11: 479.

Coffey, E. B. J., Mogilever, N. B. and Zatorre, R. J. (2017b). Speech-in-noise perception in musicians: A review. *Hearing Research* 352: 49–69.

Coffey, E. B. J., Musacchia, G. and Zatorre, R. J. (2017c). Cortical correlates of the auditory frequency-following and onset responses: EEG and fMRI evidence. *Journal of Neuroscience* 37(4): 830–838.

Coffey, E. B. J., Nicol, T., White-Schwoch, T., Chandrasekaran, B., Krizman, J., Skoe, E., Zatorre, R. J. and Kraus, N. (2019b). Evolving perspectives on the sources of the frequency-following response. *Nature Communications* 10(1): 5036.

Cohen, A., Bailey, B. and Nilsson, T. (2002). The importance of music to seniors. *Psychomusicology: A Journal of Research in Music Cognition* 18(1–2): 89.

Colasanti, A., Searle, G. E., Long, C. J., Hill, S. P., Reiley, R. R., Quelch, D., Erritzoe, D., Tziortzi, A. C., Reed, L. J., Lingford-Hughes, A. R., Waldman, A. D., Schruers, K. R. J., Matthews, P. M., Gunn, R. N., Nutt, D. J. and Rabiner, E. A. (2012). Endogenous opioid release in the human brain reward system induced by acute amphetamine administration. *Biological Psychiatry* 72(5): 371–377.

Conard, N. J., Malina, M. and Münzel, S. C. (2009). New flutes document the earliest musical tradition in southwestern Germany. *Nature* 460(7256): 737–740.

Condon, C. D. and Weinberger, N. M. (1991). Habituation produces frequency-specific plasticity of receptive fields in the auditory cortex. *Behavioral Neuroscience* 105(3): 416.

Cooper, S. J. and Turkish, S. (1989). Effects of naltrexone on food preference and concurrent behavioral responses in food-deprived rats. *Pharmacology Biochemistry and Behavior* 33(1): 17–20.

Cotter, K. N. (2019). Mental control in musical imagery: A dual component model. *Frontiers in Psychology* 10: 1904.

Cousineau, M., McDermott, J. H. and Peretz, I. (2012). The basis of musical consonance as revealed by congenital amusia. *Proceedings of the National Academy of Sciences* 109(48): 19858–19863.

Coutinho, E. and Dibben, N. (2013). Psychoacoustic cues to emotion in speech prosody and music. *Cognition and Emotion* 27(4): 658–684.

Covey, E. (2005). Neurobiological specializations in echolocating bats. *Anatomical Record Part A: Discoveries in Molecular, Cellular, and Evolutionary Biology* **287A**(1): 1103–1116.

Cowen, A. S., Fang, X., Sauter, D. and Keltner, D. (2020). What music makes us feel: At least 13 dimensions organize subjective experiences associated with music across different cultures. *Proceedings of the National Academy of Sciences* **117**(4): 1924–1934.

Cox, S. M. L., Benkelfat, C., Dagher, A., Delaney, J. S., Durand, F., McKenzie, S. A., Kolivakis, T., Casey, K. F. and Leyton, M. (2009). Striatal dopamine responses to intranasal cocaine self-administration in humans. *Biological Psychiatry* **65**(10): 846–850.

Creel, S. C., Newport, E. L. and Aslin, R. N. (2004). Distant melodies: Statistical learning of nonadjacent dependencies in tone sequences. *Journal of Experimental Psychology: Learning, Memory, and Cognition* **30**(5): 1119.

Critchley, M. and Henson, R. A. (1977). *Music and the brain: Studies in the neurology of music.* Butterworth-Heinemann.

Cui, H. (2014). From intention to action: Hierarchical sensorimotor transformation in the posterior parietal cortex. *eNeuro* **1**(1).

Culham, J. C. and Kanwisher, N. G. (2001). Neuroimaging of cognitive functions in human parietal cortex. *Current Opinion in Neurobiology* **11**(2): 157–163.

Cupchik, G. C., Phillips, K., and Hill, D. S. (2001). Shared processes in spatial rotation and musical permutation. *Brain and cognition* **46**: 373–382.

Curot, J., Busigny, T., Valton, L., Denuelle, M., Vignal, J.-P., Maillard, L., Chauvel, P., Pariente, J., Trebuchon, A., Bartolomei, F. and Barbeau, E. J. (2017). Memory scrutinized through electrical brain stimulation: A review of 80 years of experiential phenomena. *Neuroscience & Biobehavioral Reviews* **78**: 161–177.

Cusack, R., Decks, J., Aikman, G. and Carlyon, R. P. (2004). Effects of location, frequency region, and time course of selective attention on auditory scene analysis. *Journal of Experimental Psychology: Human Perception and Performance* **30**(4): 643–656.

D'Ausilio, A., Altenmüller, E., Olivetti Belardinelli, M. and Lotze, M. (2006). Cross-modal plasticity of the motor cortex while listening to a rehearsed musical piece. *European Journal of Neuroscience* **24**(3): 955–958.

Da Costa, S., Van Der Zwaag, W., Miller, L. M., Clarke, S. and Saenz, M. (2013). Tuning in to sound: Frequency-selective attentional filter in human primary auditory cortex. *Journal of Neuroscience* **33**(5): 1858–1863.

Dalboni Da Rocha, J. L., Schneider, P., Benner, J., Santoro, R., Atanasova, T., Van De Ville, D. and Golestani, N. (2020). TASH: Toolbox for the automated segmentation of Heschl's gyrus. *Scientific Reports* **10**(1): 3887.

Dalla Bella, S., Białuńska, A. and Sowiński, J. (2013). Why movement is captured by music, but less by speech: Role of temporal regularity. *PLOS ONE* **8**(8): e71945.

Dalla Bella, S., Giguère, J.-F. and Peretz, I. (2007). Singing proficiency in the general population. *Journal of the Acoustical Society of America* **121**(2): 1182–1189.

Dalla Bella, S., Peretz, I., Rousseau, L. and Gosselin, N. (2001). A developmental study of the affective value of tempo and mode in music. *Cognition* **80**(3): B1–B10.

Damasio, A., Damasio, H. and Tranel, D. (2013). Persistence of feelings and sentience after bilateral damage of the insula. *Cerebral Cortex* **23**(4): 833–846.

Damasio, A. R. (1994). *Descartes' error: Emotion, reason and the human brain.* Putnam.

Damasio, A. R., Grabowski, T. J., Bechara, A., Damasio, H., Ponto, L. L., Parvizi, J. and Hichwa, R. D. (2000). Subcortical and cortical brain activity during the feeling of self-generated emotions. *Nature Neuroscience* **3**(10): 1049–1056.

Damasio, A. R., Tranel, D., and Damasio, H. C. (1991). Somatic markers and the guidance of behavior: Theory and preliminary testing. In H. S. Levin, H. M. Eisenberg, and A. L. Benton (Eds.), Frontal lobe function and dysfunction (pp. 217–229). Oxford University Press.

Damm, L., Varoqui, D., De Cock, V. C., Dalla Bella, S. and Bardy, B. (2020). Why do we move to the beat? A multi-scale approach, from physical principles to brain dynamics. *Neuroscience & Biobehavioral Reviews* **112**: 553–584.

Daniel, R. and Pollmann, S. (2014). A universal role of the ventral striatum in reward-based learning: Evidence from human studies. *Neurobiology of Learning and Memory* **114**: 90–100.

Darwin, C., Carlyon, R. and Moore, B. (1995). Auditory grouping. In B. C. J. Moore (Ed.), *Hearing* (pp. 387–424). Academic Press.

Davis, M. H. and Johnsrude, I. S. (2003). Hierarchical processing in spoken language comprehension. *Journal of Neuroscience* **23**(8): 3423–3431.

Davis, M. H. and Johnsrude, I. S. (2007). Hearing speech sounds: Top-down influences on the interface between audition and speech perception. *Hearing Research* **229**(1): 132–147.

Daw, N. D., Gershman, S. J., Seymour, B., Dayan, P. and Dolan, R. J. (2011). Model-based influences on humans' choices and striatal prediction errors. *Neuron* **69**(6): 1204–1215.

Decharms, R. C., Blake, D. T. and Merzenich, M. M. (1998). Optimizing sound features for cortical neurons. *Science* **280**(5368): 1439–1444.

De Cheveigné, A. (1997). Harmonic fusion and pitch shifts of mistuned partials. *Journal of the Acoustical Society of America* **102**(2): 1083–1087.

Dehaene-Lambertz, G., Pallier, C., Serniclaes, W., Sprenger-Charolles, L., Jobert, A. and Dehaene, S. (2005). Neural correlates of switching from auditory to speech perception. *NeuroImage* **24**(1): 21–33.

Dehaene, S. (2011). *The number sense: How the mind creates mathematics*. Oxford University Press.

Dehaene, S. and Changeux, J.-P. (2011). Experimental and theoretical approaches to conscious processing. *Neuron* **70**(2): 200–227.

Dehaene, S. and Cohen, L. (2007). Cultural recycling of cortical maps. *Neuron* **56**(2): 384–398.

Dellacherie, D., Pfeuty, M., Hasboun, D., Lefèvre, J., Hugueville, L., Schwartz, D., Baulac, M., Adam, C. and Samson, S. (2009). The birth of musical emotion. *Annals of the New York Academy of Sciences* **1169**(1): 336–341.

De Manzano, Ö., Kuckelkorn, K. L., Ström, K. and Ullén, F. (2020). Action-perception coupling and near transfer: Listening to melodies after piano practice triggers sequence-specific representations in the auditory-motor network. *Cerebral Cortex* **30**(10): 5193–5203.

De Manzano, Ö. and Ullén, F. (2012). Activation and connectivity patterns of the presupplementary and dorsal premotor areas during free improvisation of melodies and rhythms. *NeuroImage* **63**(1): 272–280.

D'Esposito, M. and Postle, B. R. (2015). The cognitive neuroscience of working memory. *Annual Review of Psychology* **66**(1): 115–142.

Depue, R. A. and Collins, P. F. (1999). Neurobiology of the structure of personality: Dopamine, facilitation of incentive motivation, and extraversion. *Behavioral and Brain Sciences* **22**(3): 491–517.

Der-Avakian, A. and Markou, A. (2012). The neurobiology of anhedonia and other reward-related deficits. *Trends in Neurosciences* **35**(1): 68–77.

Deutsch, D. (1970). Tones and numbers: Specificity of interference in immediate memory. *Science* **168**(3939): 1604–1605.

Deutsch, D. (Ed.). (1982). *The psychology of music*. Academic press series in cognition and perception. Academic Press.

Deutsch, D., Henthorn, T. and Lapidis, R. (2011). Illusory transformation from speech to song. *Journal of the Acoustical Society of America* **129**(4): 2245–2252.

De Witte, M., Spruit, A., Van Hooren, S., Moonen, X. and Stams, G.-J. (2020). Effects of music interventions on stress-related outcomes: A systematic review and two meta-analyses. *Health Psychology Review* **14**(2): 294–324.

Dick, F., Tierney, A. T., Lutti, A., Josephs, O., Sereno, M. I. and Weiskopf, N. (2012). In vivo functional and myeloarchitectonic mapping of human primary auditory areas. *Journal of Neuroscience* **32**(46): 16095–16105.

Diedrichsen, J., King, M., Hernandez-Castillo, C., Sereno, M. and Ivry, R. B. (2019). Universal transform or multiple functionality? Understanding the contribution of the human cerebellum across task domains. *Neuron* **102**(5): 918–928.

Di Liberto, G. M., Pelofi, C., Bianco, R., Patel, P., Mehta, A. D., Herrero, J. L., De Cheveigné, A., Shamma, S. and Mesgarani, N. (2020). Cortical encoding of melodic expectations in human temporal cortex. *eLife* **9**: e51784.

Di Martino, A., Scheres, A., Margulies, D. S., Kelly, A. M. C., Uddin, L. Q., Shehzad, Z., Biswal, B., Walters, J. R., Castellanos, F. X. and Milham, M. P. (2008). Functional connectivity of human striatum: A resting state fMRI study. *Cerebral Cortex* **18**(12): 2735–2747.

Ding, N., Patel, A. D., Chen, L., Butler, H., Luo, C. and Poeppel, D. (2017). Temporal modulations in speech and music. *Neuroscience & Biobehavioral Reviews* **81**: 181–187.

Ding, N. and Simon, J. Z. (2012). Emergence of neural encoding of auditory objects while listening to competing speakers. *Proceedings of the National Academy of Sciences* **109**(29): 11854–11859.

Disbergen, N. R., Valente, G., Formisano, E. and Zatorre, R. J. (2018). Assessing top-down and bottom-up contributions to auditory stream segregation and integration with polyphonic music. *Frontiers in Neuroscience* **12**: 121.

Dorsaint-Pierre, R., Penhune, V. B., Watkins, K. E., Neelin, P., Lerch, J. P., Bouffard, M. and Zatorre, R. J. (2006). Asymmetries of the planum temporale and Heschl's gyrus: Relationship to language lateralization. *Brain* **129**(Pt 5): 1164–1176.

Douglas, K. M. and Bilkey, D. K. (2007). Amusia is associated with deficits in spatial processing. *Nature Neuroscience* **10**: 915–921.

Dowdle, L. T., Brown, T. R., George, M. S. and Hanlon, C. A. (2018). Single pulse TMS to the DLPFC, compared to a matched sham control, induces a direct, causal increase in caudate, cingulate, and thalamic bold signal. *Brain Stimulation* **11**(4): 789–796.

Dowling, W. J. (1973). The perception of interleaved melodies. *Cognitive Psychology* **5**(3): 322–337.

Dowling, W. J. (1978). Scale and contour: Two components of a theory of memory for melodies. *Psychological Review* **85**(4): 341.

Dowling, W. J. and Harwood, D. L. (1986). *Music Cognition*. Academic Press.

Drevets, W. C., Gautier, C., Price, J. C., Kupfer, D. J., Kinahan, P. E., Grace, A. A., Price, J. L and Mathis, C. A. (2001). Amphetamine-induced dopamine release in human ventral striatum correlates with euphoria. *Biological Psychiatry* **49**(2): 81–96.

Du, Y. and Zatorre, R. J. (2017). Musical training sharpens and bonds ears and tongue to hear speech better. *Proceedings of the National Academy of Sciences U S A* **114**(51): 13579–13584.

Dubé, L. and Le Bel, J. (2003). The content and structure of lay people's concept of pleasure. *Cognition and Emotion* **17**(2): 263–295.

Eckert, M. A. and Vaden, K. I. (2019). A deformation-based approach for characterizing brain asymmetries at different spatial scales of resolution. *Journal of Neuroscience Methods* **322**: 1–9.

Egermann, H., Fernando, N., Chuen, L. and McAdams, S. (2015). Music induces universal emotion-related psychophysiological responses: Comparing Canadian listeners to Congolese pygmies. *Frontiers in Psychology* **5**: 1341.

Egermann, H., Pearce, M. T., Wiggins, G. A. and McAdams, S. (2013). Probabilistic models of expectation violation predict psychophysiological emotional responses to live concert music. *Cognitive, Affective, & Behavioral Neuroscience* **13**(3): 533–553.

Eichert, N., Papp, D., Mars, R. B. and Watkins, K. E. (2020). Mapping human laryngeal motor cortex during vocalization. *Cerebral Cortex* **30**(12): 6254–6269.

Eikemo, M., Løseth, G. E., Johnstone, T., Gjerstad, J., Willoch, F. and Leknes, S. (2016). Sweet taste pleasantness is modulated by morphine and naltrexone. *Psychopharmacology* **233**(21): 3711–3723.

Ekman, P. E. and Davidson, R. J. (1994). *The nature of emotion: Fundamental questions*. Oxford University Press.

Elhilali, M. and Shamma, S. A. (2008). A cocktail party with a cortical twist: How cortical mechanisms contribute to sound segregation. *Journal of the Acoustical Society of America* **124**(6): 3751–3771.

Elie, J. E. and Theunissen, F. E. (2016). The vocal repertoire of the domesticated zebra finch: A data-driven approach to decipher the information-bearing acoustic features of communication signals. *Animal Cognition* **19**(2): 285–315.

Elliott, T. M. and Theunissen, F. E. (2009). The modulation transfer function for speech intelligibility. *PLOS Computational Biology* **5**(3): e1000302.

Engel, A., Hijmans, B. S., Cerliani, L., Bangert, M., Nanetti, L., Keller, P. E. and Keysers, C. (2014). Inter-individual differences in audio-motor learning of piano melodies and white matter fiber tract architecture. *Human Brain Mapping* **35**(5): 2483–2497.

Erb, J., Armendariz, M., De Martino, F., Goebel, R., Vanduffel, W. and Formisano, E. (2019). Homology and specificity of natural sound-encoding in human and monkey auditory cortex. *Cerebral Cortex* **29**(9): 3636–3650.

Eriksson, J., Vogel, E. K., Lansner, A., Bergström, F., and Nyberg, L. (2015). Neurocognitive architecture of working memory. *Neuron* **88**: 33–46.

Escera, C., Alho, K., Winkler, I. and Näätänen, R. (1998). Neural mechanisms of involuntary attention to acoustic novelty and change. *Journal of Cognitive Neuroscience* **10**(5): 590–604.

Eschrich, S., Münte, T. F. and Altenmüller, E. O. (2008). Unforgettable film music: The role of emotion in episodic long-term memory for music. *BMC Neuroscience* **9**(1):1–7.

Eshel, N., Bukwich, M., Rao, V., Hemmelder, V., Tian, J. and Uchida, N. (2015). Arithmetic and local circuitry underlying dopamine prediction errors. *Nature* **525**(7568): 243–246.

Etzel, J. A., Johnsen, E. L., Dickerson, J., Tranel, D. and Adolphs, R. (2006). Cardiovascular and respiratory responses during musical mood induction. *International Journal of Psychophysiology* **61**(1): 57–69.

Evers, S. and Ellger, T. (2004). The clinical spectrum of musical hallucinations. *Journal of the Neurological Sciences* **227**(1): 55–65.

Falk, S., Rathcke, T. and Dalla Bella, S. (2014). When speech sounds like music. *Journal of Experimental Psychology: Human Perception and Performance* **40**(4): 1491–1506.

Farah, M. J. (1989). The neural basis of mental imagery. *Trends in Neurosciences* **12**(10): 395–399.

Farbood, M. M., Heeger, D. J., Marcus, G., Hasson, U. and Lerner, Y. (2015). The neural processing of hierarchical structure in music and speech at different timescales. *Frontiers in Neuroscience* **9**: 157.

Farrugia, N., Jakubowski, K., Cusack, R. and Stewart, L. (2015). Tunes stuck in your brain: The frequency and affective evaluation of involuntary musical imagery correlate with cortical structure. *Consciousness and Cognition* **35**: 66–77.

Fecteau, S., Belin, P., Joanette, Y. and Armony, J. L. (2007). Amygdala responses to nonlinguistic emotional vocalizations. *NeuroImage* **36**(2): 480–487.

Feldman, R. (2017). The neurobiology of human attachments. *Trends in Cognitive Sciences* **21**(2): 80–99.

Fell, J. and Axmacher, N. (2011). The role of phase synchronization in memory processes. *Nature Reviews Neuroscience* **12**: 105–118.

Feng, L. and Wang, X. (2017). Harmonic template neurons in primate auditory cortex underlying complex sound processing. *Proceedings of the National Academy of Sciences* **114**(5): E840–E848.

Ferreri, L., Mas-Herrero, E., Cardona, G., Zatorre, R. J., Antonijoan, R. M., Valle, M., Riba, J., Ripollés, P. and Rodriguez-Fornells, A. (2021a). Dopamine modulations of reward-driven music memory consolidation. *Annals of the New York Academy of Sciences* **1502**(1): 85–98.

Ferreri, L., Mas-Herrero, E., Zatorre, R. J., Ripolles, P., Gomez-Andres, A., Alicart, H., Olive, G., Marco-Pallares, J., Antonijoan, R. M., Valle, M., Riba, J. and Rodriguez-Fornells, A. (2019). Dopamine modulates the reward experiences elicited by music. *Proceedings of the National Academy of Sciences U S A* **116**(9): 3793–3798.

Ferreri, L. and Rodriguez-Fornells, A. (2017). Music-related reward responses predict episodic memory performance. *Experimental Brain Research* **235**(12): 3721–3731.

Ferreri, L., Singer, N., Mcphee, M., Ripollés, P., Zatorre, R. J. and Mas-Herrero, E. (2021b). Engagement in music-related activities during the COVID-19 pandemic as a mirror of individual differences in musical reward and coping strategies. *Frontiers in Psychology* **12**: 673672.

Ferrier, D. (1875). Experiments on the brain of monkeys—No. I. *Proceedings of the Royal Society of London, Series B*, **23**: 409–432.

Fields, R. D. (2008). White matter in learning, cognition and psychiatric disorders. *Trends in Neurosciences* **31**(7): 361–370.

Figner, B., Knoch, D., Johnson, E. J., Krosch, A. R., Lisanby, S. H., Fehr, E. and Weber, E. U. (2010). Lateral prefrontal cortex and self-control in intertemporal choice. *Nature Neuroscience* **13**(5): 538–539.

Fink, L. K., Warrenburg, L. A., Howlin, C., Randall, W. M., Hansen, N. C. and Wald-Fuhrmann, M. (2021). Viral tunes: Changes in musical behaviours and interest in coronamusic predict socio-emotional coping during COVID-19 lockdown. *Humanities and Social Sciences Communications* **8**(1): 180.

Fiorillo, C. D., Tobler, P. N. and Schultz, W. (2003). Discrete coding of reward probability and uncertainty by dopamine neurons. *Science* **299**(5614): 1898–1902.

Fiser, J. and Aslin, R. N. (2002). Statistical learning of higher-order temporal structure from visual shape sequences. *Journal of Experimental Psychology: Learning, Memory, and Cognition* **28**(3): 458–467.

Fishman, Y. I., Reser, D. H., Arezzo, J. C. and Steinschneider, M. (2001a). Neural correlates of auditory stream segregation in primary auditory cortex of the awake monkey. *Hearing Research* **151**(1): 167–187.

Fishman, Y. I., Volkov, I. O., Noh, M. D., Garell, P. C., Bakken, H., Arezzo, J. C., Howard, M. A. and Steinschneider, M. (2001b). Consonance and dissonance of musical chords: Neural correlates in auditory cortex of monkeys and humans. *Journal of Neurophysiology* **86**(6): 2761–2788.

Flinker, A., Doyle, W. K., Mehta, A. D., Devinsky, O. and Poeppel, D. (2019). Spectrotemporal modulation provides a unifying framework for auditory cortical asymmetries. *Nature Human Behaviour* **3**(4): 393–405.

Floegel, M., Fuchs, S. and Kell, C. A. (2020). Differential contributions of the two cerebral hemispheres to temporal and spectral speech feedback control. *Nature Communications* **11**(1): 2839.

Ford, J. H., Addis, D. R. and Giovanello, K. S. (2011). Differential neural activity during search of specific and general autobiographical memories elicited by musical cues. *Neuropsychologia* **49**(9): 2514–2526.

Formisano, E., Kim, D.-S., Di Salle, F., Van De Moortele, P.-F., Ugurbil, K. and Goebel, R. (2003). Mirror-symmetric tonotopic maps in human primary auditory cortex. *Neuron* **40**(4): 859–869.

Foster, N. E. and Zatorre, R. J. (2010a). Cortical structure predicts success in performing musical transformation judgments. *NeuroImage* **53**(1): 26–36.

Foster, N. E. and Zatorre, R. J. (2010b). A role for the intraparietal sulcus in transforming musical pitch information. *Cerebral Cortex* **20**(6): 1350–1359.

Foster, N. E. V., Halpern, A. R. and Zatorre, R. J. (2013). Common parietal activation in musical mental transformations across pitch and time. *NeuroImage* **75**: 27–35.

Fouragnan, E., Retzler, C. and Philiastides, M. G. (2018). Separate neural representations of prediction error valence and surprise: Evidence from an fMRI meta-analysis. *Human Brain Mapping* **39**(7): 2887–2906.

Fraisse, P. (1948). Ii. - rythmes auditifs et rythmes visuels. *L'Année psychologique* **49**: 21–42.

Freeman, T. P., Pope, R. A., Wall, M. B., Bisby, J. A., Luijten, M., Hindocha, C., Mokrysz, C., Lawn, W., Moss, A., Bloomfield, M. A. P., Morgan, C. J. A., Nutt, D. J. and Curran, H. V. (2017). Cannabis dampens the effects of music in brain regions sensitive to reward and emotion. *International Journal of Neuropsychopharmacology* **21**(1): 21–32.

Frey, S., Campbell, J. S. W., Pike, G. B. and Petrides, M. (2008). Dissociating the human language pathways with high angular resolution diffusion fiber tractography. *Journal of Neuroscience* **28**(45): 11435.

Friederici, A. D. (2016). The neuroanatomical pathway model of language: Syntactic and semantic networks. In G. Hickok and S. L. Small (Eds.), *Neurobiology of language* (pp. 349–356). Academic Press.

Friedman, D. P., Aggleton, J. P. and Saunders, R. C. (2002). Comparison of hippocampal, amygdala, and perirhinal projections to the nucleus accumbens: Combined anterograde and retrograde tracing study in the macaque brain. *Journal of Comparative Neurology* **450**(4): 345–365.

Friston, K. (2010). The free-energy principle: A unified brain theory? *Nature Reviews Neuroscience* **11**(2): 127–138.

Friston, K. (2018). Does predictive coding have a future? *Nature Neuroscience* **21**(8): 1019–1021.

Fritz, J., Mishkin, M. and Saunders, R. C. (2005). In search of an auditory engram. *Proceedings of the National Academy of Sciences of the United States of America* **102**(26): 9359–9364.

Fritz, J., Shamma, S., Elhilali, M. and Klein, D. (2003). Rapid task-related plasticity of spectrotemporal receptive fields in primary auditory cortex. *Nature Neuroscience* **6**(11): 1216–1223.

Fritz, T., Halfpaap, J., Grahl, S., Kirkland, A. and Villringer, A. (2013a). Musical feedback during exercise machine workout enhances mood. *Frontiers in Psychology* **4**: 921.

Fritz, T., Jentschke, S., Gosselin, N., Sammler, D., Peretz, I., Turner, R., Friederici, A. D. and Koelsch, S. (2009). Universal recognition of three basic emotions in music. *Current Biology* **19**(7): 573–576.

Fritz, T. H., Bowling, D. L., Contier, O., Grant, J., Schneider, L., Lederer, A., Höer, F., Busch, Eand Villringer, A. (2018). Musical agency during physical exercise decreases pain. *Frontiers in Psychology* **8**: 2312.

Fritz, T. H., Hardikar, S., Demoucron, M., Niessen, M., Demey, M., Giot, O., Li, Y., Haynes, J.-D., Villringer, A. and Leman, M. (2013b). Musical agency reduces perceived exertion during strenuous physical performance. *Proceedings of the National Academy of Sciences* **110**(44): 17784–17789.

Frühholz, S. and Belin, P. (2018). The science of voice perception. In S. Frühholz and P Belin (Eds.), *The Oxford handbook of voice perception* (pp. 3–14). Oxford University Press.

Frühholz, S., Trost, W. and Grandjean, D. (2014). The role of the medial temporal limbic system in processing emotions in voice and music. *Progress in Neurobiology* **123**: 1–17.

Fuentemilla, L., Marco-Pallarés, J., Münte, T. F. and Grau, C. (2008). Theta EEG oscillatory activity and auditory change detection. *Brain Research* **1220**: 93–101.

Fujioka, T., Ross, B. and Trainor, L. J. (2015). Beta-band oscillations represent auditory beat and its metrical hierarchy in perception and imagery. *Journal of Neuroscience* **35**(45): 15187–15198.

Fujioka, T., Trainor, L. J., Large, E. W. and Ross, B. (2012). Internalized timing of isochronous sounds is represented in neuromagnetic beta oscillations. *Journal of Neuroscience* **32**(5): 1791–1802.

Fujioka, T., Trainor, L. J., Ross, B., Kakigi, R. and Pantev, C. (2005). Automatic encoding of polyphonic melodies in musicians and nonmusicians. *Journal of Cognitive Neuroscience* **17**(10): 1578–1592.

Furukawa, S., Xu, L. and Middlebrooks, J. C. (2000). Coding of sound-source location by ensembles of cortical neurons. *Journal of Neuroscience* **20**(3): 1216–1228.

Gaab, N., Gaser, C., Zaehle, T., Jancke, L. and Schlaug, G. (2003). Functional anatomy of pitch memory—an fMRI study with sparse temporal sampling. *NeuroImage* **19**(4): 1417–1426.

Gabard-Durnam, L. J., Hensch, T. K. and Tottenham, N. (2018). Music reveals medial prefrontal cortex sensitive period in childhood. *bioRxiv*: 412007.

Gabor, D. (1946). Theory of communication. Part 1: The analysis of information. *Journal of the Institution of Electrical Engineers-Part III: Radio and Communication Engineering* **93**(26): 429–441.

Gabrielsson, A. (1999). The performance of music. In D. Deutsch (Ed.), *The psychology of music* (pp. 501–602). Elsevier.

Gabrielsson, A. (2001). Emotion perceived and emotion felt: Same or different? *Musicae scientiae* **5**(1_ suppl): 123–147.

Gagnon, L. and Peretz, I. (2003). Mode and tempo relative contributions to "happy-sad" judgements in equitone melodies. *Cognition and Emotion* **17**(1): 25–40.

Gallivan, J. P. and Culham, J. C. (2015). Neural coding within human brain areas involved in actions. *Current Opinion in Neurobiology* **33**: 141–149.

Gallivan, J. P. and Goodale, M. A. (2018). The dorsal "action" pathway. In G. Vallar and H. B. Coslett (Eds.), *Handbook of clinical neurology* (Vol. 151, pp. 449–466).Elsevier.

Galuske, R. A. W., Schlote, W., Bratzke, H. and Singer, W. (2000). Interhemispheric asymmetries of the modular structure in human temporal cortex. *Science* **289**(5486): 1946–1949.

Gan, L., Huang, Y., Zhou, L., Qian, C. and Wu, X. (2015). Synchronization to a bouncing ball with a realistic motion trajectory. *Scientific Reports* **5**(1): 11974.

Gander, P. E., Kumar, S., Sedley, W., Nourski, K. V., Oya, H., Kovach, C. K., Kawasaki, H., Kikuchi, Y., Patterson, R. D., Howard, M. A. and Griffiths, T. D. (2019). Direct electrophysiological mapping of human pitch-related processing in auditory cortex. *NeuroImage* **202**: 116076.

Garrido, M. I., Kilner, J. M., Kiebel, S. J. and Friston, K. J. (2007). Evoked brain responses are generated by feedback loops. *Proceedings of the National Academy of Sciences* **104**(52): 20961–20966.

Gebauer, L., Kringelbach, M. L. and Vuust, P. (2012). Ever-changing cycles of musical pleasure: The role of dopamine and anticipation. *Psychomusicology: Music, Mind, and Brain* **22**(2): 152–167.

Gebel, B., Braun, C., Kaza, E., Altenmüller, E. and Lotze, M. (2013). Instrument specific brain activation in sensorimotor and auditory representation in musicians. *NeuroImage* **74**: 37–44.

Gebhardt, S., Kunkel, M. and Georgi, R. V. (2014). Emotion modulation in psychiatric patients through music. *Music Perception* **31**(5): 485–493.

Gehr, D. D., Komiya, H. and Eggermont, J. J. (2000). Neuronal responses in cat primary auditory cortex to natural and altered species-specific calls. *Hearing Research* **150**(1–2): 27–42.

Gervain, J. and Geffen, M. N. (2019). Efficient neural coding in auditory and speech perception. *Trends in Neurosciences* **42**(1): 56–65.

Geschwind, N. and Levitsky, W. (1968). Human brain: Left-right asymmetries in temporal speech region. *Science* **161**(3837): 186–187.

Giard, M. H., Perrin, F., Pernier, J. and Bouchet, P. (1990). Brain generators implicated in the processing of auditory stimulus deviance: A topographic event-related potential study. *Psychophysiology* **27**(6): 627–640.

Gibbs, R. A., Rogers, J., Katze, M. G., Bumgarner, R., Weinstock, G. M., Mardis, E. R., Remington, K. A., Strausberg, R. L., Venter, J. C. and Wilson, R. K. (2007). Evolutionary and biomedical insights from the rhesus macaque genome. *Science* **316**(5822): 222–234.

Giordano, B. L., Pernet, C., Charest, I., Belizaire, G., Zatorre, R. J. and Belin, P. (2014). Automatic domain-general processing of sound source identity in the left posterior middle frontal gyrus. *Cortex* **58**: 170–185.

Giraud, A.-L., Kleinschmidt, A., Poeppel, D., Lund, T. E., Frackowiak, R. S. J. and Laufs, H. (2007). Endogenous cortical rhythms determine cerebral specialization for speech perception and production. *Neuron* **56**(6): 1127–1134.

Giroud, J., Trébuchon, A., Schön, D., Marquis, P., Liegeois-Chauvel, C., Poeppel, D. and Morillon, B. (2020). Asymmetric sampling in human auditory cortex reveals spectral processing hierarchy. *PLoS Biology* **18**(3): e3000207.

Gläscher, J., Daw, N., Dayan, P. and O'Doherty, J. P. (2010). States versus rewards: Dissociable neural prediction error signals underlying model-based and model-free reinforcement learning. *Neuron* **66**(4): 585–595.

Gloor, P., Olivier, A., Quesney, L. F., Andermann, F. and Horowitz, S. (1982). The role of the limbic system in experiential phenomena of temporal lobe epilepsy. *Annals of Neurology* **12**(2): 129–144.

Goebl, W. and Palmer, C. (2008). Tactile feedback and timing accuracy in piano performance. *Experimental Brain Research* **186**(3): 471–479.

Gogos, A., Gavrilescu, M., Davison, S., Searle, K., Adams, J., Rossell, S. L., Bell, R., Davis, S. R. and Egan, G. F. (2010). Greater superior than inferior parietal lobule activation with increasing rotation angle during mental rotation: An fMRI study. *Neuropsychologia* **48**(2): 529–535.

Gold, B. P., Mas-Herrero, E., Zeighami, Y., Benovoy, M., Dagher, A. and Zatorre, R. J. (2019a). Musical reward prediction errors engage the nucleus accumbens and motivate learning. *Proceedings of the National Academy of Sciences U S A* **116**(8): 3310–3315.

Gold, B. P., Pearce, M. T., Mas-Herrero, E., Dagher, A. and Zatorre, R. J. (2019b). Predictability and uncertainty in the pleasure of music: A reward for learning? *Journal of Neuroscience* **39**(47): 9397–9409.

Gold, B. P., Pearce, M. T., McIntosh, A. R., Chang, C., Dagher, A. and Zatorre, R. J. (In Press). Auditory and reward circuits reflect the pleasure of learnable, naturalistic musical expectancies. *Frontiers in Neuroscience*.

Goldstein, A. (1980). Thrills in response to music and other stimuli. *Physiological Psychology* **8**(1): 126–129.

Golestani, N., Molko, N., Dehaene, S., Lebihan, D. and Pallier, C. (2006). Brain structure predicts the learning of foreign speech sounds. *Cerebral Cortex* **17**(3): 575–582.

Golestani, N. and Zatorre, R. J. (2004). Learning new sounds of speech: Reallocation of neural substrates. *NeuroImage* **21**(2): 494–506.

Gomez, P. and Danuser, B. (2007). Relationships between musical structure and psychophysiological measures of emotion. *Emotion* **7**(2): 377–387.

Gordon, C. L., Cobb, P. R. and Balasubramaniam, R. (2018). Recruitment of the motor system during music listening: An ale meta-analysis of fMRI data. *PLoS One* **13**(11): e0207213–e0207213.

Gorina-Careta, N., Kurkela, J. L. O., Hämäläinen, J., Astikainen, P. and Escera, C. (2021). Neural generators of the frequency-following response elicited to stimuli of low and high frequency: A magnetoencephalographic (MEG) study. *NeuroImage* **231**: 117866.

Gosselin, N., Samson, S., Adolphs, R., Noulhiane, M., Roy, M., Hasboun, D., Baulac, M. and Peretz, I. (2006). Emotional responses to unpleasant music correlates with damage to the parahippocampal cortex. *Brain* **129**(10): 2585–2592.

Gotts, S. J., Jo, H. J., Wallace, G. L., Saad, Z. S., Cox, R. W. and Martin, A. (2013). Two distinct forms of functional lateralization in the human brain. *Proceedings of the National Academy of Sciences* **110**(36): E3435–E3444.

Gould, S. J. and Vrba, E. S. (1982). Exaptation—a missing term in the science of form. *Paleobiology* **8**(1): 4–15.

Grahn, J. A. and Brett, M. (2007). Rhythm and beat perception in motor areas of the brain. *Journal of Cognitive Neuroscience* **19**(5): 893–906.

Grahn, J. A. and Rowe, J. B. (2009). Feeling the beat: Premotor and striatal interactions in musicians and nonmusicians during beat perception. *Journal of Neuroscience* **29**(23): 7540–7548.

Grahn, J. A. and Rowe, J. B. (2012). Finding and feeling the musical beat: Striatal dissociations between detection and prediction of regularity. *Cerebral Cortex* **23**(4): 913–921.

Granot, R., Spitz, D. H., Cherki, B. R., Loui, P., Timmers, R., Schaefer, R. S., Vuoskoski, J. K., Cárdenas-Soler, R.-N., Soares-Quadros, J. F., Li, S., Lega, C., La Rocca, S., Martínez, I. C., Tanco, M., Marchiano, M., Martínez-Castilla, P., Pérez-Acosta, G., Martínez-Ezquerro, J. D., Gutiérrez-Blasco, I. M., . . . Israel, S. (2021). "Help! I need somebody": Music as a global resource for obtaining wellbeing goals in times of crisis. *Frontiers in Psychology* 12: 1038.

Grant, M. J. (2013). The illogical logic of music torture. *Torture: Quarterly Journal on Rehabilitation of Torture Victims and Prevention of Torture* 23(2): 4–13.

Graybiel, A. M. (1973). The thalamo-cortical projection of the so-called posterior nuclear group: A study with anterograde degeneration methods in the cat. *Brain Research* **49**(2): 229–244.

Greenberg, D. M., Kosinski, M., Stillwell, D. J., Monteiro, B. L., Levitin, D. J. and Rentfrow, P. J. (2016). The song is you: Preferences for musical attribute dimensions reflect personality. *Social Psychological and Personality Science* **7**(6): 597–605.

Greenlaw, K. M., Puschmann, S. and Coffey, E. B. J. (2020). Decoding of envelope vs. fundamental frequency during complex auditory stream segregation. *Neurobiology of Language* **1**(3): 268–287.

Greer, S. M., Trujillo, A. J., Glover, G. H. and Knutson, B. (2014). Control of nucleus accumbens activity with neurofeedback. *NeuroImage* **96**: 237–244.

Grefkes, C. and Fink, G. R. (2005). The functional organization of the intraparietal sulcus in humans and monkeys. *Journal of Anatomy* **207**(1): 3–17.

Grewe, O., Nagel, F., Kopiez, R. and Altenmüller, E. (2007). Listening to music as a re-creative process: Physiological, psychological, and psychoacoustical correlates of chills and strong emotions. *Music Perception* **24**(3): 297–314.

Griffiths, T. D. (2000). Musical hallucinosis in acquired deafness: Phenomenology and brain substrate. *Brain* **123**(10): 2065–2076.

Griffiths, T. D., Büchel, C., Frackowiak, R. S. J. and Patterson, R. D. (1998). Analysis of temporal structure in sound by the human brain. *Nature Neuroscience* **1**(5): 422–427.

Griffiths, T. D., Johnsrude, I., Dean, J. L. and Green, G. G. (1999). A common neural substrate for the analysis of pitch and duration pattern in segmented sound? *NeuroReport* **10**(18): 3825–3830.

Griffiths, T. D. and Warren, J. D. (2004). What is an auditory object? *Nature Reviews Neuroscience* **5**(11): 887–892.

Griffiths, T. D., Warren, J. D., Dean, J. L. and Howard, D. (2004). "When the feeling's gone": A selective loss of musical emotion. *Journal of Neurology, Neurosurgery & Psychiatry* **75**(2): 344–345.

Grimault, S., Nolden, S., Lefebvre, C., Vachon, F., Hyde, K., Peretz, I., Zatorre, R., Robitaille, N. and Jolicoeur, P. (2014). Brain activity is related to individual differences in the number of items stored in auditory short-term memory for pitch: Evidence from magnetoencephalography. *NeuroImage* **94**: 96–106.

Gross, C. G. (1994). Hans-Lukas Teuber: A tribute. *Cerebral Cortex* **4**(5): 451–454.

Gross, J. J. and John, O. P. (2003). Individual differences in two emotion regulation processes: Implications for affect, relationships, and well-being. *Journal of Personality and Social Psychology* **85**(2): 348–362.

Gruber, M. J., Gelman, B. D. and Ranganath, C. (2014). States of curiosity modulate hippocampus-dependent learning via the dopaminergic circuit. *Neuron* **84**(2): 486–496.

Guetin, S., Florence, P., Gabelle, A., Touchon, J. and Bonté, F. (2011). Effects of music therapy on anxiety and depression in patients with alzheimer's disease: A randomized controlled trial. *Alzheimer's & Dementia* **7**(4): e49.

Gunn, R. N., Lammertsma, A. A., Hume, S. P. and Cunningham, V. J. (1997). Parametric imaging of ligand-receptor binding in PET using a simplified reference region model. *NeuroImage* **6**(4): 279–287.

Guo, S. and Koelsch, S. (2016). Effects of veridical expectations on syntax processing in music: Event-related potential evidence. *Scientific Reports* **6**(1): 1–11.

Gutschalk, A., Micheyl, C., Melcher, J. R., Rupp, A., Scherg, M. and Oxenham, A. J. (2005). Neuromagnetic correlates of streaming in human auditory cortex. *Journal of Neuroscience* **25**(22): 5382–5388.

Gutschalk, A., Patterson, R. D., Rupp, A., Uppenkamp, S. and Scherg, M. (2002). Sustained magnetic fields reveal separate sites for sound level and temporal regularity in human auditory cortex. *NeuroImage* **15**(1): 207–216.

Gyurak, A., Gross, J. J. and Etkin, A. (2011). Explicit and implicit emotion regulation: A dual-process framework. *Cognition and Emotion* **25**(3): 400–412.

Haber, S. N. (2017). Anatomy and connectivity of the reward circuit. In J.-C. Dreher and L. Tremblay (Eds.), *Decision neuroscience* (pp. 3–19). Academic Press.

Haber, S. N. and Knutson, B. (2010). The reward circuit: Linking primate anatomy and human imaging. *Neuropsychopharmacology* **35**(1): 4–26.

Hackett, T. A., Preuss, T. M. and Kaas, J. H. (2001). Architectonic identification of the core region in auditory cortex of macaques, chimpanzees, and humans. *Journal of Comparative Neurology* **441**(3): 197–222.

Hackett, T. A., Stepniewska, I. and Kaas, J. H. (1999). Prefrontal connections of the parabelt auditory cortex in macaque monkeys. *Brain Research* **817**(1): 45–58.

Halpern, A. R. (1988). Mental scanning in auditory imagery for songs. *Journal of Experimental Psychology: Learning, Memory, and Cognition* **14**(3): 434–443.

Halpern, A. R. and Bartlett, J. C. (2011). The persistence of musical memories: A descriptive study of earworms. *Music Perception* **28**(4): 425–432.

Halpern, A. R. and Zatorre, R. J. (1999). When that tune runs through your head: A PET investigation of auditory imagery for familiar melodies. *Cerebral Cortex* **9**(7): 697–704.

Halpern, A. R., Zatorre, R. J., Bouffard, M. and Johnson, J. A. (2004). Behavioral and neural correlates of perceived and imagined musical timbre. *Neuropsychologia* **42**(9): 1281–1292.

Halwani, G., Loui, P., Rueber, T. and Schlaug, G. (2011). Effects of practice and experience on the arcuate fasciculus: Comparing singers, instrumentalists, and non-musicians. *Frontiers in Psychology* **2**: 156.

Hannon, E. E., Snyder, J. S., Eerola, T. and Krumhansl, C. L. (2004). The role of melodic and temporal cues in perceiving musical meter. *Journal of Experimental Psychology: Human Perception and Performance* **30**(5): 956–974.

Hansen, N. C. and Pearce, M. T. (2014). Predictive uncertainty in auditory sequence processing. *Frontiers in Psychology* **5**: 1052.

Hansen, N. C., Vuust, P. and Pearce, M. (2016). "If you have to ask, you'll never know": Effects of specialised stylistic expertise on predictive processing of music. *PLOS ONE* **11**(10): e0163584.

Haq, I. U., Foote, K. D., Goodman, W. G., Wu, S. S., Sudhyadhom, A., Ricciuti, N., Siddiqui, M. S., Bowers, D., Jacobson, C. E., Ward, H. and Okun, M. S. (2011). Smile and laughter induction and intraoperative predictors of response to deep brain stimulation for obsessive-compulsive disorder. *NeuroImage* **54**: S247–S255.

Hargreaves, D. J. and North, A. C. (2010). Experimental aesthetics and liking for music. In P. N. Juslin and J. A. Sloboda (Eds.), *The handbook of music and emotion: Theory, research, applications* (pp. 515–546). Oxford University Press.

Harrison, P. and Pearce, M. T. (2020). Simultaneous consonance in music perception and composition. *Psychological Review* **127**(2): 216–244.

Hartmann, T. and Weisz, N. (2019). Auditory cortical generators of the frequency following response are modulated by intermodal attention. *NeuroImage* **203**: 116185.

Haslinger, B., Erhard, P., Altenmüller, E., Schroeder, U., Boecker, H. and Ceballos-Baumann, A. O. (2005). Transmodal sensorimotor networks during action observation in professional pianists. *Journal of Cognitive Neuroscience* **17**(2): 282–293.

Hasson, U., Nir, Y., Levy, I., Fuhrmann, G. and Malach, R. (2004). Intersubject synchronization of cortical activity during natural vision. *Science* **303**(5664): 1634–1640.

Hausfeld, L., Disbergen, N. R., Valente, G., Zatorre, R. J. and Formisano, E. (2021). Modulating cortical instrument representations during auditory stream segregation and integration with polyphonic music. *Frontiers in Neuroscience* **15**: 635937.

Hayashi, T., Ko, J. H., Strafella, A. P. and Dagher, A. (2013). Dorsolateral prefrontal and orbitofrontal cortex interactions during self-control of cigarette craving. *Proceedings of the National Academy of Sciences* **110**(11): 4422.

Heffner, H. E. (1987). Ferrier and the study of auditory cortex. *Archives of Neurology* **44**(2): 218–221.

Heilbron, M. and Chait, M. (2018). Great expectations: Is there evidence for predictive coding in auditory cortex? *Neuroscience* **389**: 54–73.

Heimrath, K., Kuehne, M., Heinze, H. J. and Zaehle, T. (2014). Transcranial direct current stimulation (tDCS) traces the predominance of the left auditory cortex for processing of rapidly changing acoustic information. *Neuroscience* **261**: 68–73.

Helmholtz, H. L. (1863/2009). *On the sensations of tone as a physiological basis for the theory of music.* Cambridge University Press.

Herdener, M., Esposito, F., Scheffler, K., Schneider, P., Logothetis, N. K., Uludag, K. and Kayser, C. (2013). Spatial representations of temporal and spectral sound cues in human auditory cortex. *Cortex* **49**(10): 2822–2833.

Herholz, S. C., Coffey, E. B., Pantev, C. and Zatorre, R. J. (2016). Dissociation of neural networks for predisposition and for training-related plasticity in auditory-motor learning. *Cerebral Cortex* **26**(7): 3125–3134.

Herholz, S. C., Halpern, A. R. and Zatorre, R. J. (2012). Neuronal correlates of perception, imagery, and memory for familiar tunes. *Journal of Cognitive Neuroscience* **24**(6): 1382–1397.

Herholz, S. C., Lappe, C., Knief, A. and Pantev, C. (2008). Neural basis of music imagery and the effect of musical expertise. *European Journal of Neuroscience* **28**(11): 2352–2360.

Hickok, G., Buchsbaum, B., Humphries, C. and Muftuler, T. (2003). Auditory–motor interaction revealed by fMRI: Speech, music, and working memory in area Spt. *Journal of Cognitive Neuroscience* **15**(5): 673–682.

Hickok, G. and Poeppel, D. (2007). The cortical organization of speech processing. *Nature Reviews Neuroscience* **8**(5): 393–402.

Higuchi, S., Holle, H., Roberts, N., Eickhoff, S. B. and Vogt, S. (2012). Imitation and observational learning of hand actions: Prefrontal involvement and connectivity. *NeuroImage* **59**(2): 1668–1683.

Hogan, R. E. and Kaiboriboon, K. (2004). John Hughlings-Jackson's writings on the auditory aura and localization of the auditory cortex. *Epilepsia* **45**(7): 834–837.

Holcomb, H. H., Medoff, D. R., Caudill, P. J., Zhao, Z., Lahti, A. C., Dannals, R. F. and Tamminga, C. A. (1998). Cerebral blood flow relationships associated with a difficult tone recognition task in trained normal volunteers. *Cerebral Cortex* **8**(6): 534–542.

Hollinger, A., Steele, C., Penhune, V., Zatorre, R. and Wanderley, M. (2007). FMRI-compatible electronic controllers. *Proceedings of the 7th international conference on New interfaces for musical expression* (pp. 246–249). Association for Computing Machinery.

Honing, H. (2019). *The evolving animal orchestra: In search of what makes us musical.* MIT Press.

Hoshi, E. and Tanji, J. (2006). Differential involvement of neurons in the dorsal and ventral premotor cortex during processing of visual signals for action planning. *Journal of Neurophysiology* **95**(6): 3596–3616.

Hoshi, E. and Tanji, J. (2007). Distinctions between dorsal and ventral premotor areas: Anatomical connectivity and functional properties. *Current Opinion in Neurobiology* **17**(2): 234–242.

Houde, J. F. and Chang, E. F. (2015). The cortical computations underlying feedback control in vocal production. *Current Opinion in Neurobiology* **33**: 174–181.

Hove, M. J., Fairhurst, M. T., Kotz, S. A. and Keller, P. E. (2013). Synchronizing with auditory and visual rhythms: An fMRI assessment of modality differences and modality appropriateness. *NeuroImage* **67**: 313–321.

Hove, M. J. and Risen, J. L. (2009). It's all in the timing: Interpersonal synchrony increases affiliation. *Social Cognition* **27**(6): 949–960.

Hugdahl, K., Brønnick, K., Kyllingsbrk, S., Law, I., Gade, A. and Paulson, O. B. (1999). Brain activation during dichotic presentations of consonant-vowel and musical instrument stimuli: A 15-O PET study. *Neuropsychologia* **37**(4): 431–440.

Hullett, P. W., Hamilton, L. S., Mesgarani, N., Schreiner, C. E. and Chang, E. F. (2016). Human superior temporal gyrus organization of spectrotemporal modulation tuning derived from speech stimuli. *Journal of Neuroscience* **36**(6): 2014–2020.

Hunter, P. G. and Schellenberg, E. G. (2011). Interactive effects of personality and frequency of exposure on liking for music. *Personality and Individual Differences* **50**(2): 175–179.

Huron, D. (2001). Tone and voice: A derivation of the rules of voice-leading from perceptual principles. *Music Perception* **19**: 1–64.

Huron, D. B. (2006). *Sweet anticipation: Music and the psychology of expectation.* MIT Press.

Hutsler, J. and Galuske, R. a. W. (2003). Hemispheric asymmetries in cerebral cortical networks. *Trends in Neurosciences* **26**(8): 429–435.

Hutsler, J. J. and Gazzaniga, M. S. (1996). Acetylcholinesterase staining in human auditory and language cortices: Regional variation of structural features. *Cerebral Cortex* **6**: 260–270.

Hutsler, J. J. (2003). The specialized structure of human language cortex: Pyramidal cell size asymmetries within auditory and language-associated regions of the temporal lobes. *Brain and Language* **86**(2): 226–242.

Hyde, K. L., Lerch, J., Norton, A., Forgeard, M., Winner, E., Evans, A. C. and Schlaug, G. (2009). Musical training shapes structural brain development. *Journal of Neuroscience* **29**(10): 3019–3025.

Hyde, K. L., Lerch, J. P., Zatorre, R. J., Griffiths, T. D., Evans, A. C. and Peretz, I. (2007). Cortical thickness in congenital amusia: When less is better than more. *Journal of Neuroscience* **27**(47): 13028–13032.

Hyde, K. L. and Peretz, I. (2004). Brains that are out of tune but in time. *Psychological Science* **15**(5): 356–360.

Hyde, K. L., Peretz, I. and Zatorre, R. J. (2008). Evidence for the role of the right auditory cortex in fine pitch resolution. *Neuropsychologia* **46**(2): 632–639.

Hyde, K. L., Zatorre, R. J., Griffiths, T. D., Lerch, J. P. and Peretz, I. (2006). Morphometry of the amusic brain: A two-site study. *Brain* **129**(Pt 10): 2562–2570.

Hyde, K. L., Zatorre, R. J. and Peretz, I. (2011). Functional MRI evidence of an abnormal neural network for pitch processing in congenital amusia. *Cerebral Cortex* **21**(2): 292–299.

Ilie, G. and Thompson, W. F. (2006). A comparison of acoustic cues in music and speech for three dimensions of affect. *Music Perception* **23**(4): 319–330.

Irvine, D. R. (2012). *The auditory brainstem: A review of the structure and function of auditory brainstem processing mechanisms.* Springer Science & Business Media.

Iturria-Medina, Y., Pérez Fernández, A., Morris, D. M., Canales-Rodríguez, E. J., Haroon, H. A., García Pentón, L., Augath, M., Galán García, L., Logothetis, N., Parker, G. J. M. and Melie-García, L. (2010). Brain hemispheric structural efficiency and interconnectivity rightward asymmetry in human and non-human primates. *Cerebral Cortex* **21**(1): 56–67.

Iversen, J. R., Repp, B. H. and Patel, A. D. (2009). Top-down control of rhythm perception modulates early auditory responses. *Annals of the New York Academy of Sciences* **1169**(1): 58–73.

Ivry, R. B., Spencer, R. M., Zelaznik, H. N. and Diedrichsen, J. (2002). The cerebellum and event timing. *Annals of the New York Academy of Sciences* **978**(1): 302–317.

Jacoby, N., Margulis, E. H., Clayton, M., Hannon, E., Honing, H., Iversen, J., Klein, T. R., Mehr, S. A., Pearson, L., Peretz, I., Perlman, M., Polak, R., Ravignani, A., Savage, P. E., Steingo, G., Stevens, C. J., Trainor, L., Trehub, S., Veal, M. and Wald-Fuhrmann, M. (2020). Cross-cultural work in music cognition: Challenges, insights, and recommendations. *Music Perception* **37**(3): 185–195.

Jacoby, N. and McDermott, J. H. (2017). Integer ratio priors on musical rhythm revealed cross-culturally by iterated reproduction. *Current Biology* **27**: 359–370.

Jacoby, N., Undurraga, E. A., McPherson, M. J., Valdés, J., Ossandón, T. and McDermott, J. H. (2019). Universal and non-universal features of musical pitch perception revealed by singing. *Current Biology* **29**(19): 3229–3243.

Jakobovits, L. A. (1966). Studies of fads: I. The "hit parade". *Psychological Reports* **18**(2): 443–450.

Jakubowski, K., Farrugia, N., Halpern, A. R., Sankarpandi, S. K. and Stewart, L. (2015). The speed of our mental soundtracks: Tracking the tempo of involuntary musical imagery in everyday life. *Memory & Cognition* **43**(8): 1229–1242.

James, W. (1890). *The principles of psychology* (Vol. ii). Henry Holt and Company.

Jamison, H. L., Watkins, K. E., Bishop, D. V. and Matthews, P. M. (2006). Hemispheric specialization for processing auditory nonspeech stimuli. *Cerebral Cortex* **16**(9): 1266–1275.

Janata, P. (2009). The neural architecture of music-evoked autobiographical memories. *Cerebral Cortex* **19**(11): 2579–2594.

Janata, P., Tomic, S. T. and Haberman, J. M. (2012). Sensorimotor coupling in music and the psychology of the groove. *Journal of Experimental Psychology: General* **141**(1): 54.

Janata, P., Tomic, S. T. and Rakowski, S. K. (2007). Characterisation of music-evoked autobiographical memories. *Memory* **15**(8): 845–860.

Jäncke, L., Loose, R., Lutz, K., Specht, K. and Shah, N. J. (2000). Cortical activations during paced finger-tapping applying visual and auditory pacing stimuli. *Cognitive Brain Research* **10**(1): 51–66.

Jäncke, L. and Steinmetz, H. (1993). Auditory lateralization and planum temporale asymmetry. *NeuroReport* **5**: 169–172.

Janssen, S. M. J., Chessa, A. G. and Murre, J. M. J. (2007). Temporal distribution of favourite books, movies, and records: Differential encoding and re-sampling. *Memory* **15**(7): 755–767.

Jasmin, K., Lima, C. F. and Scott, S. K. (2019). Understanding rostral–caudal auditory cortex contributions to auditory perception. *Nature Reviews Neuroscience* **20**(7): 425–434.

Jeannerod, M. (2001). Neural simulation of action: A unifying mechanism for motor cognition. *NeuroImage* **14**(1): S103–S109.

Johansen-Berg, H., Della-Maggiore, V., Behrens, T. E., Smith, S. M. and Paus, T. (2007). Integrity of white matter in the corpus callosum correlates with bimanual co-ordination skills. *NeuroImage* **36**: T16–T21.

John, O. P. and Gross, J. J. (2007). Individual differences in emotion regulation. In J. J. Gross (Ed.), *Handbook of emotion regulation* (pp. 351–372). Guilford Press.

Johnsrude, I. S., Penhune, V. B. and Zatorre, R. J. (2000). Functional specificity in the right human auditory cortex for perceiving pitch direction. *Brain* **123** (Pt 1): 155–163.

Jones, M. R. (1987). Dynamic pattern structure in music: Recent theory and research. *Perception & Psychophysics* **41**(6): 621–634.

Jung, Y. and Hong, S. (2003). Corticostriatal connections of the superior temporal regions in the macaque monkey. *Korean Journal of Biological Sciences* **7**(4): 317–325.

Juslin, P. N. (2019). *Musical emotions explained: Unlocking the secrets of musical affect*. Oxford University Press.

Juslin, P. N. and Laukka, P. (2003). Communication of emotions in vocal expression and music performance: Different channels, same code? *Psychological Bulletin* **129**(5): 770–814.

Juslin, P. N. and Laukka, P. (2004). Expression, perception, and induction of musical emotions: A review and a questionnaire study of everyday listening. *Journal of New Music Research* **33**(3): 217–238.

Juslin, P. N. and Sloboda, J. A. (2001). *Music and emotion: Theory and research*. Oxford University Press.

Juslin, P. N. and Västfjäll, D. (2008). Emotional responses to music: The need to consider underlying mechanisms. *Behavioral and Brain Sciences* **31**(5): 559–575.

Kaas, J. H. and Hackett, T. A. (2000). Subdivisions of auditory cortex and processing streams in primates. *Proceedings of the National Academy of Sciences* **97**(22): 11793–11799.

Kallinen, K. and Ravaja, N. (2006). Emotion perceived and emotion felt: Same and different. *Musicae Scientiae* **10**(2): 191–213.

Kampe, K. K. W., Frith, C. D., Dolan, R. J. and Frith, U. (2001). Reward value of attractiveness and gaze. *Nature* **413**(6856): 589–589.

Kang, M. J., Hsu, M., Krajbich, I. M., Loewenstein, G., Mcclure, S. M., Wang, J. T.-Y. and Camerer, C. F. (2009). The wick in the candle of learning: Epistemic curiosity activates reward circuitry and enhances memory. *Psychological Science* **20**(8): 963–973.

Kanwisher, N. and Yovel, G. (2006). The fusiform face area: A cortical region specialized for the perception of faces. *Philosophical Transactions of the Royal Society B: Biological Sciences* **361**(1476): 2109–2128.

Kearney, E. and Guenther, F. H. (2019). Articulating: The neural mechanisms of speech production. *Language, Cognition and Neuroscience* **34**(9): 1214–1229.

Kell, A. J. E. and McDermott, J. H. (2019). Invariance to background noise as a signature of non-primary auditory cortex. *Nature Communications* **10**(1): 3958.

Keller, J., Young, C. B., Kelley, E., Prater, K., Levitin, D. J. and Menon, V. (2013). Trait anhedonia is associated with reduced reactivity and connectivity of mesolimbic and paralimbic reward pathways. *Journal of Psychiatric Research* **47**(10): 1319–1328.

Keller, P. E. (2012). Mental imagery in music performance: Underlying mechanisms and potential benefits. *Annals of the New York Academy of Sciences* **1252**(1): 206–213.

Kello, C. T., Dalla Bella, S., Médé, B. and Balasubramaniam, R. (2017). Hierarchical temporal structure in music, speech and animal vocalizations: Jazz is like a conversation, humpbacks sing like hermit thrushes. *Journal of the Royal Society Interface* **14**(135): 20170231.

Keynan, J. N., Cohen, A., Jackont, G., Green, N., Goldway, N., Davidov, A., Meir-Hasson, Y., Raz, G., Intrator, N., Fruchter, E., Ginat, K., Laska, E., Cavazza, M. and Hendler, T. (2019). Electrical fingerprint of the amygdala guides neurofeedback training for stress resilience. *Nature Human Behaviour* **3**(1): 63–73.

Kidd, C. and Hayden, B. Y. (2015). The psychology and neuroscience of curiosity. *Neuron* **88**(3): 449–460.

Kidd, G., Mason, C. R., Richards, V. M., Gallun, F. J. and Durlach, N. I. (2008). Informational masking. In W. A. Yost, A. N. Popper, and R. R. Fay (Eds.), *Auditory perception of sound sources* (pp. 143–189). Springer.

Kivy, P. (1990). *Music alone: Philosophical reflections on the purely musical experience.* Cornell University Press.

Kleber, B., Birbaumer, N., Veit, R., Trevorrow, T. and Lotze, M. (2007). Overt and imagined singing of an Italian aria. *NeuroImage* **36**(3): 889–900.

Kleber, B., Friberg, A., Zeitouni, A. and Zatorre, R. (2017). Experience-dependent modulation of right anterior insula and sensorimotor regions as a function of noise-masked auditory feedback in singers and nonsingers. *NeuroImage* **147**: 97–110.

Kleber, B., Veit, R., Birbaumer, N., Gruzelier, J. and Lotze, M. (2009). The brain of opera singers: Experience-dependent changes in functional activation. *Cerebral Cortex* **20**(5): 1144–1152.

Kleber, B., Veit, R., Moll, C. V., Gaser, C., Birbaumer, N. and Lotze, M. (2016). Voxel-based morphometry in opera singers: Increased gray-matter volume in right somatosensory and auditory cortices. *NeuroImage* **133**: 477–483.

Kleber, B., Zeitouni, A. G., Friberg, A. and Zatorre, R. J. (2013). Experience-dependent modulation of feedback integration during singing: Role of the right anterior insula. *Journal of Neuroscience* **33**(14): 6070–6080.

Klein, C., Liem, F., Hänggi, J., Elmer, S. and Jäncke, L. (2016). The "silent" imprint of musical training. *Human Brain Mapping* **37**(2): 536–546.

Klein, M. E. and Zatorre, R. J. (2015). Representations of invariant musical categories are decodable by pattern analysis of locally distributed bold responses in superior temporal and intraparietal sulci. *Cerebral Cortex* **25**(7): 1947–1957.

Knecht, S., Breitenstein, C., Bushuven, S., Wailke, S., Kamping, S., Flöel, A., Zwitserlood, Pand Ringelstein, E. B. (2004). Levodopa: Faster and better word learning in normal humans. *Annals of Neurology* **56**(1): 20–26.

Knutson, B., Adams, C. M., Fong, G. W. and Hommer, D. (2001). Anticipation of increasing monetary reward selectively recruits nucleus accumbens. *Journal of Neuroscience* **21**(16): RC159.

Knutson, B. and Bossaerts, P. (2007). Neural antecedents of financial decisions. *Journal of Neuroscience* **27**(31): 8174–8177.

Knutson, B., Fong, G. W., Bennett, S. M., Adams, C. M. and Hommer, D. (2003). A region of mesial prefrontal cortex tracks monetarily rewarding outcomes: Characterization with rapid event-related fMRI. *NeuroImage* **18**(2): 263–272.

Knutson, B. and Gibbs, S. E. B. (2007). Linking nucleus accumbens dopamine and blood oxygenation. *Psychopharmacology* **191**(3): 813–822.

Knutson, B., Westdorp, A., Kaiser, E. and Hommer, D. (2000). FMRI visualization of brain activity during a monetary incentive delay task. *NeuroImage* **12**(1): 20–27.

Ko, J. H., Monchi, O., Ptito, A., Bloomfield, P., Houle, S. and Strafella, A. P. (2008). Theta burst stimulation-induced inhibition of dorsolateral prefrontal cortex reveals hemispheric asymmetry in striatal dopamine release during a set-shifting task – a TMS–[11C]raclopride PET study. *European Journal of Neuroscience* **28**(10): 2147–2155.

Kober, H., Mende-Siedlecki, P., Kross, E. F., Weber, J., Mischel, W., Hart, C. L. and Ochsner, K. N. (2010). Prefrontal–striatal pathway underlies cognitive regulation of craving. *Proceedings of the National Academy of Sciences* **107**(33): 14811–14816.

Koelsch, S. (2014). Brain correlates of music-evoked emotions. *Nature Reviews Neuroscience* **15**(3): 170–180.

Koelsch, S. (2018). Investigating the neural encoding of emotion with music. *Neuron* **98**(6): 1075–1079.

Koelsch, S. (2020). A coordinate-based meta-analysis of music-evoked emotions. *NeuroImage* **223**: 117350.

Koelsch, S. and Friederici, A. D. (2003). Toward the neural basis of processing structure in music: Comparative results of different neurophysiological investigation methods. *Annals of the New York Academy of Sciences, the Neurosciences and Music* **999**: 15–28.

Koelsch, S., Fritz, T., Dy, V. C., Muller, K. and Friederici, A. D. (2006). Investigating emotion with music: An fMRI study. *Human Brain Mapping* **27**(3): 239–250.

Koelsch, S., Gunter, T. C., V Cramon, D. Y., Zysset, S., Lohmann, G. and Friederici, A. D. (2002a). Bach speaks: A cortical "language-network" serves the processing of music. *NeuroImage* **17**(2): 956–966.

Koelsch, S., Kilches, S., Steinbeis, N. and Schelinski, S. (2008). Effects of unexpected chords and of performer's expression on brain responses and electrodermal activity. *PLoS One* **3**(7): e2631.

Koelsch, S., Schmidt, B.-H. and Kansok, J. (2002b). Effects of musical expertise on the early right anterior negativity: An event-related brain potential study. *Psychophysiology* **39**(5): 657–663.

Koelsch, S., Schröger, E. and Tervaniemi, M. (1999). Superior pre-attentive auditory processing in musicians. *NeuroReport* **10**(6): 1309–1313.

Koelsch, S., Schulze, K., Sammler, D., Fritz, T., Müller, K. and Gruber, O. (2009). Functional architecture of verbal and tonal working memory: An fMRI study. *Human Brain Mapping* **30**(3): 859–873.

Koelsch, S., Vuust, P. and Friston, K. (2019). Predictive processes and the peculiar case of music. *Trends Cogn Sci* **23**(1): 63–77.

Koenig, W., Dunn, H. K. and Lacy, L. Y. (1946). The sound spectrograph. *Journal of the Acoustical Society of America* **18**(1): 19–49.

Kohler, E., Keysers, C., Umiltà, M. A., Fogassi, L., Gallese, V. and Rizzolatti, G. (2002). Hearing sounds, understanding actions: Action representation in mirror neurons. *Science* **297**(5582): 846–848.

Kondo, H. M. and Kashino, M. (2009). Involvement of the thalamocortical loop in the spontaneous switching of percepts in auditory streaming. *Journal of Neuroscience* **29**(40): 12695–12701.

Konečni, V. J. (2003). Review of the book *Music and emotion: Theory and research*, by P. N. Juslin, & J. A. Sloboda. *Music Perception* **20**: 332–341.

Kong, X.-Z., Mathias, S. R., Guadalupe, T., Glahn, D. C., Franke, B., Crivello, F., Tzourio-Mazoyer, N., Fisher, S. E., Thompson, P. M. and Francks, C. (2018). Mapping cortical brain asymmetry in 17,141 healthy individuals worldwide via the enigma consortium. *Proceedings of the National Academy of Sciences* **115**(22): E5154–E5163.

Koob, G. F. and Volkow, N. D. (2010). Neurocircuitry of addiction. *Neuropsychopharmacology* **35**(1): 217–238.

Kornysheva, K. and Diedrichsen, J. (2014). Human premotor areas parse sequences into their spatial and temporal features. *eLife* **3**: e03043.

Kostopoulos, P. and Petrides, M. (2016). Selective memory retrieval of auditory what and auditory where involves the ventrolateral prefrontal cortex. *Proceedings of the National Academy of Sciences* **113**(7): 1919–1924.

Kraus, N., Anderson, S. and White-Schwoch, T. (2017). The frequency-following response: A window into human communication. In N. Kraus, S. Anderson, T. White-Schwoch, R. Fay, A. Popper (Eds.), *Springer Handbook of Auditory Research*, vol. 61 (pp. 1–15). Springer, Cham.

Kraus, N., Strait, D. and Parbery-Clark, A. (2012). Cognitive factors shape brain networks for auditory skills: Spotlight on auditory working memory. *Annals of the New York Academy of Sciences* **1252**(1): 100–107.

Krause, V., Pollok, B. and Schnitzler, A. (2010). Perception in action: The impact of sensory information on sensorimotor synchronization in musicians and non-musicians. *Acta Psychologia (Amst)* **133**(1): 28–37.

Kravitz, D. J., Saleem, K. S., Baker, C. I., Ungerleider, L. G. and Mishkin, M. (2013). The ventral visual pathway: An expanded neural framework for the processing of object quality. *Trends in Cognitive Sciences* **17**(1): 26–49.

Kringelbach, M. L. (2005). The human orbitofrontal cortex: Linking reward to hedonic experience. *Nature Reviews Neuroscience* **6**(9): 691–702.

Kringelbach, M. L., O'Doherty, J., Rolls, E. T. and Andrews, C. (2003). Activation of the human orbitofrontal cortex to a liquid food stimulus is correlated with its subjective pleasantness. *Cerebral Cortex* **13**(10): 1064–1071.

Kringelbach, M. L. and Rolls, E. T. (2004). The functional neuroanatomy of the human orbitofrontal cortex: Evidence from neuroimaging and neuropsychology. *Progress in Neurobiology* **72**(5): 341–372.

Krishnan, A., Bidelman, G. M., Smalt, C. J., Ananthakrishnan, S. and Gandour, J. T. (2012). Relationship between brainstem, cortical and behavioral measures relevant to pitch salience in humans. *Neuropsychologia* **50**(12): 2849–2859.

Krishnan, A., Xu, Y., Gandour, J. and Cariani, P. (2005). Encoding of pitch in the human brainstem is sensitive to language experience. *Cognitive Brain Research* **25**(1): 161–168.

Krizman, J. and Kraus, N. (2019). Analyzing the FFR: A tutorial for decoding the richness of auditory function. *Hearing Research* **382**: 107779.

Krumbholz, K., Patterson, R. D., Seither-Preisler, A., Lammertmann, C. and Lütkenhöner, B. (2003). Neuromagnetic evidence for a pitch processing center in Heschl's gyrus. *Cerebral Cortex* **13**(7): 765–772.

Krumhansl, C. (1990). *Cognitive foundations of musical pitch.* Oxford University Press.

Krumhansl, C. L. (1997). An exploratory study of musical emotions and psychophysiology. *Canadian Journal of Experimental Psychology/Revue canadienne de psychologie expérimentale* **51**(4): 336–353.

Krumhansl, C. L. (2017). Listening niches across a century of popular music. *Frontiers in Psychology* **8**: 431.

Krumhansl, C. L., Sandell, G. J. and Sergeant, D. C. (1987). The perception of tone hierarchies and mirror forms in twelve-tone serial music. *Music Perception:* **5**(1): 31–77.

Krumhansl, C. L. and Zupnick, J. A. (2013). Cascading reminiscence bumps in popular music. *Psychological Science* **24**(10): 2057–2068.

Kühn, S. and Gallinat, J. (2012). The neural correlates of subjective pleasantness. *NeuroImage* **61**(1): 289–294.

Kumar, S., Joseph, S., Gander, P. E., Barascud, N., Halpern, A. R. and Griffiths, T. D. (2016). A brain system for auditory working memory. *Journal of Neuroscience* **36**(16): 4492–4505.

Kumar, S. and Schönwiesner, M. (2012). Mapping human pitch representation in a distributed system using depth-electrode recordings and modeling. *Journal of Neuroscience* **32**(39): 13348–13351.

Kumar, S., Sedley, W., Barnes, G. R., Teki, S., Friston, K. J. and Griffiths, T. D. (2014). A brain basis for musical hallucinations. *Cortex* **52**: 86–97.

Kumar, S., Sedley, W., Nourski, K. V., Kawasaki, H., Oya, H., Patterson, R. D., Howard Iii, M. A., Friston, K. J. and Griffiths, T. D. (2011). Predictive coding and pitch processing in the auditory cortex. *Journal of Cognitive Neuroscience* **23**(10): 3084–3094.

Kumar, S., Von Kriegstein, K., Friston, K. and Griffiths, T. D. (2012). Features versus feelings: Dissociable representations of the acoustic features and valence of aversive sounds. *Journal of Neuroscience* **32**(41): 14184–14192.

Kung, S. J., Chen, J. L., Zatorre, R. J. and Penhune, V. B. (2013). Interacting cortical and basal ganglia networks underlying finding and tapping to the musical beat. *Journal of Cognitive Neuroscience* **25**(3): 401–420.

Labbé, E., Schmidt, N., Babin, J. and Pharr, M. (2007). Coping with stress: The effectiveness of different types of music. *Applied Psychophysiology and Biofeedback* **32**(3): 163–168.

Laeng, B., Garvija, L., Løseth, G., Eikemo, M., Ernst, G. and Leknes, S. (2021). 'Defrosting'music chills with naltrexone: The role of endogenous opioids for the intensity of musical pleasure. *Consciousness and Cognition* **90**: 103105.

Lahav, A., Saltzman, E. and Schlaug, G. (2007). Action representation of sound: Audiomotor recognition network while listening to newly acquired actions. *Journal of Neuroscience* **27**(2): 308–314.

Laje, R. and Buonomano, D. V. (2013). Robust timing and motor patterns by taming chaos in recurrent neural networks. *Nature Neuroscience* **16**(7): 925–933.

Landemard, A., Bimbard, C., Demené, C., Shamma, S., Norman-Haignere, S. and Boubenec, Y. (2021). Distinct higher-order representations of natural sounds in human and ferret auditory cortex. *eLife* **10**: e65566.

Lane, H. and Tranel, B. (1971). The Lombard sign and the role of hearing in speech. *Journal of Speech and Hearing Research* **14**(4): 677–709.

Large, E. W. (2000). On synchronizing movements to music. *Human Movement Science* **19**(4): 527–566.

Large, E. W. and Palmer, C. (2002). Perceiving temporal regularity in music. *Cognitive Science* **26**(1): 1–37.

Large, E. W. and Snyder, J. S. (2009). Pulse and meter as neural resonance. *Annals of the New York Academy of Sciences* **1169**(1): 46–57.

Larson, C. R., Burnett, T. A., Kiran, S. and Hain, T. C. (1999). Effects of pitch-shift velocity on voice f0 responses. *Journal of the Acoustical Society of America* **107**(1): 559–564.

Leaver, A. M. and Rauschecker, J. P. (2010). Cortical representation of natural complex sounds: Effects of acoustic features and auditory object category. *Journal of Neuroscience* **30**(22): 7604–7612.

Lebel, C. and Beaulieu, C. (2009). Lateralization of the arcuate fasciculus from childhood to adulthood and its relation to cognitive abilities in children. *Human Brain Mapping* **30**(11): 3563–3573.

LeDoux, J. (1998). *The emotional brain: The mysterious underpinnings of emotional life.* Simon and Schuster.

LeDoux, J. E., Sakaguchi, A. and Reis, D. J. (1984). Subcortical efferent projections of the medial geniculate nucleus mediate emotional responses conditioned to acoustic stimuli. *Journal of Neuroscience* **4**(3): 683–698.

Lee, D., Seo, H. and Jung, M. W. (2012). Neural basis of reinforcement learning and decision making. *Annual Review of Neuroscience* **35**(1): 287–308.

Lee, H. and Noppeney, U. (2011). Long-term music training tunes how the brain temporally binds signals from multiple senses. *Proceedings of the National Academy of Sciences* **108**(51): E1441–E1450.

Lee, M. C., Wagner, H. N., Tanada, S., Frost, J. J., Bice, A. N. and Dannals, R. F. (1988). Duration of occupancy of opiate receptors by naltrexone. *Journal of Nuclear Medicine* **29**(7): 1207–1211.

Lee, Y.-S., Janata, P., Frost, C., Hanke, M. and Granger, R. (2011). Investigation of melodic contour processing in the brain using multivariate pattern-based fMRI. *NeuroImage* **57**(1): 293–300.

Lehmann, A. and Schönwiesner, M. (2014). Selective attention modulates human auditory brainstem responses: Relative contributions of frequency and spatial cues. *PloS One* **9**(1): e85442.

Lerner, Y., Honey, C. J., Silbert, L. J. and Hasson, U. (2011). Topographic mapping of a hierarchy of temporal receptive windows using a narrated story. *Journal of Neuroscience* **31**(8): 2906–2915.

Lerner, Y., Papo, D., Zhdanov, A., Belozersky, L. and Hendler, T. (2009). Eyes wide shut: Amygdala mediates eyes-closed effect on emotional experience with music. *PLoS One* **4**(7): e6230.

Leung, Y. and Dean, R. T. (2018). Learning unfamiliar pitch intervals: A novel paradigm for demonstrating the learning of statistical associations between musical pitches. *PLoS One* **13**(8): e0203026.

Lévêque, Y., Fauvel, B., Groussard, M., Caclin, A., Albouy, P., Platel, H. and Tillmann, B. (2016). Altered intrinsic connectivity of the auditory cortex in congenital amusia. *Journal of Neurophysiology* **116**(1): 88–97.

Lewis, J. W., Brefczynski, J. A., Phinney, R. E., Janik, J. J. and Deyoe, E. A. (2005). Distinct cortical pathways for processing tool versus animal sounds. *Journal of Neuroscience* **25**(21): 5148–5158.

Lewis, J. W. and Van Essen, D. C. (2000). Corticocortical connections of visual, sensorimotor, and multimodal processing areas in the parietal lobe of the macaque monkey. *Journal of Comparative Neurology* **428**(1): 112–137.

Ley, A., Vroomen, J., Hausfeld, L., Valente, G., De Weerd, P. and Formisano, E. (2012). Learning of new sound categories shapes neural response patterns in human auditory cortex. *Journal of Neuroscience* **32**(38): 13273–13280.

Leyton, M., Boileau, I., Benkelfat, C., Diksic, M., Baker, G. and Dagher, A. (2002). Amphetamine-induced increases in extracellular dopamine, drug wanting, and novelty seeking: A PET/[11c]raclopride study in healthy men. *Neuropsychopharmacology* **27**(6): 1027–1035.

Li, Q., Wang, X., Wang, S., Xie, Y., Li, X., Xie, Y. and Li, S. (2018). Musical training induces functional and structural auditory-motor network plasticity in young adults. *Human Brain Mapping* **39**(5): 2098–2110.

Lidow, M. S., Goldman-Rakic, P. S., Gallager, D. W. and Rakic, P. (1991). Distribution of dopaminergic receptors in the primate cerebral cortex: Quantitative autoradiographic analysis using [^{3}H]raclopride, [^{3}H] spiperone and [^{3}H]sch23390. *Neuroscience* **40**(3): 657–671.

Liebenthal, E., Binder, J. R., Piorkowski, R. L. and Remez, R. E. (2003). Short-term reorganization of auditory analysis induced by phonetic experience. *Journal of Cognitive Neuroscience* **15**(4): 549–558.

Liégeois-Chauvel, C., De Graaf, J. B., Laguitton, V. and Chauvel, P. (1999). Specialization of left auditory cortex for speech perception in man depends on temporal coding. *Cerebral Cortex* **9**(5): 484–496.

Liégeois-Chauvel, C., Peretz, I., Babaï, M., Laguitton, V. and Chauvel, P. (1998). Contribution of different cortical areas in the temporal lobes to music processing. *Brain* **121**(10): 1853–1867.

Liggins, J., Pihl, R. O., Benkelfat, C. and Leyton, M. (2012). The dopamine augmenter l-dopa does not affect positive mood in healthy human volunteers. *PLoS One* **7**(1): e28370.

Ligneul, R., Mermillod, M. and Morisseau, T. (2018). From relief to surprise: Dual control of epistemic curiosity in the human brain. *NeuroImage* **181**: 490–500.

Liikkanen, L. A. (2012). Musical activities predispose to involuntary musical imagery. *Psychology of Music* **40**(2): 236–256.

Liikkanen, L. A. and Jakubowski, K. (2020). Involuntary musical imagery as a component of ordinary music cognition: A review of empirical evidence. *Psychonomic Bulletin & Review.* **27**: 1195–1217.

Lin, W.-J., Horner, A. J. and Burgess, N. (2016). Ventromedial prefrontal cortex, adding value to autobiographical memories. *Scientific Reports* **6**(1): 28630.

Linden, J. F., Liu, R. C., Sahani, M., Schreiner, C. E. and Merzenich, M. M. (2003). Spectrotemporal structure of receptive fields in areas AI and AAF of mouse auditory cortex. *Journal of Neurophysiology* **90**(4): 2660–2675.

Lindquist, K. A., Satpute, A. B., Wager, T. D., Weber, J. and Barrett, L. F. (2016). The brain basis of positive and negative affect: Evidence from a meta-analysis of the human neuroimaging literature. *Cerebral Cortex* **26**(5): 1910–1922.

Litman, J. (2005). Curiosity and the pleasures of learning: Wanting and liking new information. *Cognition and Emotion* **19**(6): 793–814.

Liu, B.-H., Wu, G. K., Arbuckle, R., Tao, H. W. and Zhang, L. I. (2007). Defining cortical frequency tuning with recurrent excitatory circuitry. *Nature Neuroscience* **10**(12): 1594–1600.

Liu, H., Stufflebeam, S. M., Sepulcre, J., Hedden, T. and Buckner, R. L. (2009). Evidence from intrinsic activity that asymmetry of the human brain is controlled by multiple factors. *Proceedings of the National Academy of Sciences* **106**(48): 20499–20503.

Loh, K. K., Petrides, M., Hopkins, W. D., Procyk, E. and Amiez, C. (2017). Cognitive control of vocalizations in the primate ventrolateral-dorsomedial frontal (vlf-dmf) brain network. *Neuroscience & Biobehavioral Reviews* **82**: 32–44.

London, J. (2012). *Hearing in time: Psychological aspects of musical meter.* Oxford University Press.

Lonsdale, A. J. and North, A. C. (2011). Why do we listen to music? A uses and gratifications analysis. *British Journal of Psychology* **102**(1): 108–134.

López-Barroso, D., Catani, M., Ripollés, P., Dell'Acqua, F., Rodríguez-Fornells, A. and de Diego-Balaguer, R. (2013). Word learning is mediated by the left arcuate fasciculus. *Proceedings of the National Academy of Sciences* **110**(32): 13168.

Loui, P., Alsop, D. and Schlaug, G. (2009). Tone deafness: A new disconnection syndrome? *Journal of Neuroscience* **29**(33): 10215–10220.

Loui, P., Li, H. C. and Schlaug, G. (2011). White matter integrity in right hemisphere predicts pitch-related grammar learning. *NeuroImage* **55**(2): 500–507.

Loui, P., Raine, L. B., Chaddock-Heyman, L., Kramer, A. F. and Hillman, C. H. (2019). Musical instrument practice predicts white matter microstructure and cognitive abilities in childhood. *Frontiers in Psychology* **10**: 1198.

Loui, P., Wessel, D. L. and Kam, C. L. H. (2010). Humans rapidly learn grammatical structure in a new musical scale. *Music Perception* **27**(5): 377–388.

Lu, K., Xu, Y., Yin, P., Oxenham, A. J., Fritz, J. B. and Shamma, S. A. (2017). Temporal coherence structure rapidly shapes neuronal interactions. *Nature Communications* **8**(1): 13900.

Luciana, M., Wahlstrom, D., Porter, J. N. and Collins, P. F. (2012). Dopaminergic modulation of incentive motivation in adolescence: Age-related changes in signaling, individual differences, and implications for the development of self-regulation. *Developmental Ppsychology* **48**(3): 844–861.

Lumaca, M., Trusbak Haumann, N., Brattico, E., Grube, M. and Vuust, P. (2019). Weighting of neural prediction error by rhythmic complexity: A predictive coding account using mismatch negativity. *European Journal of Neuroscience* **49**(12): 1597–1609.

Luna, B., Marek, S., Larsen, B., Tervo-Clemmens, B. and Chahal, R. (2015). An integrative model of the maturation of cognitive control. *Annual Review of Neuroscience* **38**(1): 151–170.

Luo, H., Husain, F. T., Horwitz, B. and Poeppel, D. (2005). Discrimination and categorization of speech and non-speech sounds in an MEG delayed-match-to-sample study. *NeuroImage* **28**(1): 59–71.

Luria, A. R., Tsvetkova, L. S. and Futer, D. S. (1965). Aphasia in a composer. *Journal of the Neurological Sciences* **2**(3): 288–292.

MacInnes, J. J., Dickerson, K. C., Chen, N.-K. and Adcock, R. A. (2016). Cognitive neurostimulation: Learning to volitionally sustain ventral tegmental area activation. *Neuron* **89**(6): 1331–1342.

Madison, G. (2006). Experiencing groove induced by music: Consistency and phenomenology. *Music perception* **24**(2): 201–208.

Maeder, P. P., Meuli, R. A., Adriani, M., Bellmann, A., Fornari, E., Thiran, J.-P., Pittet, A. and Clarke, S. (2001). Distinct pathways involved in sound recognition and localization: A human fMRI study. *NeuroImage* **14**(4): 802–816.

Maess, B., Koelsch, S., Gunter, T. C. and Friederici, A. D. (2001). Musical syntax is processed in Broca's area: An MEG study. *Nature Neuroscience* **4**(5): 540–545.

Mallik, A., Chanda, M. L. and Levitin, D. J. (2017). Anhedonia to music and mu-opioids: Evidence from the administration of naltrexone. *Scientific Reports* **7**(1): 41952.

Malmierca, M. S. (2015). Auditory system. In G. Paxinos (Ed.), *The rat nervous system* (pp. 865–946). Elsevier.

Malmierca, M. S., Anderson, L. A., and Antunes, F. M. (2015). The cortical modulation of stimulus-specific adaptation in the auditory midbrain and thalamus: a potential neuronal correlate for predictive coding. *Frontiers in Systems Neuroscience* 9: 19.

Malmierca, M. S., Cristaudo, S., Pérez-González, D. and Covey, E. (2009). Stimulus-specific adaptation in the inferior colliculus of the anesthetized rat. *Journal of Neuroscience* 29(17): 5483–5493.

Malmierca, M. S. and Hackett, T. A. (2010). Structural organization of the ascending auditory pathway. In A. Rees and A. R. Palmer (Eds.), *The Auditory Brain* (pp. 9–41). Oxford University Press.

Manley, G. A., Gummer, A. W., Popper, A. N. and Fay, R. R. (2017). Understanding the cochlea. Springer.

Manning, F. and Schutz, M. (2013). "Moving to the beat" improves timing perception. *Psychonomic Bulletin & Review* 20(6): 1133–1139.

Manning, L. and Thomas-Antérion, C. (2011). Marc Dax and the discovery of the lateralisation of language in the left cerebral hemisphere. *Revue Neurologique* 167(12): 868–872.

Mäntysalo, S. and Näätänen, R. (1987). The duration of a neuronal trace of an auditory stimulus as indicated by event-related potentials. *Biological Psychology* 24(3): 183–195.

Maratos, A., Crawford, M. J. and Procter, S. (2011). Music therapy for depression: It seems to work, but how? *The British Journal of Psychiatry* 199(2): 92–93.

Margulis, E. H. (2014). *On repeat: How music plays the mind*. Oxford University Press.

Martin, S., Mikutta, C., Leonard, M. K., Hungate, D., Koelsch, S., Shamma, S., Chang, E. F., Millan, J. D. R., Knight, R. T. and Pasley, B. N. (2018). Neural encoding of auditory features during music perception and imagery. *Cerebral Cortex* 28(12): 4222–4233.

Martínez-Molina, N., Mas-Herrero, E., Rodríguez-Fornells, A., Zatorre, R. J. and Marco-Pallarés, J. (2016). Neural correlates of specific musical anhedonia. *Proceedings of the National Academy of Sciences U S A* 113(46): E7337–E7345.

Martínez-Molina, N., Mas-Herrero, E., Rodríguez-Fornells, A., Zatorre, R. J. and Marco-Pallarés, J. (2019). White matter microstructure reflects individual differences in music reward sensitivity. *Journal of Neuroscience* 39(25): 5018–5027.

Marvin, C. B. and Shohamy, D. (2016). Curiosity and reward: Valence predicts choice and information prediction errors enhance learning. *Journal of Experimental Psychology: General* 145(3): 266–272.

Masataka, N. (2006). Preference for consonance over dissonance by hearing newborns of deaf parents and of hearing parents. *Developmental Science* 9(1): 46–50.

Mas-Herrero, E., Dagher, A., Farrés-Franch, M. and Zatorre, R. J. (2021a). Unraveling the temporal dynamics of reward signals in music-induced pleasure with TMS. *The Journal of Neuroscience* 41(17): 3889–3899.

Mas-Herrero, E., Dagher, A. and Zatorre, R. J. (2018a). Modulating musical reward sensitivity up and down with transcranial magnetic stimulation. *Nature Human Behaviour* 2(1): 27–32.

Mas-Herrero, E., Ferreri, L., Cardona, G., Zatorre, R. J., Pla-Juncà, F., Antonijoan, R. M., . . . Rodriguez-Fornells, A. (2023). The role of opioid transmission in music-induced pleasure. *Annals of the New York Academy of Sciences* 1520: 105–114.

Mas-Herrero, E., Karhulahti, M., Marco-Pallarés, J., Zatorre, R. J. and Rodríguez-Fornells, A. (2018b). The impact of visual art and emotional sounds in specific musical anhedonia. *Progress in Brain Research* 237: 399–413.

Mas-Herrero, E., Maini, L., Sescousse, G. and Zatorre, R. J. (2021b). Common and distinct neural correlates of music and food-induced pleasure: A coordinate-based meta-analysis of neuroimaging studies. *Neuroscience & Biobehavioral Reviews* 123: 61–71.

Mas-Herrero, E., Marco-Pallarés, J., Lorenzo-Seva, U., Zatorre, R. J. and Rodríguez-Fornells, A. (2013). Individual differences in music reward experiences. *Music Perception* 31(2): 118–138.

Mas-Herrero, E., Singer, N., Ferreri, L., McPhee, M., Zatorre, R. J., and Ripollés, P. (2023). Music engagement is negatively correlated with depressive symptoms during the COVID-19 pandemic via reward-related mechanisms. *Annals of the New York Academy of Sciences* 1519: 186–198.

Mas-Herrero, E., Zatorre, R. J., Rodríguez-Fornells, A. and Marco-Pallarés, J. (2014). Dissociation between musical and monetary reward responses in specific musical anhedonia. *Current Biology* 24(6): 699–704.

Massoudi, R., Van Wanrooij, M. M., Versnel, H. and Van Opstal, A. J. (2015). Spectrotemporal response properties of core auditory cortex neurons in awake monkey. *PLoS One* 10(2): e0116118.

Mathias, B., Palmer, C., Perrin, F. and Tillmann, B. (2014). Sensorimotor learning enhances expectations during auditory perception. *Cerebral Cortex* **25**(8): 2238–2254.

Mathias, B., Tillmann, B. and Palmer, C. (2016). Sensory, cognitive, and sensorimotor learning effects in recognition memory for music. *Journal of Cognitive Neuroscience* **28**(8): 1111–1126.

Matsushita, R., Andoh, J. and Zatorre, R. J. (2015). Polarity-specific transcranial direct current stimulation disrupts auditory pitch learning. *Frontiers in Neuroscience* **9**: 174.

Matsushita, R., Puschmann, S., Baillet, S. and Zatorre, R. J. (2021). Inhibitory effect of tDCS on auditory evoked response: Simultaneous MEG-tDCS reveals causal role of right auditory cortex in pitch learning. *NeuroImage* **233**: 117915.

Matthews, T. E., Witek, M. A. G., Heggli, O. A., Penhune, V. B. and Vuust, P. (2019). The sensation of groove is affected by the interaction of rhythmic and harmonic complexity. *PLoS One* **14**(1): e0204539.

Matthews, T. E., Witek, M. A. G., Lund, T., Vuust, P. and Penhune, V. B. (2020). The sensation of groove engages motor and reward networks. *NeuroImage* **214**: 116768.

McAdams, S., Winsberg, S., Donnadieu, S., De Soete, G. and Krimphoff, J. (1995). Perceptual scaling of synthesized musical timbres: Common dimensions, specificities, and latent subject classes. *Psychological Research* **58**(3): 177–192.

McDermott, J. H., Lehr, A. J. and Oxenham, A. J. (2010). Individual differences reveal the basis of consonance. *Current Biology* **20**(11): 1035–1041.

McDermott, J. H. and Oxenham, A. J. (2008). Spectral completion of partially masked sounds. *Proceedings of the National Academy of Sciences* **105**(15): 5939–5944.

McDermott, J. H., Schultz, A. F., Undurraga, E. A. and Godoy, R. A. (2016). Indifference to dissonance in native amazonians reveals cultural variation in music perception. *Nature* **535**(7613): 547–550.

McGettigan, C. and Scott, S. K. (2012). Cortical asymmetries in speech perception: What's wrong, what's right and what's left? *Trends in Cognitive Sciences* **16**(5): 269–276.

Mehlhorn, K., Newell, B. R., Todd, P. M., Lee, M. D., Morgan, K., Braithwaite, V. A., Hausmann, D., Fiedler, K. and Gonzalez, C. (2015). Unpacking the exploration–exploitation tradeoff: A synthesis of human and animal literatures. *Decision* **2**(3): 191–215.

Mehr, S. A. and Krasnow, M. M. (2017). Parent-offspring conflict and the evolution of infant-directed song. *Evolution and Human Behavior* **38**: 674–684.

Mehr, S. A., Singh, M., Knox, D., Ketter, D. M., Pickens-Jones, D., Atwood, S., Lucas, C., Jacoby, N., Egner, A. A., Hopkins, E. J., Howard, R. M., Hartshorne, J. K., Jennings, M. V., Simson, J., Bainbridge, C. M., Pinker, S., O'Donnell, T. J., Krasnow, M. M. and Glowacki, L. (2019). Universality and diversity in human song. *Science* **366**(6468): eaax0868.

Mehr, S. A., Singh, M., York, H., Glowacki, L. and Krasnow, M. M. (2018). Form and function in human song. *Current Biology* **28**(3): 356–368.

Mellers, B., Schwartz, A. and Ritov, I. (1999). Emotion-based choice. *Journal of Experimental Psychology: General* **128**(3): 332–345.

Mencke, I., Omigie, D., Wald-Fuhrmann, M. and Brattico, E. (2019). Atonal music: Can uncertainty lead to pleasure? *Frontiers in Neuroscience* **12**: 979.

Menon, V. and Levitin, D. J. (2005). The rewards of music listening: Response and physiological connectivity of the mesolimbic system. *NeuroImage* **28**(1): 175–184.

Mesgarani, N. and Chang, E. F. (2012). Selective cortical representation of attended speaker in multi-talker speech perception. *Nature* **485**(7397): 233–236.

Meyer, K., Kaplan, J. T., Essex, R., Webber, C., Damasio, H. and Damasio, A. (2010). Predicting visual stimuli on the basis of activity in auditory cortices. *Nature Neuroscience* **13**(6): 667–668.

Meyer, L. (1956). *Emotion and meaning in music.* University of Chicago Press.

Meyer, M., Liem, F., Hirsiger, S., Jäncke, L. and Hänggi, J. (2013). Cortical surface area and cortical thickness demonstrate differential structural asymmetry in auditory-related areas of the human cortex. *Cerebral Cortex* **24**(10): 2541–2552.

Meyer, T. and Olson, C. R. (2011). Statistical learning of visual transitions in monkey inferotemporal cortex. *Proceedings of the National Academy of Sciences* **108**(48): 19401–19406.

Michalak, J., Troje, N. F., Fischer, J., Vollmar, P., Heidenreich, T. and Schulte, D. (2009). Embodiment of sadness and depression—gait patterns associated with dysphoric mood. *Psychosomatic Medicine* **71**(5): 580–587.

Micheyl, C., Delhommeau, K., Perrot, X. and Oxenham, A. J. (2006). Influence of musical and psychoacoustical training on pitch discrimination. *Hearing Research* **219**(1): 36–47.

Micheyl, C., Tian, B., Carlyon, R. P. and Rauschecker, J. P. (2005). Perceptual organization of tone sequences in the auditory cortex of awake macaques. *Neuron* **48**(1): 139–148.

Middlebrooks, J. C. and Onsan, Z. A. (2012). Stream segregation with high spatial acuity. *Journal of the Acoustical Society of America* **132**(6): 3896–3911.

Miller, S. E., Schlauch, R. S. and Watson, P. J. (2010). The effects of fundamental frequency contour manipulations on speech intelligibility in background noise. *Journal of the Acoustical Society of America* **128**(1): 435–443.

Milner, B. (1962). Laterality effects in audition. In V. B. Mountcastle (Ed.), *Interhemispheric relations and cerebral dominance* (pp. 177–195). Johns Hopkins Press.

Milner, B., Taylor, L. and Sperry, R. W. (1968). Lateralized suppression of dichotically presented digits after commissural section in man. *Science* **161**(3837): 184–185.

Milner, D. and Goodale, M. (2006). *The visual brain in action*. Oxford University Press.

Mishkin, M., Ungerleider, L. G. and Macko, K. A. (1983). Object vision and spatial vision: Two cortical pathways. *Trends in Neurosciences* **6**: 414–417.

Mišić, B., Betzel, R. F., Griffa, A., De Reus, M. A., He, Y., Zuo, X. N., Van Den Heuvel, M. P., Hagmann, P., Sporns, O. and Zatorre, R. J. (2018). Network-based asymmetry of the human auditory system. *Cerebral Cortex* **28**(7): 2655–2664.

Mišić, B., Betzel, R. F., Nematzadeh, A., Goni, J., Griffa, A., Hagmann, P., Flammini, A., Ahn, Y.-Y. and Sporns, O. (2015). Cooperative and competitive spreading dynamics on the human connectome. *Neuron* **86**(6): 1518–1529.

Mitterschiffthaler, M. T., Fu, C. H. Y., Dalton, J. A., Andrew, C. M. and Williams, S. C. R. (2007). A functional MRI study of happy and sad affective states induced by classical music. *Human Brain Mapping* **28**(11): 1150–1162.

Moerel, M., De Martino, F. and Formisano, E. (2012). Processing of natural sounds in human auditory cortex: Tonotopy, spectral tuning, and relation to voice sensitivity. *Journal of Neuroscience* **32**(41): 14205–14216.

Moerel, M., De Martino, F., Santoro, R., Ugurbil, K., Goebel, R., Yacoub, E. and Formisano, E. (2013). Processing of natural sounds: Characterization of multipeak spectral tuning in human auditory cortex. *Journal of Neuroscience* **33**(29): 11888–11898.

Moerel, M., De Martino, F., Santoro, R., Yacoub, E. and Formisano, E. (2015). Representation of pitch chroma by multi-peak spectral tuning in human auditory cortex. *NeuroImage* **106**: 161–169.

Moerel, M., De Martino, F., Uğurbil, K., Yacoub, E. and Formisano, E. (2019). Processing complexity increases in superficial layers of human primary auditory cortex. *Scientific Reports* **9**(1): 5502.

Molholm, S., Martinez, A., Ritter, W., Javitt, D. C. and Foxe, J. J. (2004). The neural circuitry of pre-attentive auditory change-detection: An fMRI study of pitch and duration mismatch negativity generators. *Cerebral Cortex* **15**(5): 545–551.

Molholm, S., Sehatpour, P., Mehta, A. D., Shpaner, M., Gomez-Ramirez, M., Ortigue, S., Dyke, J. P., Schwartz, T. H. and Foxe, J. J. (2006). Audio-visual multisensory integration in superior parietal lobule revealed by human intracranial recordings. *Journal of Neurophysiology* **96**(2): 721–729.

Molnar-Szakacs, I. and Overy, K. (2006). Music and mirror neurons: From motion to 'e'motion. *Social Cognitive and Affective Neuroscience* **1**(3): 235–241.

Montag, C., Reuter, M. and Axmacher, N. (2011). How one's favorite song activates the reward circuitry of the brain: Personality matters! *Behavioural Brain Research* **225**(2): 511–514.

Moore, B. C., Glasberg, B. R. and Peters, R. W. (1986). Thresholds for hearing mistuned partials as separate tones in harmonic complexes. *Journal of the Acoustical Society of America* **80**(2): 479–483.

Moore, B. C. and Gockel, H. (2002). Factors influencing sequential stream segregation. *Acta Acustica United with Acustica* **88**(3): 320–333.

Morawetz, C., Bode, S., Derntl, B. and Heekeren, H. R. (2017). The effect of strategies, goals and stimulus material on the neural mechanisms of emotion regulation: A meta-analysis of fMRI studies. *Neuroscience & Biobehavioral Reviews* **72**: 111–128.

Morillon, B., Arnal, L. H., Schroeder, C. E. and Keitel, A. (2019). Prominence of delta oscillatory rhythms in the motor cortex and their relevance for auditory and speech perception. *Neuroscience & Biobehavioral Reviews* **107**: 136–142.

Morillon, B. and Baillet, S. (2017). Motor origin of temporal predictions in auditory attention. *Proceedings of the National Academy of Sciences* **114**(42): E8913–E8921.

Morillon, B., Schroeder, C. E. and Wyart, V. (2014). Motor contributions to the temporal precision of auditory attention. *Nature Communications* **5**(1): 5255.

Moser, D., Baker, J. M., Sanchez, C. E., Rorden, C. and Fridriksson, J. (2009). Temporal order processing of syllables in the left parietal lobe. *The Journal of Neuroscience* **29**(40): 12568–12573.

Möttönen, R., Calvert, G. A., Jääskeläinen, I. P., Matthews, P. M., Thesen, T., Tuomainen, J. and Sams, M. (2006). Perceiving identical sounds as speech or non-speech modulates activity in the left posterior superior temporal sulcus. *NeuroImage* **30**(2): 563–569.

Mueller, K., Fritz, T., Mildner, T., Richter, M., Schulze, K., Lepsien, J., Schroeter, M. L. and Möller, H. E. (2015). Investigating the dynamics of the brain response to music: A central role of the ventral striatum/nucleus accumbens. *NeuroImage* **116**: 68–79.

Munoz-Lopez, M., Mohedamo-Moriano, A. and Insausti, R. (2010). Anatomical pathways for auditory memory in primates. *Frontiers in Neuroanatomy* **4**: 129.

Murray, M. M., Camen, C., Gonzalez Andino, S. L., Bovet, P. and Clarke, S. (2006). Rapid brain discrimination of sounds of objects. *The Journal of Neuroscience* **26**(4): 1293–1302.

Musacchia, G., Sams, M., Skoe, E. and Kraus, N. (2007). Musicians have enhanced subcortical auditory and audiovisual processing of speech and music. *Proceedings of the National Academy of Sciences* **104**(40): 15894–15898.

Näätänen, R., Kujala, T. and Light, G. (2019). *Mismatch negativity: A window to the brain.* Oxford University Press.

Nan, Y., Sun, Y. and Peretz, I. (2010). Congenital amusia in speakers of a tone language: Association with lexical tone agnosia. *Brain* **133**(9): 2635–2642.

Narmour, E. (2000). Music expectation by cognitive rule-mapping. *Music Perception* **17**(3): 329–398.

Nilsson, M., Soli, S. D. and Sullivan, J. A. (1994). Development of the hearing in noise test for the measurement of speech reception thresholds in quiet and in noise. *Journal of the Acoustical Society of America* **95**(2): 1085–1099.

Norman-Haignere, S., Kanwisher, N. and McDermott, J. H. (2013). Cortical pitch regions in humans respond primarily to resolved harmonics and are located in specific tonotopic regions of anterior auditory cortex. *Journal of Neuroscience* **33**(50): 19451–19469.

Norman-Haignere, S., Kanwisher, N. G. and McDermott, J. H. (2015). Distinct cortical pathways for music and speech revealed by hypothesis-free voxel decomposition. *Neuron* **88**(6): 1281–1296.

Norman-Haignere, S. V., Albouy, P., Caclin, A., McDermott, J. H., Kanwisher, N. G. and Tillmann, B. (2016). Pitch-responsive cortical regions in congenital amusia. *Journal of Neuroscience* **36**(10): 2986–2994.

Norman-Haignere, S. V., Feather, J., Boebinger, D., Brunner, P., Ritaccio, A., McDermott, J. H., Schalk, G. and Kanwisher, N. (2022). A neural population selective for song in human auditory cortex. *Current Biology* **32**(7): 1470–1484.

Norman-Haignere, S. V., Feather, J., Boebinger, D., Brunner, P., Ritaccio, A., McDermott, J. H., . . . Kanwisher, N. (2022). A neural population selective for song in human auditory cortex. *Current Biology* **32**(7): 1470–1484.

North, A. C. and Hargreaves, D. J. (1997). Liking, arousal potential, and the emotions expressed by music. *Scandinavian Journal of Psychology* **38**(1): 45–53.

Novembre, G. and Keller, P. E. (2014). A conceptual review on action-perception coupling in the musicians' brain: What is it good for? *Frontiers in Human Neuroscience* **8**: 603.

Nozaradan, S., Schwartze, M., Obermeier, C. and Kotz, S. A. (2017). Specific contributions of basal ganglia and cerebellum to the neural tracking of rhythm. *Cortex* **95**: 156–168.

Nucifora, P. G., Verma, R., Melhem, E. R., Gur, R. E. and Gur, R. C. (2005). Leftward asymmetry in relative fiber density of the arcuate fasciculus. *NeuroReport* **16**(8): 791–794.

Nummenmaa, L., Putkinen, V. and Sams, M. (2021). Social pleasures of music. *Current Opinion in Behavioral Sciences* **39**: 196–202.

Nummenmaa, L., Saanijoki, T., Tuominen, L., Hirvonen, J., Tuulari, J. J., Nuutila, P. and Kalliokoski, K. (2018). M-opioid receptor system mediates reward processing in humans. *Nature Communications* **9**(1): 1500.

Nusbaum, E. C. and Silvia, P. J. (2011). Shivers and timbres: Personality and the experience of chills from music. *Social Psychological and Personality Science* **2**(2): 199–204.

Obleser, J., Eisner, F. and Kotz, S. A. (2008). Bilateral speech comprehension reflects differential sensitivity to spectral and temporal features. *Journal of Neuroscience* **28**(32): 8116–8123.

Ochsner, K. N., Ray, R. D., Cooper, J. C., Robertson, E. R., Chopra, S., Gabrieli, J. D. E. and Gross, J. J. (2004). For better or for worse: Neural systems supporting the cognitive down- and up-regulation of negative emotion. *NeuroImage* **23**(2): 483–499.

Ochsner, K. N., Silvers, J. A. and Buhle, J. T. (2012). Functional imaging studies of emotion regulation: A synthetic review and evolving model of the cognitive control of emotion. *Annals of the New York Academy of Sciences* **1251**(1): E1–E24.

Ocklenburg, S., Friedrich, P., Fraenz, C., Schlüter, C., Beste, C., Güntürkün, O. and Genç, E. (2018). Neurite architecture of the planum temporale predicts neurophysiological processing of auditory speech. *Science Advances* **4**(7): eaar6830.

Ocklenburg, S., Schlaffke, L., Hugdahl, K. and Westerhausen, R. (2014). From structure to function in the lateralized brain: How structural properties of the arcuate and uncinate fasciculus are associated with dichotic listening performance. *Neuroscience Letters* **580**: 32–36.

O'Doherty, J., Kringelbach, M. L., Rolls, E. T., Hornak, J. and Andrews, C. (2001). Abstract reward and punishment representations in the human orbitofrontal cortex. *Nature Neuroscience* **4**(1): 95–102.

Ohira, H., Nomura, M., Ichikawa, N., Isowa, T., Iidaka, T., Sato, A., Fukuyama, S., Nakajima, T. and Yamada, J. (2006). Association of neural and physiological responses during voluntary emotion suppression. *NeuroImage* **29**(3): 721–733.

Okamoto, H. and Kakigi, R. (2015). Hemispheric asymmetry of auditory mismatch negativity elicited by spectral and temporal deviants: A magnetoencephalographic study. *Brain Topography* **28**(3): 471–478.

Oldham, S., Murawski, C., Fornito, A., Youssef, G., Yücel, M. and Lorenzetti, V. (2018). The anticipation and outcome phases of reward and loss processing: A neuroimaging meta-analysis of the monetary incentive delay task. *Human Brain Mapping* **39**(8): 3398–3418.

Olds, J. and Milner, P. (1954). Positive reinforcement produced by electrical stimulation of septal area and other regions of rat brain. *Journal of Comparative and Physiological Psychology* **47**(6): 419–427.

Olive, M. F., Koenig, H. N., Nannini, M. A. and Hodge, C. W. (2001). Stimulation of endorphin neurotransmission in the nucleus accumbens by ethanol, cocaine, and amphetamine. *Journal of Neuroscience* **21**(23): RC184–RC184.

Omigie, D., Pearce, M., Lehongre, K., Hasboun, D., Navarro, V., Adam, C. and Samson, S. (2019). Intracranial recordings and computational modeling of music reveal the time course of prediction error signaling in frontal and temporal cortices. *Journal of Cognitive Neuroscience* **31**(6): 855–873.

Omigie, D., Pearce, M. T. and Stewart, L. (2012). Tracking of pitch probabilities in congenital amusia. *Neuropsychologia* **50**(7): 1483–1493.

Öngür, D. and Price, J. L. (2000). The organization of networks within the orbital and medial prefrontal cortex of rats, monkeys and humans. *Cerebral Cortex* **10**(3): 206–219.

Opitz, B., Rinne, T., Mecklinger, A., Von Cramon, D. Y. and Schröger, E. (2002). Differential contribution of frontal and temporal cortices to auditory change detection: fMRI and ERP results. *NeuroImage* **15**(1): 167–174.

Oram, N. and Cuddy, L. L. (1995). Responsiveness of western adults to pitch-distributional information in melodic sequences. *Psychological Research* **57**(2): 103–118.

Orpella, J., Mas-Herrero, E., Ripollés, P., Marco-Pallarés, J. and de Diego-Balaguer, R. (2021). Language statistical learning responds to reinforcement learning principles rooted in the striatum. *PLoS Biology* **19**(9): e3001119.

Ortiz-Rios, M., Azevedo, F. a. C., Kuśmierek, P., Balla, D. Z., Munk, M. H., Keliris, G. A., Logothetis, N. K. and Rauschecker, J. P. (2017). Widespread and opponent fMRI signals represent sound location in macaque auditory cortex. *Neuron* **93**(4): 971–983.e974.

O'Sullivan, J. A., Shamma, S. A. and Lalor, E. C. (2015). Evidence for neural computations of temporal coherence in an auditory scene and their enhancement during active listening. *Journal of Neuroscience* 35(18): 7256–7263.

Oudeyer, P. Y., Gottlieb, J. and Lopes, M. (2016). Intrinsic motivation, curiosity, and learning: Theory and applications in educational technologies. In B. Studer and S. Knecht (Eds.), *Progress in brain research* (Vol. 229, pp. 257–284). Elsevier.

Overath, T., Kumar, S., Von Kriegstein, K. and Griffiths, T. D. (2008). Encoding of spectral correlation over time in auditory cortex. *Journal of Neuroscience* 28(49): 13268–13273.

Overath, T., McDermott, J. H., Zarate, J. M. and Poeppel, D. (2015). The cortical analysis of speech-specific temporal structure revealed by responses to sound quilts. *Nature Neuroscience* 18(6): 903–911.

Owen, A. M., Mcmillan, K. M., Laird, A. R. and Bullmore, E. (2005). N-back working memory paradigm: A meta-analysis of normative functional neuroimaging studies. *Human Brain Mapping* 25(1): 46–59.

Özdemir, E., Norton, A. and Schlaug, G. (2006). Shared and distinct neural correlates of singing and speaking. *NeuroImage* 33(2): 628–635.

Paavilainen, P. (2013). The mismatch-negativity (MMN) component of the auditory event-related potential to violations of abstract regularities: A review. *International Journal of Psychophysiology* 88(2): 109–123.

Padoa-Schioppa, C. and Assad, J. A. (2006). Neurons in the orbitofrontal cortex encode economic value. *Nature* 441(7090): 223–226.

Padoa-Schioppa, C. and Conen, K. E. (2017). Orbitofrontal cortex: A neural circuit for economic decisions. *Neuron* 96(4): 736–754.

Palmer, C. (1997). Music performance. *Annual Review of Psychology* 48(1): 115–138.

Palomar-García, M. A., Zatorre, R. J., Ventura-Campos, N., Bueicheku, E. and Avila, C. (2017). Modulation of functional connectivity in auditory-motor networks in musicians compared with nonmusicians. *Cerebral Cortex* 27(5): 2768–2778.

Panksepp, J. (1995). The emotional sources of "chills" induced by music. *Music Perception* 13(2): 171–207.

Panksepp, J., Lane, R. D., Solms, M. and Smith, R. (2017). Reconciling cognitive and affective neuroscience perspectives on the brain basis of emotional experience. *Neuroscience & Biobehavioral Reviews* 76: 187–215.

Paquette, S., Fujii, S., Li, H. C. and Schlaug, G. (2017). The cerebellum's contribution to beat interval discrimination. *NeuroImage* 163: 177–182.

Paquette, S., Peretz, I. and Belin, P. (2013). The "musical emotional bursts": A validated set of musical affect bursts to investigate auditory affective processing. *Frontiers in psychology* 4: 509.

Paquette, S., Takerkart, S., Saget, S., Peretz, I. and Belin, P. (2018). Cross-classification of musical and vocal emotions in the auditory cortex. *Annals of the New York Academy of Sciences* 1423: 329–337.

Parbery-Clark, A., Skoe, E. and Kraus, N. (2009). Musical experience limits the degradative effects of background noise on the neural processing of sound. *Journal of Neuroscience* 29(45): 14100–14107.

Park, M., Hennig-Fast, K., Bao, Y., Carl, P., Pöppel, E., Welker, L., Reiser, M., Meindl, T. and Gutyrchik, E. (2013). Personality traits modulate neural responses to emotions expressed in music. *Brain Research* 1523: 68–76.

Parkin, B. L., Ekhtiari, H. and Walsh, V. F. (2015). Non-invasive human brain stimulation in cognitive neuroscience: A primer. *Neuron* 87(5): 932–945.

Parras, G. G., Nieto-Diego, J., Carbajal, G. V., Valdés-Baizabal, C., Escera, C. and Malmierca, M. S. (2017). Neurons along the auditory pathway exhibit a hierarchical organization of prediction error. *Nature Communications* 8(1): 2148.

Parsons, L. M., Sergent, J., Hodges, D. A. and Fox, P. T. (2005). The brain basis of piano performance. *Neuropsychologia* 43(2): 199–215.

Passamonti, L., Terracciano, A., Riccelli, R., Donzuso, G., Cerasa, A., Vaccaro, M. G., Novellino, F., Fera, F. and Quattrone, A. (2015). Increased functional connectivity within mesocortical networks in open people. *NeuroImage* 104: 301–309.

Patel, A. D. (2010). *Music, language, and the brain*. Oxford University Press.

Patel, A. D. (2011). Why would musical training benefit the neural encoding of speech? The OPERA hypothesis. *Frontiers in Psychology* 2: 142.

Patel, A. D. (2021). Vocal learning as a preadaptation for the evolution of human beat perception and synchronization. *Philosophical Transactions of the Royal Society B: Biological Sciences* **376**(1835): 20200326.

Patel, A. D., Gibson, E., Ratner, J., Besson, M. and Holcomb, P. J. (1998). Processing syntactic relations in language and music: An event-related potential study. *Journal of Cognitive Neuroscience* **10**(6): 717–733.

Patel, A. D. and Iversen, J. R. (2014). The evolutionary neuroscience of musical beat perception: The action simulation for auditory prediction (asap) hypothesis. *Frontiers in Systems Neuroscience* **8**: 57–57.

Patterson, R. D., Uppenkamp, S., Johnsrude, I. S. and Griffiths, T. D. (2002). The processing of temporal pitch and melody information in auditory cortex. *Neuron* **36**(4): 767–776.

Paulus, M. P. and Stein, M. B. (2006). An insular view of anxiety. *Biological Psychiatry* **60**(4): 383–387.

Pearce, E., Launay, J. and Dunbar, R. I. (2015). The ice-breaker effect: Singing mediates fast social bonding. *Royal Society Open Science* **2**(10): 150221.

Pearce, M. T. (2018). Statistical learning and probabilistic prediction in music cognition: Mechanisms of stylistic enculturation. *Annals of the New York Academy of Sciences* **1423**(1): 378.

Pearce, M. T. and Halpern, A. R. (2015). Age-related patterns in emotions evoked by music. *Psychology of Aesthetics, Creativity, and the Arts* **9**(3): 248–253.

Peciña, S., Smith, K. S. and Berridge, K. C. (2006). Hedonic hot spots in the brain. *The Neuroscientist* **12**(6): 500–511.

Pelletier, C. L. (2004). The effect of music on decreasing arousal due to stress: A meta-analysis. *Journal of Music Therapy* **41**(3): 192–214.

Penfield, W. (1958). Some cechanisms of consciousness discovered during electrical stimulation of the brain. *Proceedings of the National Academy of Sciences* **44**(2): 51–66.

Penfield, W. and Perot, P. (1963). The brain's record of auditory and visual experience: A final summary and discussion. *Brain* **86**(4): 595–696.

Penhune, V. B., Zatorre, R., MacDonald, J. and Evans, A. (1996). Interhemispheric anatomical differences in human primary auditory cortex: Probabilistic mapping and volume measurement from magnetic resonance scans. *Cerebral Cortex* **6**(5): 661–672.

Penhune, V. B., Cismaru, R., Dorsaint-Pierre, R., Petitto, L.-A. and Zatorre, R. J. (2003). The morphometry of auditory cortex in the congenitally deaf measured using MRI. *NeuroImage* **20**(2): 1215–1225.

Penhune, V. B., Zatorre, R. J. and Feindel, W. H. (1999). The role of auditory cortex in retention of rhythmic patterns as studied in patients with temporal lobe removals including Heschl's gyrus. *Neuropsychologia* **37**(3): 315–331.

Pereira, C. S., Teixeira, J., Figueiredo, P., Xavier, J., Castro, S. L. and Brattico, E. (2011). Music and emotions in the brain: Familiarity matters. *PloS One* **6**(11): e27241.

Peretz, I., Ayotte, J., Zatorre, R. J., Mehler, J., Ahad, P., Penhune, V. B. and Jutras, B. (2002). Congenital amusia: A disorder of fine-grained pitch discrimination. *Neuron* **33**(2): 185–191.

Peretz, I., Brattico, E., Järvenpää, M. and Tervaniemi, M. (2009). The amusic brain: In tune, out of key, and unaware. *Brain* **132**(5): 1277–1286.

Peretz, I., Champod, A. S. and Hyde, K. (2003). Varieties of musical disorders: The Montreal battery of evaluation of amusia. *Annals of the New York Academy of Sciences* **999**(1): 58–75.

Peretz, I. and Coltheart, M. (2003). Modularity of music processing. *Nature Neuroscience* **6**(7): 688–691.

Peretz, I., Cummings, S. and Dubé, M.-P. (2007). The genetics of congenital amusia (tone deafness): A family-aggregation study. *The American Journal of Human Genetics* **81**(3): 582–588.

Peretz, I., Gaudreau, D. and Bonnel, A.-M. (1998). Exposure effects on music preference and recognition. *Memory & Cognition* **26**(5): 884–902.

Peretz, I. and Kolinsky, R. (1993). Boundaries of separability between melody and rhythm in music discrimination: A neuropsychological perspective. *Quarterly Journal of Experimental Psychology Section A* **46**(2): 301–325.

Peretz, I., Saffran, J., Schön, D. and Gosselin, N. (2012). Statistical learning of speech, not music, in congenital amusia. *Annals of the New York Academy of Sciences* **1252**(1): 361–366.

Peretz, I., Vuvan, D., Lagrois, M.-É. and Armony, J. L. (2015). Neural overlap in processing music and speech. *Philosophical Transactions of the Royal Society B: Biological Sciences* **370**(1664): 20140090.

Pernet, C. R., Mcaleer, P., Latinus, M., Gorgolewski, K. J., Charest, I., Bestelmeyer, P. E. G., Watson, R. H., Fleming, D., Crabbe, F., Valdes-Sosa, M. and Belin, P. (2015). The human voice areas: Spatial organization and inter-individual variability in temporal and extra-temporal cortices. *NeuroImage* 119: 164–174.

Pesnot Lerousseau, J., Trébuchon, A., Morillon, B. and Schön, D. (2021). Frequency selectivity of persistent cortical oscillatory responses to auditory rhythmic stimulation. *Journal of Neuroscience* 41(38): 7991–8006.

Pessiglione, M., Seymour, B., Flandin, G., Dolan, R. J. and Frith, C. D. (2006). Dopamine-dependent prediction errors underpin reward-seeking behaviour in humans. *Nature* 442(7106): 1042–1045.

Petacchi, A., Laird, A. R., Fox, P. T. and Bower, J. M. (2005). Cerebellum and auditory function: An ALE meta-analysis of functional neuroimaging studies. *Human Brain Mapping* 25(1): 118–128.

Petkov, C. I., Kayser, C., Steudel, T., Whittingstall, K., Augath, M. and Logothetis, N. K. (2008). A voice region in the monkey brain. *Nature Neuroscience* 11(3): 367–374.

Petrides, M. (2000). Middorsolateral and midventrolateral prefrontal cortex: Two levels of executive control for the processing of mnemonic information. In S. Monsell and J. Driver (Eds.), Control of Cognitive Processes (pp. 535–548). MIT Press.

Petrides, M. (2005). Lateral prefrontal cortex: Architectonic and functional organization. *Philosophical Transactions of the Royal Society B: Biological Sciences* 360(1456): 781–795.

Petrides, M. (2014). Neuroanatomy of language regions of the human brain. Academic Press.

Petrides, M. and Pandya, D. N. (1988). Association fiber pathways to the frontal cortex from the superior temporal region in the rhesus monkey. *Journal of Comparative Neurology* 273(1): 52–66.

Petrides, M. and Pandya, D. N. (2009). Distinct parietal and temporal pathways to the homologues of broca's area in the monkey. *PLoS Biology* 7(8): e1000170.

Pfordresher, P. Q. and Halpern, A. R. (2013). Auditory imagery and the poor-pitch singer. *Psychonomic Bulletin & Review* 20(4): 747–753.

Pfordresher, P. Q., Halpern, A. R. and Greenspon, E. B. (2015). A mechanism for sensorimotor translation in singing: The multi-modal imagery association (MMIA) model. *Music Perception* 32(3): 242–253.

Pfordresher, P. Q., Mantell, J. T., Brown, S., Zivadinov, R. and Cox, J. L. (2014). Brain responses to altered auditory feedback during musical keyboard production: An fMRI study. *Brain Research* 1556: 28–37.

Phelps, E. A. and Ledoux, J. E. (2005). Contributions of the amygdala to emotion processing: From animal models to human behavior. *Neuron* 48(2): 175–187.

Phillips, D. P. and Farmer, M. E. (1990). Acquired word deafness, and the temporal grain of sound representation in the primary auditory cortex. *Behavioural Brain Research* 40(2): 85–94.

Phillips-Silver, J. and Trainor, L. J. (2005). Feeling the beat: Movement influences infant rhythm perception. *Science* 308(5727): 1430.

Pinho, A. L., De Manzano, Ö., Fransson, P., Eriksson, H. and Ullén, F. (2014). Connecting to create: Expertise in musical improvisation is associated with increased functional connectivity between premotor and prefrontal areas. *The Journal of Neuroscience* 34(18): 6156–6163.

Plack, C. J., Oxenham, A. J. and Fay, R. R. (2006). *Pitch: Neural coding and perception*. Springer Science & Business Media.

Plantinga, J. and Trainor, L. J. (2005). Memory for melody: Infants use a relative pitch code. *Cognition* 98: 1–11.

Poeppel, D. (2003). The analysis of speech in different temporal integration windows: Cerebral lateralization as 'asymmetric sampling in time'. *Speech Communication* 41(1): 245–255.

Poeppel, D. and Assaneo, M. F. (2020). Speech rhythms and their neural foundations. *Nature Reviews Neuroscience* 21(6): 322–334.

Pogarell, O., Koch, W., Pöpperl, G., Tatsch, K., Jakob, F., Mulert, C., Grossheinrich, N., Rupprecht, R., Möller, H.-J. and Hegerl, U. (2007). Acute prefrontal rTMS increases striatal dopamine to a similar degree as d-amphetamine. *Psychiatry Research: Neuroimaging* 156(3): 251–255.

Popescu, T., Neuser, M. P., Neuwirth, M., Bravo, F., Mende, W., Boneh, O., Moss, F. C. and Rohrmeier, M. (2019). The pleasantness of sensory dissonance is mediated by musical style and expertise. *Scientific Reports* 9(1): 1070.

Poppa, T. and Bechara, A. (2018). The somatic marker hypothesis: Revisiting the role of the 'body-loop' in decision-making. *Current Opinion in Behavioral Sciences* 19: 61–66.

Povel, D.-J. and Essens, P. (1985). Perception of temporal patterns. *Music Perception* **2**(4): 411–440.

Pressnitzer, D., Sayles, M., Micheyl, C. and Winter, I. M. (2008). Perceptual organization of sound begins in the auditory periphery. *Current Biology* **18**(15): 1124–1128.

Pressnitzer, D., Suied, C. and Shamma, S. (2011). Auditory scene analysis: The sweet music of ambiguity. *Frontiers in Human Neuroscience* **5**: 158.

Preuschoff, K., Bossaerts, P. and Quartz, S. R. (2006). Neural differentiation of expected reward and risk in human subcortical structures. *Neuron* **51**(3): 381–390.

Pribram, K. H. (1982). Localization and distribution of function in the brain. In J. Orbach (Ed.), *Neuropsychology after Lashley* (pp. 273–296). Lawrence Erlbaum.

Puschmann, S., Baillet, S. and Zatorre, R. J. (2019). Musicians at the cocktail party: Neural substrates of musical training during selective listening in multispeaker situations. *Cerebral Cortex* **29**(8): 3253–3265.

Puschmann, S., Uppenkamp, S., Kollmeier, B. and Thiel, C. M. (2010). Dichotic pitch activates pitch processing centre in Heschl's gyrus. *NeuroImage* **49**(2): 1641–1649.

Putkinen, V., Tervaniemi, M., Saarikivi, K., Ojala, P. and Huotilainen, M. (2014). Enhanced development of auditory change detection in musically trained school-aged children: A longitudinal event-related potential study. *Developmental Science* **17**(2): 282–297.

Quinci, M. A., Belden, A., Goutama, V., Gong, D., Hanser, S., Donovan, N. J., . . . Loui, P. (2022). Longitudinal changes in auditory and reward systems following receptive music-based intervention in older adults. *Scientific Reports* **12**: 11517.

Quiroga-Martinez, D. R., Hansen, N. C., Højlund, A., Pearce, M. T., Brattico, E. and Vuust, P. (2019). Reduced prediction error responses in high-as compared to low-uncertainty musical contexts. *Cortex* **120**: 181–200.

Ragert, M., Fairhurst, M. T. and Keller, P. E. (2014). Segregation and integration of auditory streams when listening to multi-part music. *PloS One* **9**(1): e84085.

Rakowski, A. (1990). Intonation variants of musical intervals in isolation and in musical contexts. *Psychology of Music* **18**(1): 60–72.

Rauschecker, J. P. (2018). Where, when, and how: Are they all sensorimotor? Towards a unified view of the dorsal pathway in vision and audition. *Cortex* **98**: 262–268.

Rauschecker, J. P. and Scott, S. K. (2009). Maps and streams in the auditory cortex: Nonhuman primates illuminate human speech processing. *Nature Neuroscience* **12**(6): 718–724.

Recanzone, G. H., Guard, D. C. and Phan, M. L. (2000). Frequency and intensity response properties of single neurons in the auditory cortex of the behaving macaque monkey. *Journal of Neurophysiology* **83**(4): 2315–2331.

Reddy, L. and Kanwisher, N. (2006). Coding of visual objects in the ventral stream. *Current Opinion in Neurobiology* **16**(4): 408–414.

Reese, N. B., Garcia-Rill, E. and Skinner, R. D. (1995). The pedunculopontine nucleus—auditory input, arousal and pathophysiology. *Progress in Neurobiology* **47**(2): 105–133.

Reetzke, R., Xie, Z., Llanos, F. and Chandrasekaran, B. (2018). Tracing the trajectory of sensory plasticity across different stages of speech learning in adulthood. *Current Biology* **28**(9): 1419–1427.e1414.

Regenbogen, C., Seubert, J., Johansson, E., Finkelmeyer, A., Andersson, P. and Lundström, J. N. (2018). The intraparietal sulcus governs multisensory integration of audiovisual information based on task difficulty. *Human Brain Mapping* **39**(3): 1313–1326.

Regev, M., Halpern, A. R., Owen, A. M., Patel, A. D. and Zatorre, R. J. (2021). Mapping specific mental content during musical imagery. *Cerebral Cortex* **31**(8): 3622–3640.

Remez, R. E., Rubin, P. E., Pisoni, D. B. and Carrell, T. D. (1981). Speech perception without traditional speech cues. *Science* **212**(4497): 947–949.

Rentfrow, P. J. and Gosling, S. D. (2003). The do re mi's of everyday life: The structure and personality correlates of music preferences. *Journal of Personality and Social Psychology* **84**(6): 1236–1256.

Repp, B. H. (2010). Sensorimotor synchronization and perception of timing: Effects of music training and task experience. *Human Movement Science* **29**(2): 200–213.

Repp, B. H., Iversen, J. R. and Patel, A. D. (2008). Tracking an imposed beat within a metrical grid. *Music Perception:* **26**(1): 1–18.

Repp, B. H. and Penel, A. (2002). Auditory dominance in temporal processing: New evidence from synchronization with simultaneous visual and auditory sequences. *Journal of Experimental Psychology: Human Perception and Performance* **28**(5): 1085–1099.

Repp, B. H. and Su, Y.-H. (2013). Sensorimotor synchronization: A review of recent research (2006–2012). *Psychonomic Bulletin & Review* **20**(3): 403–452.

Reynolds, S. M. and Berridge, K. C. (2008). Emotional environments retune the valence of appetitive versus fearful functions in nucleus accumbens. *Nature Neuroscience* **11**(4): 423–425.

Ribeiro, F. S., Lessa, J. P. A., Delmolin, G. and Santos, F. H. (2021). Music listening in times of covid-19 outbreak: A Brazilian study. *Frontiers in Psychology* **12**: 1471.

Richard, J. M. and Berridge, K. C. (2013). Prefrontal cortex modulates desire and dread generated by nucleus accumbens glutamate disruption. *Biological Psychiatry* **73**(4): 360–370.

Riecke, L., Peters, J. C., Valente, G., Kemper, V. G., Formisano, E. and Sorger, B. (2017). Frequency-selective attention in auditory scenes recruits frequency representations throughout human superior temporal cortex. *Cerebral Cortex* **27**(5): 3002–3014.

Rimmele, J. M., Morillon, B., Poeppel, D. and Arnal, L. H. (2018). Proactive sensing of periodic and aperiodic auditory patterns. *Trends in Cognitive Sciences* **22**(10): 870–882.

Ripollés, P., Ferreri, L., Mas-Herrero, E., Alicart, H., Gómez-Andrés, A., Marco-Pallarés, J., Antonijoan, R. M., Noesselt, T., Valle, M. and Riba, J. (2018). Intrinsically regulated learning is modulated by synaptic dopamine signaling. *Elife* **7**: e38113.

Ripollés, P., Marco-Pallarés, J., Alicart, H., Tempelmann, C., Rodriguez-Fornells, A. and Noesselt, T. (2016). Intrinsic monitoring of learning success facilitates memory encoding via the activation of the SN/VTA-hippocampal loop. *Elife* **5**: e17441.

Ripollés, P., Marco-Pallarés, J., Hielscher, U., Mestres-Missé, A., Tempelmann, C., Heinze, H.-J., Rodríguez-Fornells, A. and Noesselt, T. (2014). The role of reward in word learning and its implications for language acquisition. *Current Biology* **24**(21): 2606–2611.

Rissman, J., Chow, T. E., Reggente, N. and Wagner, A. D. (2016). Decoding fMRI signatures of real-world autobiographical memory retrieval. *Journal of Cognitive Neuroscience* **28**(4): 604–620.

Ritsner, M. S. (2014). *Anhedonia: A comprehensive handbook* (Vol. I, pp. 19–54). Springer.

Rizzolatti, G. and Craighero, L. (2004). The mirror-neuron system. *Annual Review of Neuroscience* **27**(1): 169–192.

Roberts, L. E., Eggermont, J. J., Caspary, D. M., Shore, S. E., Melcher, J. R. and Kaltenbach, J. A. (2010). Ringing ears: The neuroscience of tinnitus. *The Journal of Neuroscience* **30**(45): 14972–14979.

Rocchi, F., Oya, H., Balezeau, F., Billig, A. J., Kocsis, Z., Jenison, R. L., Nourski, K. V., Kovach, C. K., Steinschneider, M., Kikuchi, Y., Rhone, A. E., Dlouhy, B. J., Kawasaki, H., Adolphs, R., Greenlee, J. D. W., Griffiths, T. D., Howard, M. A. and Petkov, C. I. (2021). Common fronto-temporal effective connectivity in humans and monkeys. *Neuron* **109**(5): 852–868.

Rodríguez, F. A., Read, H. L. and Escabí, M. A. (2010). Spectral and temporal modulation tradeoff in the inferior colliculus. *Journal of Neurophysiology* **103**(2): 887–903.

Rodríguez-Rey, R., Garrido-Hernansaiz, H. and Collado, S. (2020). Psychological impact and associated factors during the initial stage of the coronavirus (COVID-19) pandemic among the general population in Spain. *Frontiers in Psychology* **11**: 1540.

Rohrmeier, M. and Rebuschat, P. (2012). Implicit learning and acquisition of music. *Topics in Cognitive Science* **4**(4): 525–553.

Rohrmeier, M. A. and Koelsch, S. (2012). Predictive information processing in music cognition. A critical review. *International Journal of Psychophysiology* **83**(2): 164–175.

Romanski, L. M., Bates, J. F. and Goldman-Rakic, P. S. (1999a). Auditory belt and parabelt projections to the prefrontal cortex in the rhesus monkey. *Journal of Comparative Neurology* **403**(2): 141–157.

Romanski, L. M., Tian, B., Fritz, J., Mishkin, M., Goldman-Rakic, P. S. and Rauschecker, J. P. (1999b). Dual streams of auditory afferents target multiple domains in the primate prefrontal cortex. *Nature Neuroscience* **2**(12): 1131–1136.

Rømer Thomsen, K., Whybrow, P. C. and Kringelbach, M. L. (2015). Reconceptualizing anhedonia: Novel perspectives on balancing the pleasure networks in the human brain. *Frontiers in Behavioral Neuroscience* **9**: 49.

Ross, D., Choi, J. and Purves, D. (2007). Musical intervals in speech. *Proceedings of the National Academy of Sciences* **104**(23): 9852–9857.

Ross, J. M., Iversen, J. R. and Balasubramaniam, R. (2018). The role of posterior parietal cortex in beat-based timing perception: A continuous theta burst stimulation study. *Journal of Cognitive Neuroscience* **30**(5): 634–643.

Rudner, M., Rönnberg, J. and Hugdahl, K. (2005). Reversing spoken items—mind twisting not tongue twisting. *Brain and Language* **92**(1): 78–90.

Rusconi, E., Kwan, B., Giordano, B. L., Umiltà, C. and Butterworth, B. (2006). Spatial representation of pitch height: The SMARC effect. *Cognition* **99**(2): 113–129.

Russell, J. A. (2003). Core affect and the psychological construction of emotion. *Psychological Review* **110**(1): 145–172.

Saarikallio, S. (2011). Music as emotional self-regulation throughout adulthood. *Psychology of Music* **39**(3): 307–327.

Saarikallio, S. H. (2008). Music in mood regulation: Initial scale development. *Musicae Scientiae* **12**(2): 291–309.

Saarikallio, S. H., Maksimainen, J. P. and Randall, W. M. (2019). Relaxed and connected: Insights into the emotional–motivational constituents of musical pleasure. *Psychology of Music* **47**(5): 644–662.

Saarinen, J., Paavilainen, P., Schöger, E., Tervaniemi, M. and Näätänen, R. (1992). Representation of abstract attributes of auditory stimuli in the human brain. *NeuroReport* **3**(12): 1149–1151.

Sachs, M. E., Ellis, R. J., Schlaug, G. and Loui, P. (2016). Brain connectivity reflects human aesthetic responses to music. *Social Cognitive and Affective Neuroscience* **11**(6): 884–891.

Sachs, M. E., Habibi, A., Damasio, A. and Kaplan, J. T. (2018). Decoding the neural signatures of emotions expressed through sound. *NeuroImage* **174**: 1–10.

Saetveit, J., Lewis, D. and Seashore, C. E. (1940). Revision of the Seashore measures of musical talent. *University of Iowa Studies: Series of Aims & Progress of Research* **65**: 62–62.

Saffran, J. R. (2020). Statistical language learning in infancy. *Child Development Perspectives* **14**(1): 49–54.

Saffran, J. R., Aslin, R. N. and Newport, E. L. (1996). Statistical learning by 8-month-old infants. *Science* **274**(5294): 1926–1928.

Saffran, J. R., Johnson, E. K., Aslin, R. N. and Newport, E. L. (1999). Statistical learning of tone sequences by human infants and adults. *Cognition* **70**(1): 27–52.

Saldaña, E., Feliciano, M. and Mugnaini, E. (1996). Distribution of descending projections from primary auditory neocortex to inferior colliculus mimics the topography of intracollicular projections. *Journal of Comparative Neurology* **371**(1): 15–40.

Salimpoor, V. N., Benovoy, M., Larcher, K., Dagher, A. and Zatorre, R. J. (2011). Anatomically distinct dopamine release during anticipation and experience of peak emotion to music. *Nature Neuroscience* **14**(2): 257–262.

Salimpoor, V. N., Benovoy, M., Longo, G., Cooperstock, J. R. and Zatorre, R. J. (2009). The rewarding aspects of music listening are related to degree of emotional arousal. *PLoS One* **4**(10): e7487.

Salimpoor, V. N., Van Den Bosch, I., Kovacevic, N., Mcintosh, A. R., Dagher, A. and Zatorre, R. J. (2013). Interactions between the nucleus accumbens and auditory cortices predict music reward value. *Science* **340**(6129): 216–219.

Sammler, D., Grosbras, M.-H., Anwander, A., Bestelmeyer, Patricia e. G. and Belin, P. (2015). Dorsal and ventral pathways for prosody. *Current Biology* **25**(23): 3079–3085.

Samson, S. and Zatorre, R. J. (1988). Melodic and harmonic discrimination following unilateral cerebral excision. *Brain and Cognition* **7**(3): 348–360.

Samson, S. and Zatorre, R. J. (1992). Learning and retention of melodic and verbal information after unilateral temporal lobectomy. *Neuropsychologia* **30**(9): 815–826.

Sander, D., Grafman, J. and Zalla, T. (2003). The human amygdala: An evolved system for relevance detection. *Reviews in the Neurosciences* **14**(4): 303–316.

Santoro, R., Moerel, M., De Martino, F., Goebel, R., Ugurbil, K., Yacoub, E. and Formisano, E. (2014). Encoding of natural sounds at multiple spectral and temporal resolutions in the human auditory cortex. *PLoS Computational Biology* **10**(1): e1003412.

Särkämö, T., Tervaniemi, M., Laitinen, S., Forsblom, A., Soinila, S., Mikkonen, M., Autti, T., Silvennoinen, H. M., Erkkilä, J. and Laine, M. (2008). Music listening enhances cognitive recovery and mood after middle cerebral artery stroke. *Brain* **131**(3): 866–876.

Sataloff, R. T., Spiegel, J. R. and Hawkshaw, M. (1993). Voice disorders. *Medical Clinics of North America* **77**(3): 551–570.

Sato, S., Mcbride, J., Pfordresher, P., Tierney, A., Six, J., Fujii, S. and Savage, P. E. (2020). Automatic acoustic analyses quantify variation in pitch structure within and between human music, speech, and bird song. *PsyArXiv*.

Satoh, M., Nakase, T., Nagata, K. and Tomimoto, H. (2011). Musical anhedonia: Selective loss of emotional experience in listening to music. *Neurocase* **17**(5): 410–417.

Savage, P. E., Brown, S., Sakai, E. and Currie, T. E. (2015). Statistical universals reveal the structures and functions of human music. *Proceedings of the National Academy of Sciences* **112**(29): 8987–8992.

Schäfer, T. and Mehlhorn, C. (2017). Can personality traits predict musical style preferences? A meta-analysis. *Personality and Individual Differences* **116**: 265–273.

Schäfer, T. and Sedlmeier, P. (2009). From the functions of music to music preference. *Psychology of Music* **37**(3): 279–300.

Schellenberg, E. G., Peretz, I. and Vieillard, S. (2008). Liking for happy-and sad-sounding music: Effects of exposure. *Cognition & Emotion* **22**(2): 218–237.

Scherer, K. R. (1995). Expression of emotion in voice and music. *Journal of Voice* **9**(3): 235–248.

Schlaepfer, T. E., Cohen, M. X., Frick, C., Kosel, M., Brodesser, D., Axmacher, N., Joe, A. Y., Kreft, M., Lenartz, D. and Sturm, V. (2008). Deep brain stimulation to reward circuitry alleviates anhedonia in refractory major depression. *Neuropsychopharmacology* **33**(2): 368–377.

Schlaug, G., Jäncke, L., Huang, Y., Staiger, J. F. and Steinmetz, H. (1995). Increased corpus callosum size in musicians. *Neuropsychologia* **33**(8): 1047–1055.

Schmitz, J., Fraenz, C., Schlüter, C., Friedrich, P., Jung, R. E., Güntürkün, O., Genç, E. and Ocklenburg, S. (2019). Hemispheric asymmetries in cortical gray matter microstructure identified by neurite orientation dispersion and density imaging. *NeuroImage* **189**: 667–675.

Schneider, P., Scherg, M., Dosch, H. G., Specht, H. J., Gutschalk, A. and Rupp, A. (2002). Morphology of Heschl's gyrus reflects enhanced activation in the auditory cortex of musicians. *Nature Neuroscience* **5**(7): 688–694.

Schneider, P., Sluming, V., Roberts, N., Scherg, M., Goebel, R., Specht, H. J., Dosch, H. G., Bleeck, S., Stippich, C. and Rupp, A. (2005). Structural and functional asymmetry of lateral Heschl's gyrus reflects pitch perception preference. *Nature Neuroscience* **8**(9): 1241–1247.

Schneider, S., Peters, J., Bromberg, U., Brassen, S., Miedl, S. F., Banaschewski, T., Barker, G. J., Conrod, P., Flor, H. and Garavan, H. (2012). Risk taking and the adolescent reward system: A potential common link to substance abuse. *American Journal of Psychiatry* **169**(1): 39–46.

Schönwiesner, M., Dechent, P., Voit, D., Petkov, C. I. and Krumbholz, K. (2014). Parcellation of human and monkey core auditory cortex with fMRI pattern classification and objective detection of tonotopic gradient reversals. *Cerebral Cortex* **25**(10): 3278–3289.

Schönwiesner, M., Novitski, N., Pakarinen, S., Carlson, S., Tervaniemi, M. and Näätänen, R. (2007). Heschl's gyrus, posterior superior temporal gyrus, and mid-ventrolateral prefrontal cortex have different roles in the detection of acoustic changes. *Journal of Neurophysiology* **97**(3): 2075–2082.

Schönwiesner, M., Rübsamen, R. and Von Cramon, D. Y. (2005). Hemispheric asymmetry for spectral and temporal processing in the human antero-lateral auditory belt cortex. *European Journal of Neuroscience* **22**(6): 1521–1528.

Schönwiesner, M. and Zatorre, R. J. (2008). Depth electrode recordings show double dissociation between pitch processing in lateral Heschl's gyrus and sound onset processing in medial Heschl's gyrus. *Exp Brain Res* **187**(1): 97–105.

Schönwiesner, M. and Zatorre, R. J. (2009). Spectro-temporal modulation transfer function of single voxels in the human auditory cortex measured with high-resolution fMRI. *Proceedings of the National Academy of Sciences U S A* **106**(34): 14611–14616.

Schott, B. H., Minuzzi, L., Krebs, R. M., Elmenhorst, D., Lang, M., Winz, O. H., Seidenbecher, C. I., Coenen, H. H., Heinze, H.-J., Zilles, K., Düzel, E. and Bauer, A. (2008). Mesolimbic functional magnetic resonance imaging activations during reward anticipation correlate with reward-related ventral striatal dopamine release. *Journal of Neuroscience* **28**(52): 14311–14319.

Schröger, E., Bendixen, A., Trujillo-Barreto, N. J. and Roeber, U. (2007). Processing of abstract rule violations in audition. *PLoS One* **2**(11): e1131.

Schulkind, M. D., Hennis, L. K. and Rubin, D. C. (1999). Music, emotion, and autobiographical memory: They're playing your song. *Memory & Cognition* **27**(6): 948–955.

Schultz, W. (2016). Dopamine reward prediction-error signalling: A two-component response. *Nature Reviews Neuroscience* **17**(3): 183–195.

Schultz, W. (2017). Reward prediction error. *Current Biology* **27**(10): R369–R371.

Schultz, W., Dayan, P. and Montague, P. R. (1997). A neural substrate of prediction and reward. *Science* **275**(5306): 1593–1599.

Schulze, K., Zysset, S., Mueller, K., Friederici, A. D. and Koelsch, S. (2011). Neuroarchitecture of verbal and tonal working memory in nonmusicians and musicians. *Human Brain Mapping* **32**(5): 771–783.

Schwartenbeck, P., Fitzgerald, T., Dolan, R. and Friston, K. (2013). Exploration, novelty, surprise, and free energy minimization. *Frontiers in Psychology* **4**: 710.

Schwartze, M. and Kotz, S. A. (2013). A dual-pathway neural architecture for specific temporal prediction. *Neuroscience & Biobehavioral Reviews* **37**(10, Part 2): 2587–2596.

Scott, B. H., Saleem, K. S., Kikuchi, Y., Fukushima, M., Mishkin, M. and Saunders, R. C. (2017). Thalamic connections of the core auditory cortex and rostral supratemporal plane in the macaque monkey. *Journal of Comparative Neurology* **525**(16): 3488–3513.

Segado, M., Hollinger, A., Thibodeau, J., Penhune, V. and Zatorre, R. J. (2018). Partially overlapping brain networks for singing and cello playing. *Frontiers in Neuroscience* **12**: 351.

Segado, M., Zatorre, R. J. and Penhune, V. B. (2021). Effector-independent brain network for auditory-motor integration: FMRI evidence from singing and cello playing. *NeuroImage* **237**: 118128.

Segarra, P., Poy, R., López, R. and Moltó, J. (2014). Characterizing Carver and White's BIS/BAS subscales using the five factor model of personality. *Personality and Individual Differences* **61–62**: 18–23.

Seger, C. A., Peterson, E. J., Cincotta, C. M., Lopez-Paniagua, D. and Anderson, C. W. (2010). Dissociating the contributions of independent corticostriatal systems to visual categorization learning through the use of reinforcement learning modeling and granger causality modeling. *NeuroImage* **50**(2): 644–656.

Seger, C. A., Spiering, B. J., Sares, A. G., Quraini, S. I., Alpeter, C., David, J. and Thaut, M. H. (2013). Corticostriatal contributions to musical expectancy perception. *Journal of Cognitive Neuroscience* **25**(7): 1062–1077.

Seldon, H. L. (1981a). Structure of human auditory cortex. I. Cytoarchitectonics and dendritic distributions. *Brain Research* **229**: 277–294.

Seldon, H. L. (1981b). Structure of human auditory cortex. II. Axon distributions and morphological correlates of speech perception. *Brain Research* **229**: 295–310.

Semmes, J. (1968). Hemispheric specialization: A possible clue to mechanism. *Neuropsychologia* **6**(1): 11–26.

Sescousse, G., Caldú, X., Segura, B. and Dreher, J.-C. (2013). Processing of primary and secondary rewards: A quantitative meta-analysis and review of human functional neuroimaging studies. *Neuroscience & Biobehavioral Reviews* **37**(4): 681–696.

Sestieri, C., Di Matteo, R., Ferretti, A., Del Gratta, C., Caulo, M., Tartaro, A., Olivetti Belardinelli, M. and Romani, G. L. (2006). "What" versus "where" in the audiovisual domain: An fMRI study. *NeuroImage* **33**(2): 672–680.

Shamma, S. (2001). On the role of space and time in auditory processing. *Trends in Cognitive Sciences* **5**(8): 340–348.

Shannon, R. V. (2016). Is birdsong more like speech or music? *Trends in Cognitive Sciences* **20**(4): 245–247.

Shannon, R. V., Zeng, F.-G., Kamath, V., Wygonski, J. and Ekelid, M. (1995). Speech recognition with primarily temporal cues. *Science* **270**(5234): 303–304.

Shany, O., Singer, N., Gold, B. P., Jacoby, N., Tarrasch, R., Hendler, T. and Granot, R. (2019). Surprise-related activation in the nucleus accumbens interacts with music-induced pleasantness. *Social Cognitive and Affective Neuroscience* **14**(4): 459–470.

Shapleske, J., Rossell, S. L., Woodruff, P. and David, A. (1999). The planum temporale: A systematic, quantitative review of its structural, functional and clinical significance. *Brain Research Reviews* **29**(1): 26–49.

Shepard, R. N. (1978). The mental image. *American Psychologist* **33**(2): 125.

Shepard, R. N. (1982). Geometrical approximations to the structure of musical pitch. *Psychological Review* **89**(4): 305.

Shifriss, R., Bodner, E. and Palgi, Y. (2015). When you're down and troubled: Views on the regulatory power of music. *Psychology of Music* **43**(6): 793–807.

Shin, J. C. and Ivry, R. B. (2002). Concurrent learning of temporal and spatial sequences. *Journal of Experimental Psychology: Learning, Memory, and Cognition* **28**(3): 445–457.

Shohamy, D. and Adcock, R. A. (2010). Dopamine and adaptive memory. *Trends in Cognitive Sciences* **14**(10): 464–472.

Sienkiewicz-Jarosz, H., Scinska, A., Swiecicki, L., Lipczynska-Lojkowska, W., Kuran, W., Ryglewicz, D., Kolaczkowski, M., Samochowiec, J. and Bienkowski, P. (2013). Sweet liking in patients with Parkinson's disease. *Journal of the Neurological Sciences* **329**(1–2): 17–22.

Sievers, B., Polansky, L., Casey, M. and Wheatley, T. (2013). Music and movement share a dynamic structure that supports universal expressions of emotion. *Proceedings of the National Academy of Sciences* **110**(1): 70–75.

Sigalovsky, I. S., Fischl, B. and Melcher, J. R. (2006). Mapping an intrinsic MR property of gray matter in auditory cortex of living humans: A possible marker for primary cortex and hemispheric differences. *NeuroImage* **32**(4): 1524–1537.

Signoret, J. L., Van Eeckhout, P., Poncet, M. and Castaigne, P. (1987). [Aphasia without amusia in a blind organist. Verbal alexia-agraphia without musical alexia-agraphia in Braille]. *Revue neurologique* **143**(3): 172–181.

Sihvonen, A. J., Särkämö, T., Leo, V., Tervaniemi, M., Altenmüller, E. and Soinila, S. (2017). Music-based interventions in neurological rehabilitation. *The Lancet Neurology* **16**(8): 648–660.

Sihvonen, A. J., Särkämö, T., Rodríguez-Fornells, A., Ripollés, P., Münte, T. F. and Soinila, S. (2019). Neural architectures of music–insights from acquired amusia. *Neuroscience & Biobehavioral Reviews* **107**: 104–114.

Simon, J. J., Walther, S., Fiebach, C. J., Friederich, H.-C., Stippich, C., Weisbrod, M. and Kaiser, S. (2010). Neural reward processing is modulated by approach- and avoidance-related personality traits. *NeuroImage* **49**(2): 1868–1874.

Simoncelli, E. P. and Olshausen, B. A. (2001). Natural image statistics and neural representation. *Annual Review of Neuroscience* **24**(1): 1193–1216.

Simonyan, K. and Horwitz, B. (2011). Laryngeal motor cortex and control of speech in humans. *Neuroscientist* **17**(2): 197–208.

Singer, N., Poker, G., Dunsky, N., Nemni, S., Balter, S., Doron, M., . . . Hendler, T. (2023). Development and validation of an fMRI-informed EEG model of reward-related ventral striatum activation. *NeuroImage* 120183.

Singh, N. C. and Theunissen, F. E. (2003). Modulation spectra of natural sounds and ethological theories of auditory processing. *Journal of the Acoustical Society of America* **114**(6): 3394–3411.

Skov, M. and Nadal, M. (2020). A farewell to art: Aesthetics as a topic in psychology and neuroscience. *Perspectives on Psychological Science* **15**(3): 630–642.

Sloboda, J. A. (1991). Music structure and emotional response: Some empirical findings. *Psychology of Music* **19**(2): 110–120.

Small, D. M., Jones-Gotman, M. and Dagher, A. (2003). Feeding-induced dopamine release in dorsal striatum correlates with meal pleasantness ratings in healthy human volunteers. *NeuroImage* **19**(4): 1709–1715.

Small, D. M., Zatorre, R. J., Dagher, A., Evans, A. C. and Jones-Gotman, M. (2001). Changes in brain activity related to eating chocolate: From pleasure to aversion. *Brain* **124**(Pt 9): 1720–1733.

Smillie, L. D. (2008). What is reinforcement sensitivity? Neuroscience paradigms for approach-avoidance process theories of personality. *European Journal of Personality* **22**(5): 359–384.

Smith, E. C. and Lewicki, M. S. (2006). Efficient auditory coding. *Nature* **439**(7079): 978–982.

Smith, G. (1995). Dopamine and food reward. *Progress in Psychobiology, Physiology, Psychology* **16**: 83–144.

Smith, J. (1983). Reproduction and representation of musical rhythms: The effects of musical skill. In D. Rogers and J.A. Sloboda (Eds.), *The acquisition of symbolic skills* (pp. 273–282). Springer.

Smolewska, K. A., Mccabe, S. B. and Woody, E. Z. (2006). A psychometric evaluation of the highly sensitive person scale: The components of sensory-processing sensitivity and their relation to the bis/bas and "big five". *Personality and Individual Differences* **40**(6): 1269–1279.

Snyder, J. S. and Elhilali, M. (2017). Recent advances in exploring the neural underpinnings of auditory scene perception. *Annals of the New York Academy of Sciences* **1396**(1): 39–55.

Song, J. H., Skoe, E., Wong, P. C. and Kraus, N. (2008). Plasticity in the adult human auditory brainstem following short-term linguistic training. *Journal of Cognitive Neuroscience* **20**(10): 1892–1902.

Sperry, R. W., Gazzaniga, M. S. and Bogen, J. E. (1969). Interhemispheric relationships: The neocortical commissures; syndromes of hemisphere disconnection. *Handbook of Clinical Neurology* **4**(273–290).

Spreng, R. N. and Grady, C. L. (2010). Patterns of brain activity supporting autobiographical memory, prospection, and theory of mind, and their relationship to the default mode network. *Journal of Cognitive Neuroscience* **22**(6): 1112–1123.

Staeren, N., Renvall, H., De Martino, F., Goebel, R. and Formisano, E. (2009). Sound categories are represented as distributed patterns in the human auditory cortex. *Current Biology* **19**(6): 498–502.

Stark, E. A., Vuust, P. and Kringelbach, M. L. (2018). Music, dance, and other art forms: New insights into the links between hedonia (pleasure) and eudaimonia (well-being). *Progress in Brain Research* **237**: 129–152.

Steele, C. J., Bailey, J. A., Zatorre, R. J. and Penhune, V. B. (2013). Early musical training and white-matter plasticity in the corpus callosum: Evidence for a sensitive period. *Journal of Neuroscience* **33**(3): 1282–1290.

Stefanacci, L. and Amaral, D. G. (2002). Some observations on cortical inputs to the macaque monkey amygdala: An anterograde tracing study. *Journal of Comparative Neurology* **451**(4): 301–323.

Steinbeis, N., Koelsch, S. and Sloboda, J. A. (2006). The role of harmonic expectancy violations in musical emotions: Evidence from subjective, physiological, and neural responses. *Journal of Cognitive Neuroscience* **18**(8): 1380–1393.

Steinmetzger, K. and Rosen, S. (2015). The role of periodicity in perceiving speech in quiet and in background noise. *Journal of the Acoustical Society of America* **138**(6): 3586–3599.

Stephan, M. A., Lega, C. and Penhune, V. B. (2018). Auditory prediction cues motor preparation in the absence of movements. *NeuroImage* **174**: 288–296.

Stewart, L., Henson, R., Kampe, K., Walsh, V., Turner, R. and Frith, U. (2003). Brain changes after learning to read and play music. *NeuroImage* **20**(1): 71–83.

Stewart, L., Von Kriegstein, K., Warren, J. D. and Griffiths, T. D. (2006). Music and the brain: Disorders of musical listening. *Brain* **129**(10): 2533–2553.

Strafella, A. P., Paus, T., Barrett, J. and Dagher, A. (2001). Repetitive transcranial magnetic stimulation of the human prefrontal cortex induces dopamine release in the caudate nucleus. *Journal of Neuroscience* **21**(15): RC157.

Strait, D. L. and Kraus, N. (2011). Can you hear me now? Musical training shapes functional brain networks for selective auditory attention and hearing speech in noise. *Frontiers in Psychology* **2**: 113.

Su, Y.-H. and Pöppel, E. (2012). Body movement enhances the extraction of temporal structures in auditory sequences. *Psychological Research* **76**(3): 373–382.

Summerfield, Cand De Lange, F. P. (2014). Expectation in perceptual decision making: Neural and computational mechanisms. *Nature Reviews Neuroscience* **15**(11): 745–756.

Tabas, A., Mihai, G., Kiebel, S., Trampel, R. and Von Kriegstein, K. (2020). Abstract rules drive adaptation in the subcortical sensory pathway. *eLife* **9**: e64501.

Takaya, S., Kuperberg, G., Liu, H., Greve, D., Makris, N. and Stufflebeam, S. (2015). Asymmetric projections of the arcuate fasciculus to the temporal cortex underlie lateralized language function in the human brain. *Frontiers in Neuroanatomy* **9**: 119.

Tallal, P., Miller, S. and Fitch, R. H. (1993). Neurobiological basis of speech: A case for the preeminence of temporal processing. *Annals-New York Academy of Sciences* **682**: 27–27–47.

Tanabe, H. C., Honda, M. and Sadato, N. (2005). Functionally segregated neural substrates for arbitrary audiovisual paired-association learning. *Journal of Neuroscience* **25**(27): 6409–6418.

Tardif, E. and Clarke, S. (2001). Intrinsic connectivity of human auditory areas: A tracing study with DII. *European Journal of Neuroscience* **13**(5): 1045–1050.

Tarr, B., Launay, J. and Dunbar, R. I. (2016). Silent disco: Dancing in synchrony leads to elevated pain thresholds and social closeness. *Evolution and Human Behavior* **37**(5): 343–349.

Teki, S., Barascud, N., Picard, S., Payne, C., Griffiths, T. D. and Chait, M. (2016). Neural correlates of auditory figure-ground segregation based on temporal coherence. *Cerebral Cortex* **26**(9): 3669–3680.

Temperley, D. (2010). Modeling common-practice rhythm. *Music Perception* **27**(5): 355–376.

Tervaniemi, M., Medvedev, S. V., Alho, K., Pakhomov, S. V., Roudas, M. S., Van Zuijen, T. L. and Näätänen, R. (2000). Lateralized automatic auditory processing of phonetic versus musical information: A PET study. *Human Brain Mapping* **10**(2): 74–79.

Tervaniemi, M., Rytkönen, M., Schröger, E., Ilmoniemi, R. J. and Näätänen, R. (2001). Superior formation of cortical memory traces for melodic patterns in musicians. *Learning & Memory* **8**(5): 295–300.

Tervaniemi, M., Szameitat, A. J., Kruck, S., Schröger, E., Alter, K., De Baene, W. and Friederici, A. D. (2006). From air oscillations to music and speech: Functional magnetic resonance imaging evidence for fine-tuned neural networks in audition. *Journal of Neuroscience* **26**(34): 8647–8652.

Thayer, R. E., Newman, J. R. and Mcclain, T. M. (1994). Self-regulation of mood: Strategies for changing a bad mood, raising energy, and reducing tension. *Journal of Personality and Social Psychology* **67**(5): 910–925.

Thompson, P. M., Lee, A. D., Dutton, R. A., Geaga, J. A., Hayashi, K. M., Eckert, M. A., Bellugi, U., Galaburda, A. M., Korenberg, J. R., Mills, D. L., Toga, A. W. and Reiss, A. L. (2005). Abnormal cortical complexity and thickness profiles mapped in williams syndrome. *The Journal of Neuroscience* **25**(16): 4146–4158.

Thompson, W. F., Geeves, A. M. and Olsen, K. N. (2019). Who enjoys listening to violent music and why? *Psychology of Popular Media Culture* **8**(3): 218–232.

Thompson, W. F., Marin, M. M. and Stewart, L. (2012). Reduced sensitivity to emotional prosody in congenital amusia rekindles the musical protolanguage hypothesis. *Proceedings of the National Academy of Sciences* **109**(46): 19027–19032.

Tian, B., Reser, D., Durham, A., Kustov, A. and Rauschecker, J. P. (2001). Functional specialization in rhesus monkey auditory cortex. *Science* **292**(5515): 290–293.

Tierney, A., Dick, F., Deutsch, D. and Sereno, M. (2013). Speech versus song: Multiple pitch-sensitive areas revealed by a naturally occurring musical illusion. *Cerebral Cortex* **23**(2): 249–254.

Tillmann, B., Albouy, P. and Caclin, A. (2015). Congenital amusias. In M. J. Aminoff, F. Boller, and D. F. Swaab (Eds.), *Handbook of clinical neurology* (Vol. 129, pp. 589–605). Elsevier.

Tillmann, B., Bharucha, J. J. and Bigand, E. (2000). Implicit learning of tonality: A self-organizing approach. *Psychological Review* **107**(4): 885–913.

Tillmann, B. and Bigand, E. (2010). Musical structure processing after repeated listening: Schematic expectations resist veridical expectations. *Musicae Scientiae* **14**(2_suppl): 33–47.

Tillmann, B., Koelsch, S., Escoffier, N., Bigand, E., Lalitte, P., Friederici, A. D. and Von Cramon, D. Y. (2006). Cognitive priming in sung and instrumental music: Activation of inferior frontal cortex. *NeuroImage* **31**(4): 1771–1782.

Tillmann, B., Lévêque, Y., Fornoni, L., Albouy, P. and Caclin, A. (2016). Impaired short-term memory for pitch in congenital amusia. *Brain Research* **1640**: 251–263.

Toiviainen, P. and Snyder, J. S. (2003). Tapping to Bach: Resonance-based modeling of pulse. *Music Perception:* **21**(1): 43–80.

Trainor, L. J. (1996). Effects of harmonics on relative pitch discrimination in a musical context. *Perception & Psychophysics* **58**(5): 704–712.

Trainor, L. J., Tsang, C. D. and Cheung, V. H. (2002). Preference for sensory consonance in 2-and 4-month-old infants. *Music Perception* **20**(2): 187–194.

Tramo, M. J., Shah, G. D. and Braida, L. D. (2002). Functional role of auditory cortex in frequency processing and pitch perception. *Journal of Neurophysiology* **87**(1): 122–139.

Treadway, M. T. and Zald, D. H. (2011). Reconsidering anhedonia in depression: Lessons from translational neuroscience. *Neuroscience & Biobehavioral Reviews* **35**(3): 537–555.

Trevor, C., Arnal, L. H. and Frühholz, S. (2020). Terrifying film music mimics alarming acoustic feature of human screams. *Journal of the Acoustical Society of America* **147**(6): EL540–EL545.

Trost, W., Frühholz, S., Cochrane, T., Cojan, Y. and Vuilleumier, P. (2015). Temporal dynamics of musical emotions examined through intersubject synchrony of brain activity. *Social Cognitive and Affective Neuroscience* **10**(12): 1705–1721.

Trost, W., Frühholz, S., Schön, D., Labbé, C., Pichon, S., Grandjean, D. and Vuilleumier, P. (2014). Getting the beat: Entrainment of brain activity by musical rhythm and pleasantness. *NeuroImage* **103**: 55–64.

Tsang, C. D., Friendly, R. H. and Trainor, L. J. (2011). Singing development as a sensorimotor interaction problem. *Psychomusicology: Music, Mind and Brain* **21**(1–2): 31–44.

Tzourio-Mazoyer, N., Crivello, F. and Mazoyer, B. (2018). Is the planum temporale surface area a marker of hemispheric or regional language lateralization? *Brain Structure and Function* **223**(3): 1217–1228.

Uhlig, M., Fairhurst, M. T. and Keller, P. E. (2013). The importance of integration and top-down salience when listening to complex multi-part musical stimuli. *NeuroImage* **77**: 52–61.

Ulanovsky, N., Las, L. and Nelken, I. (2003). Processing of low-probability sounds by cortical neurons. *Nature Neuroscience* **6**(4): 391–398.

Urošević, S., Collins, P., Muetzel, R., Lim, K. and Luciana, M. (2012). Longitudinal changes in behavioral approach system sensitivity and brain structures involved in reward processing during adolescence. *Developmental Psychology* **48**(5): 1488–1500.

Urry, H. L., Van Reekum, C. M., Johnstone, T., Kalin, N. H., Thurow, M. E., Schaefer, H. S., Jackson, C. A., Frye, C. J., Greischar, L. L., Alexander, A. L. and Davidson, R. J. (2006). Amygdala and ventromedial pre-frontal cortex are inversely coupled during regulation of negative affect and predict the diurnal pattern of cortisol secretion among older adults. *Journal of Neuroscience* **26**(16): 4415–4425.

Van Der Heijden, K., Rauschecker, J. P., De Gelder, B. and Formisano, E. (2019). Cortical mechanisms of spatial hearing. *Nature Reviews Neuroscience* **20**(10): 609–623.

van Egmond, R. and Povel, D. J. (1996). Perceived similarity of exact and inexact transpositions. *Acta Psychologica* **92**: 283–295.

Vanneste, S., Song, J.-J. and De Ridder, D. (2013). Tinnitus and musical hallucinosis: The same but more. *NeuroImage* **82**: 373–383.

Van Vugt, F., Hartmann, K., Altenmüller, E., Mohammadi, B. and Margulies, D. (2021). The impact of early musical training on striatal functional connectivity. *NeuroImage* **238**: 118251.

Vaquero, L., Ramos-Escobar, N., Cucurell, D., François, C., Putkinen, V., Segura, E., Huotilainen, M., Penhune, V. B. and Rodríguez-Fornells, A. (2021). Arcuate fasciculus architecture is associated with in-dividual differences in pre-attentive detection of unpredicted music changes. *NeuroImage* **229**: 117759.

Vaquero, L., Ramos-Escobar, N., François, C., Penhune, V. B. and Rodríguez-Fornells, A. (2018). White-matter structural connectivity predicts short-term melody and rhythm learning in non-musicians. *NeuroImage* **181**: 252–262.

Vella, E. J. and Mills, G. (2016). Personality, uses of music, and music preference: The influence of openness to experience and extraversion. *Psychology of Music* **45**(3): 338–354.

Venezia, J. H., Richards, V. M. and Hickok, G. (2021). Speech-driven spectrotemporal receptive fields be-yond the auditory cortex. *Hearing Research* **408**: 108307.

Venezia, J. H., Thurman, S. M., Richards, V. M. and Hickok, G. (2019). Hierarchy of speech-driven spectrotemporal receptive fields in human auditory cortex. *NeuroImage* **186**: 647–666.

Verosky, N. J. and Morgan, E. (2021). Pitches that wire together fire together: Scale degree associations across time predict melodic expectations. *Cognitive Science* **45**(10): e13037.

Vieillard, S., Peretz, I., Gosselin, N., Khalfa, S., Gagnon, L. and Bouchard, B. (2008). Happy, sad, scary and peaceful musical excerpts for research on emotions. *Cognition & Emotion* **22**(4): 720–752.

Voisin, J., Bidet-Caulet, A., Bertrand, O. and Fonlupt, P. (2006). Listening in silence activates auditory areas: A functional magnetic resonance imaging study. *Journal of Neuroscience* **26**(1): 273–278.

Von Economo, C. and Horn, L. (1930). Über windungsrelief, maße und rindenarchitektonik der supratemporalfläche, ihre individuellen und ihre seitenunterschiede. *Zeitschrift für die gesamte Neurologie und Psychiatrie* **130**(1): 678–757.

Von Ehrenfels, C. (1937). On gestalt-qualities. *Psychological Review* **44**(6): 521–524.

Von Helmholtz, H. (1925). *Helmholtz's treatise on physiological optics*. Optical Society of America.

Von Kriegstein, K., Eger, E., Kleinschmidt, A. and Giraud, A. L. (2003). Modulation of neural responses to speech by directing attention to voices or verbal content. *Cognitive Brain Research* **17**(1): 48–55.

Vuoskoski, J. K. and Eerola, T. (2011). The role of mood and personality in the perception of emotions rep-resented by music. *Cortex* **47**(9): 1099–1106.

Vuust, P., Gebauer, L. K. and Witek, M. A. (2014). Neural underpinnings of music: The polyrhythmic brain. In H. Merchant and V. de Lafuente (Eds.), *Neurobiology of interval timing* (pp. 339–356). Springer.

Vuust, P., Ostergaard, L., Pallesen, K. J., Bailey, C. and Roepstorff, A. (2009). Predictive coding of music-brain responses to rhythmic incongruity. *Cortex* **45**(1): 80–92.

Vuust, P., Pallesen, K. J., Bailey, C., Van Zuijen, T. L., Gjedde, A., Roepstorff, A. and Østergaard, L. (2005). To musicians, the message is in the meter: Pre-attentive neuronal responses to incongruent rhythm are left-lateralized in musicians. *NeuroImage* **24**(2): 560–564.

Vuvan, D. T., Paquette, S., Mignault Goulet, G., Royal, I., Felezeu, M. and Peretz, I. (2018a). The Montreal protocol for identification of amusia. *Behavior Research Methods* **50**(2): 662–672.

Vuvan, D. T., Zendel, B. R. and Peretz, I. (2018b). Random feedback makes listeners tone-deaf. *Scientific Reports* **8**(1): 7283.

Wager, T. D., Davidson, M. L., Hughes, B. L., Lindquist, M. A. and Ochsner, K. N. (2008). Prefrontal-subcortical pathways mediating successful emotion regulation. *Neuron* **59**(6): 1037–1050.

Wallmark, Z., Deblieck, C. and Iacoboni, M. (2018). Neurophysiological effects of trait empathy in music listening. *Frontiers in Behavioral Neuroscience* **12**: 66.

Walworth, D. D. (2003). The effect of preferred music genre selection versus preferred song selection on experimentally induced anxiety levels. *Journal of Music Therapy* **40**(1): 2–14.

Wang, C., Pan, R., Wan, X., Tan, Y., Xu, L., Ho, C. S. and Ho, R. C. (2020). Immediate psychological responses and associated factors during the initial stage of the 2019 coronavirus disease (Covid-19) epidemic among the general population in China. *International Journal of Environmental Research and Public Health* **17**(5): 1729.

Wang, X., Merzenich, M. M., Beitel, R. and Schreiner, C. E. (1995). Representation of a species-specific vocalization in the primary auditory cortex of the common marmoset: Temporal and spectral characteristics. *Journal of Neurophysiology* **74**(6): 2685–2706.

Warren, J. D. and Griffiths, T. D. (2003). Distinct mechanisms for processing spatial sequences and pitch sequences in the human auditory brain. *The Journal of Neuroscience* **23**(13): 5799–5804.

Warren, J. D., Uppenkamp, S., Patterson, R. D. and Griffiths, T. D. (2003). Separating pitch chroma and pitch height in the human brain. *Proceedings of the National Academy of Sciences* **100**(17): 10038–10042.

Warren, J. E., Sauter, D. A., Eisner, F., Wiland, J., Dresner, M. A., Wise, R. J. S., Rosen, S. and Scott, S. K. (2006). Positive emotions preferentially engage an auditory–motor "mirror" system. *Journal of Neuroscience* **26**(50): 13067–13075.

Warren, R. M. and Warren, R. P. (1968). *Helmholtz on perception: Its physiology and development.* John Wiley and Sons.

Warrier, C., Wong, P., Penhune, V., Zatorre, R., Parrish, T., Abrams, D. and Kraus, N. (2009). Relating structure to function: Heschl's gyrus and acoustic processing. *Journal of Neuroscience* **29**(1): 61–69.

Warrier, C. M. and Zatorre, R. J. (2002). Influence of tonal context and timbral variation on perception of pitch. *Perception & Psychophysics* **64**(2): 198–207.

Warrier, C. M. and Zatorre, R. J. (2004). Right temporal cortex is critical for utilization of melodic contextual cues in a pitch constancy task. *Brain* **127**(Pt 7): 1616–1625.

Watabe-Uchida, M., Eshel, N. and Uchida, N. (2017). Neural circuitry of reward prediction error. *Annual Review of Neuroscience* **40**(1): 373–394.

Watkins, K. E., Strafella, A. P. and Paus, T. (2003). Seeing and hearing speech excites the motor system involved in speech production. *Neuropsychologia* **41**(8): 989–994.

Watson, D., Wiese, D., Vaidya, J. and Tellegen, A. (1999). The two general activation systems of affect: Structural findings, evolutionary considerations, and psychobiological evidence. *Journal of Personality and Social Psychology* **76**(5): 820–838.

Weiss, M. W. and Bidelman, G. M. (2015). Listening to the brainstem: Musicianship enhances intelligibility of subcortical representations for speech. *The Journal of Neuroscience* **35**(4): 1687–1691.

Welch, G. F., Howard, D. M. and Nix, J., Eds. (2019). *The Oxford handbook of singing.* Oxford University Press.

Werner, H. (1940). Musical "micro-scales" and "micro-melodies". *The Journal of Psychology* **10**(1): 149–156.

Westbury, C. F., Zatorre, R. J. and Evans, A. C. (1999). Quantifying variability in the planum temporale: A probability map. *Cerebral Cortex* **9**(4): 392–405.

Wheeler, M. E., Petersen, S. E. and Buckner, R. L. (2000). Memory's echo: Vivid remembering reactivates sensory-specific cortex. *Proceedings of the National Academy of Sciences* **97**(20): 11125–11129.

Whiteford, K. L. and Oxenham, A. J. (2018). Learning for pitch and melody discrimination in congenital amusia. *Cortex* **103**: 164–178.

Whitfield, I. C. (1985). The role of auditory cortex in behavior. In A. Peters and E. G. Jones (Eds.), *Association and auditory cortices* (pp. 329–349). Springer US.

Wiestler, T. and Diedrichsen, J. (2013). Skill learning strengthens cortical representations of motor sequences. *eLife* **2**: e00801.

Williamson, V. J. and Stewart, L. (2010). Memory for pitch in congenital amusia: Beyond a fine-grained pitch discrimination problem. *Memory* **18**(6): 657–669.

Wilson, E. C., Melcher, J. R., Micheyl, C., Gutschalk, A. and Oxenham, A. J. (2007a). Cortical fMRI activation to sequences of tones alternating in frequency: Relationship to perceived rate and streaming. *Journal of Neurophysiology* **97**(3): 2230–2238.

Wilson, M. and Cook, P. F. (2016). Rhythmic entrainment: Why humans want to, fireflies can't help it, pet birds try, and sea lions have to be bribed. *Psychonomic Bulletin & Review* **23**(6): 1647–1659.

Wilson, S. M., Molnar-Szakacs, I. and Iacoboni, M. (2007b). Beyond superior temporal cortex: Intersubject correlations in narrative speech comprehension. *Cerebral Cortex* **18**(1): 230–242.

Wiltermuth, S. S. and Heath, C. (2009). Synchrony and cooperation. *Psychological Science* **20**(1): 1–5.

Winer, J. A. (2005). Decoding the auditory corticofugal systems. *Hearing Research* **207**(1–2): 1–9.

Wise, R. A. (1980). The dopamine synapse and the notion of 'pleasure centers' in the brain. *Trends in Neurosciences* **3**(4): 91–95.

Witek, M. a. G., Clarke, E. F., Wallentin, M., Kringelbach, M. L. and Vuust, P. (2014). Syncopation, body-movement and pleasure in groove music. *PLoS One* **9**(4): e94446.

Witteman, J., Van Ijzendoorn, M. H., Van De Velde, D., Van Heuven, V. J. and Schiller, N. O. (2011). The nature of hemispheric specialization for linguistic and emotional prosodic perception: A meta-analysis of the lesion literature. *Neuropsychologia* **49**(13): 3722–3738.

Wittgenstein, L. (1966). *Lectures and conversations on aesthetics, psychology, and religious belief.* University of California Press.

Wittmann, B. C., Daw, N. D., Seymour, B. and Dolan, R. J. (2008). Striatal activity underlies novelty-based choice in humans. *Neuron* **58**(6): 967–973.

Wollman, I., Fritz, C. and Poitevineau, J. (2014). Influence of vibrotactile feedback on some perceptual features of violins. *The Journal of the Acoustical Society of America* **136**(2): 910–921.

Wollman, I., Penhune, V., Segado, M., Carpentier, T. and Zatorre, R. J. (2018). Neural network retuning and neural predictors of learning success associated with cello training. *Proceedings of the National Academy of Sciences U S A* **115**(26): E6056–E6064.

Wong, P. C. M., Skoe, E., Russo, N. M., Dees, T. and Kraus, N. (2007). Musical experience shapes human brainstem encoding of linguistic pitch patterns. *Nature Neuroscience* **10**(4): 420–422.

Woolgar, A., Duncan, J., Manes, F. and Fedorenko, E. (2018). The multiple-demand system but not the language system supports fluid intelligence. *Nature Human Behaviour* **2**(3): 200–204.

Woolley, S. M. N., Fremouw, T. E., Hsu, A. and Theunissen, F. E. (2005). Tuning for spectro-temporal modulations as a mechanism for auditory discrimination of natural sounds. *Nature Neuroscience* **8**(10): 1371–1379.

Worden, F. G. and Marsh, J. T. (1968). Frequency-following (microphonic-like) neural responses evoked by sound. *Electroencephalography and Clinical Neurophysiology* **25**(1): 42–52.

Wright, A. A., Rivera, J. J., Hulse, S. H., Shyan, M. and Neiworth, J. J. (2000). Music perception and octave generalization in rhesus monkeys. *Journal of Experimental Psychology: General* **129**(3): 291–307.

Wundt, W. M. (1904). *Principles of physiological psychology.* Sonnenschein.

Xu, G., Zhang, L., Shu, H., Wang, X. and Li, P. (2013). Access to lexical meaning in pitch-flattened chinese sentences: An fMRI study. *Neuropsychologia* **51**(3): 550–556.

Yamagishi, S., Otsuka, S., Furukawa, S. and Kashino, M. (2016). Subcortical correlates of auditory perceptual organization in humans. *Hearing Research* **339**: 104–111.

Yeterian, E. H. and Pandya, D. N. (1998). Corticostriatal connections of the superior temporal region in rhesus monkeys. *Journal of Comparative Neurology* **399**(3): 384–402.

Zacks, J. M. (2008). Neuroimaging studies of mental rotation: A meta-analysis and review. *Journal of Cognitive Neuroscience* **20**(1): 1–19.

Zald, D. H. and Kim, S. W. (1996). Anatomy and function of the orbital frontal cortex: I. Anatomy, neurocircuitry, and obsessive-compulsive disorder. *Journal of Neuropsychiatry and Clinical Neurosciences* **8**(2): 125–138.

Zald, D. H. and Pardo, J. V. (2002). The neural correlates of aversive auditory stimulation. *NeuroImage* **16**(3, Part A): 746–753.

Zamorano, A. M., Cifre, I., Montoya, P., Riquelme, I. and Kleber, B. (2017). Insula-based networks in professional musicians: Evidence for increased functional connectivity during resting state fMRI. *Human Brain Mapping* **38**(10): 4834–4849.

Zangenehpour, S. and Zatorre, R. J. (2010). Crossmodal recruitment of primary visual cortex following brief exposure to bimodal audiovisual stimuli. *Neuropsychologia* **48**(2): 591–600.

Zarate, J. M., Wood, S. and Zatorre, R. J. (2010). Neural networks involved in voluntary and involuntary vocal pitch regulation in experienced singers. *Neuropsychologia* **48**(2): 607–618.

Zarate, J. M. and Zatorre, R. J. (2008). Experience-dependent neural substrates involved in vocal pitch regulation during singing. *NeuroImage* **40**(4): 1871–1887.

Zatorre, R. J. (1985). Discrimination and recognition of tonal melodies after unilateral cerebral excisions. *Neuropsychologia* **23**(1): 31–41.

Zatorre, R. J. (1988). Pitch perception of complex tones and human temporal-lobe function. *The Journal of the Acoustical Society of America* **84**(2): 566–572.

Zatorre, R. J. and Baum, S. R. (2012). Musical melody and speech intonation: Singing a different tune. *PLoS Biology* **10**(7): e1001372.

Zatorre, R. J. and Belin, P. (2001). Spectral and temporal processing in human auditory cortex. *Cerebral Cortex* **11**(10): 946–953.

Zatorre, R. J., Belin, P. and Penhune, V. B. (2002a). Structure and function of auditory cortex: Music and speech. *Trends in Cognitive Sciences* **6**(1): 37–46.

Zatorre, R. J., Bouffard, M., Ahad, P. and Belin, P. (2002b). Where is 'where' in the human auditory cortex? *Nature Neuroscience* **5**(9): 905–909.

Zatorre, R. J., Bouffard, M. and Belin, P. (2004). Sensitivity to auditory object features in human temporal neocortex. *Journal of Neuroscience* **24**(14): 3637–3642.

Zatorre, R. J., Chen, J. L. and Penhune, V. B. (2007). When the brain plays music: Auditory-motor interactions in music perception and production. *Nature Reviews Neuroscience* **8**(7): 547–558.

Zatorre, R. J., Delhommeau, K. and Zarate, J. M. (2012). Modulation of auditory cortex response to pitch variation following training with microtonal melodies. *Frontiers in Psychology* **3**: 544.

Zatorre, R. J., Evans, A. C. and Meyer, E. (1994). Neural mechanisms underlying melodic perception and memory for pitch. *Journal of Neuroscience* **14**(4): 1908–1919.

Zatorre, R. J., Evans, A. C., Meyer, E. and Gjedde, A. (1992). Lateralization of phonetic and pitch discrimination in speech processing. *Science* **256**(5058): 846–849.

Zatorre, R. J. and Gandour, J. T. (2008). Neural specializations for speech and pitch: Moving beyond the dichotomies. *Philosophical Transactions of the Royal Society London B Biological Science* **363**(1493): 1087–1104.

Zatorre, R. J. and Halpern, A. R. (1993). Effect of unilateral temporal-lobe excision on perception and imagery of songs. *Neuropsychologia* **31**(3): 221–232.

Zatorre, R. J., Halpern, A. R. and Bouffard, M. (2010). Mental reversal of imagined melodies: A role for the posterior parietal cortex. *Journal of Cognitive Neuroscience* **22**(4): 775–789.

Zatorre, R. J., Halpern, A. R., Perry, D. W., Meyer, E. and Evans, A. C. (1996). Hearing in the mind's ear: A pet investigation of musical imagery and perception. *Journal of Cognitive Neuroscience* **8**(1): 29–46.

Zatorre, R. J. and Samson, S. (1991). Role of the right temporal neocortex in retention of pitch in auditory short-term memory. *Brain* **114**(6): 2403–2417.

Zendel, B. R., West, G. L., Belleville, S. and Peretz, I. (2019). Musical training improves the ability to understand speech-in-noise in older adults. *Neurobiology of Aging* **81**: 102–115.

Zentner, M. and Eerola, T. (2010). Rhythmic engagement with music in infancy. *Proceedings of the National Academy of Sciences* **107**(13): 5768–5773.

Zentner, M., Grandjean, D. and Scherer, K. R. (2008). Emotions evoked by the sound of music: Characterization, classification, and measurement. *Emotion* **8**(4): 494–521.

Zetzsche, T., Meisenzahl, E. M., Preuss, U. W., Holder, J. J., Kathmann, N., Leinsinger, G., Hahn, K., Hegerl, U. and Möller, H.-J. (2001). In-vivo analysis of the human planum temporale (pt): Does the definition of pt borders influence the results with regard to cerebral asymmetry and correlation with handedness? *Psychiatry Research: Neuroimaging* **107**(2): 99–115.

Zhou, J., Gardner, M. P. H. and Schoenbaum, G. (2021). Is the core function of orbitofrontal cortex to signal values or make predictions? *Current Opinion in Behavioral Sciences* **41**: 1–9.

Index

For the benefit of digital users, indexed terms that span two pages (e.g., 52–53) may, on occasion, appear on only one of those pages.

Figures are indicated by *f* following the page number